Neuromorphic Devices for Brain-Inspired Computing

Neuromorphic Devices for Brain-Inspired Computing

Artificial Intelligence, Perception, and Robotics

Edited by Qing Wan and Yi Shi

WILEY-VCH

Editors

Professor Qing Wan
Nanjing University
Electronic Science & Engineering
No. 163, Xianlin Avenue
Qixia District
210093 Nanjing
China

Professor Yi Shi
Nanjing University
School of Electronic Science and Engineering
No. 163, Xianlin Da Dao
Qixia District
210023 Nanjing
China

Cover Image: © vs148/Shutterstock

Library of Congress Card No.: applied for

British Library Cataloguing-in-Publication Data
A catalogue record for this book is available from the British Library.

Bibliographic information published by the Deutsche Nationalbibliothek
The Deutsche Nationalbibliothek lists this publication in the Deutsche Nationalbibliografie; detailed bibliographic data are available on the Internet at <http://dnb.d-nb.de>.

Print ISBN: 978-3-527-34979-1
ePDF ISBN: 978-3-527-83529-4
ePub ISBN: 978-3-527-83530-0
oBook ISBN: 978-3-527-83531-7

Typesetting Straive, Chennai, India
Printing and Binding CPI Group (UK) Ltd, Croydon, CR0 4YY

Printed on acid-free paper

C117192_011221

Contents

Preface *xi*

1 **Two-Terminal Neuromorphic Memristors** *1*
Hui-Kai He, He-Ming Huang, and Rui Yang
1.1 Memristive Devices *1*
1.1.1 Memristive Device Structure and Materials *1*
1.1.1.1 Memristive Device Structure *1*
1.1.1.2 Memristive Materials *2*
1.1.2 Resistive Switching Behavior *4*
1.1.2.1 Volatile Resistive Switching *5*
1.1.2.2 Nonvolatile Resistive Switching *7*
1.2 Resistive Switching Mechanisms *7*
1.2.1 Filamentary-Type Resistive Switching *8*
1.2.1.1 Cation Migration-Related Filaments *8*
1.2.1.2 Anion Migration-Related Filaments *11*
1.2.2 Interface-Type Resistive Switching *14*
1.3 Memristive Bioinspired Devices *17*
1.3.1 Memristive Synapses *18*
1.3.1.1 Short-Term Memristive Synapses *20*
1.3.1.2 Long-Term Memristive Synapses *23*
1.3.2 Memristive Neurons *27*
1.3.2.1 Bioplausible Memristive Neurons *31*
1.3.2.2 Biophysical Memristive Neurons *34*
1.4 Memristive Neural Networks *36*
1.4.1 Memristive ANN Computing *36*
1.4.2 Memristive SNN Computing *39*
1.5 Summary and Outlook *41*
References *42*

2 **Spintronic Neuromorphic Devices** *47*
Deming Zhang, Sai Li, Xinran Wang, Ao Du, Wang Kang, Lang Zeng, and Weisheng Zhao
2.1 Introduction *47*
2.2 Magnetic Tunnel Junction for Neuromorphic Computing *48*

2.2.1 Device Structure and the Write/Read Operation *48*
2.2.1.1 Binary Magnetic Tunnel Junction *48*
2.2.1.2 Multi-level Spintronic Memristor *49*
2.2.2 Working as the Synaptic Device *51*
2.2.2.1 Stochastic Binary Synaptic Device *51*
2.2.2.2 Analog-Like Synaptic Device *52*
2.2.3 Working as the Neural Device *54*
2.2.3.1 Spiking Neural Device *54*
2.2.3.2 Artificial Neural Device *57*
2.2.4 All-Spin Neural Network *61*
2.2.4.1 All-Spin Artificial Neural Network with Compound Scheme *61*
2.2.4.2 All-Spin Spiking Neural Network with Spintronic Memristor *62*
2.2.5 Summary and Outlook *62*
2.3 Skyrmion-Based Neuromorphic Computing *62*
2.3.1 The Introduction of Skyrmions *62*
2.3.2 Skyrmion-Based Synapse Devices *64*
2.3.3 Skyrmion-Based Neuron Devices *65*
2.3.4 Skyrmion-Based Reservoir Computing *67*
2.3.5 Skyrmion-Based Stochastic Computing *68*
2.3.6 Challenges and Perspectives *69*
2.4 Spin Torque Oscillators for Neuromorphic Computing *72*
2.4.1 Introduction to Spin Torque Oscillator *72*
2.4.2 Associative Memory Based on Injection Locking of STO *73*
2.4.3 Reservoir Computing Based on STO *74*
2.4.4 Recurrent Neural Network based on Delayed Feedback of STO *75*
2.4.5 Neuromorphic Computing Based on the Synchronization of STO *77*
2.4.6 Problems and Perspectives *77*
2.5 Conclusion and Outlook *80*
References *80*

3 Multiterminal Neuromorphic Devices with Cognitive Behaviors *91*
Li Qiang Zhu, Jia Cheng Cai, Zheng Yu Ren, Wen Xiong, and Qing Wan
3.1 Introduction *91*
3.2 Multiterminal Neuromorphic Memristors *92*
3.2.1 Memristor-Based Neuromorphic Devices *92*
3.2.2 Multiterminal Memristor for Neuromorphic System *96*
3.2.2.1 Synaptic Competition and Cooperation on Multiterminal Memristor *96*
3.2.2.2 Heterosynaptic Plasticity on Multiterminal Memristor *97*
3.2.2.3 Multiterminal Memtransistor *98*
3.3 Multiterminal Neuromorphic Transistors *99*
3.3.1 Neuron Transistors *99*
3.3.2 Neuromorphic Devices for Chemical Biosensor Applications *101*
3.3.2.1 Chemical Biosensors *101*
3.3.2.2 Neuron Transistors for Chemical Biosensor *102*

3.3.2.3 Multi-gate Neuromorphic Transistor for pH Sensor *103*
3.3.3 Dendritic Algorithm on Multiterminal Neuromorphic Transistors *103*
3.3.3.1 EGT-Based Neuromorphic Transistors *103*
3.3.3.2 Multi-gate Neuromorphic Transistor *105*
3.3.3.3 Dendrite Neuron and Dendritic Algorithm *106*
3.3.3.4 Multi-gate Neuromorphic Transistor for Pattern Recognition *109*
3.4 Neuromorphic Transistors for Perception Learning Activities *109*
3.4.1 Artificial Tactile Device *109*
3.4.2 Artificial Vision Device *112*
3.4.3 Artificial Auditory Device *113*
3.5 Conclusion and Outlook *114*
Acknowledgments *116*
References *116*

4 Neuromorphic Devices Based on Chalcogenide Materials *125*
Jia Chen, Yifan Lu, Zhe Yang, Yi Li, and Xiangshui Miao
4.1 Introduction *125*
4.2 Ovonic Memory Switching (OMS) and Threshold Switching (OTS) in Chalcogenide Materials *126*
4.3 Artificial Synapses Based on MS Behaviors *129*
4.4 Artificial Neurons Based on TS Effects *136*
4.5 Hardware Neural Networks *139*
4.6 Summary and Outlook *143*
References *144*

5 Neuromorphic Devices Based on Organic Materials *149*
Junyao Zhang and Jia Huang
5.1 Introduction *149*
5.2 Two-Terminal Organic Neuromorphic Devices *149*
5.2.1 Metal Filament Conducting-Based Memristors *150*
5.2.2 Redox Reaction-Based Memristors *152*
5.2.3 Ion Migration-Based Memristors *153*
5.2.4 Charge Trapping-Based Memristors *155*
5.3 Three-Terminal Organic Neuromorphic Devices *157*
5.3.1 Floating-Gate Transistors *157*
5.3.2 Electrolyte-Gate Transistors *158*
5.3.3 Ferroelectric-Gate Transistors *162*
5.3.4 Optoelectronic Transistors *164*
5.4 Innovative Applications of Organic Neuromorphic Devices for Bionic Perception Systems *165*
5.4.1 Artificial Visualization Systems *166*
5.4.2 Tactile-Perception Systems *167*
5.5 Summary and Outlook *169*
References *170*

6 Neuromorphic Computing Systems with Emerging Devices *173*
Qiumeng Wei, Jianshi Tang, Bin Gao, Xinyi Li, He Qian, and Huaqiang Wu
6.1 Introduction *173*
6.1.1 Background Introduction *173*
6.1.2 The Motivation for Neuromorphic Computing *174*
6.1.3 Progress and Challenges of CMOS-Based Neuromorphic Computing *174*
6.1.4 Principles of Memristors *176*
6.2 DNNs Based on Synaptic Devices *178*
6.2.1 Device Performance Requirements *178*
6.2.2 Array Demonstrations *179*
6.2.3 Chip and System Implementations *183*
6.2.4 Architecture and Algorithm Optimization *185*
6.2.4.1 Quantization *185*
6.2.4.2 Nonideal Analog Switching Characteristics *186*
6.2.4.3 Synaptic Array Size *186*
6.3 SNNs Based on Neuromorphic Devices *187*
6.3.1 Learning Rules and Memory Principles *188*
6.3.1.1 Synapses and Neurons in SNNs *188*
6.3.1.2 Learning Algorithms and Benchmarks for SNNs *188*
6.3.2 SNNs with Synaptic Devices and Neuronal Devices *189*
6.3.2.1 Synaptic Devices with Plasticity *189*
6.3.2.2 Neuronal Devices *191*
6.3.3 SNN Implementations with Synaptic Arrays *192*
6.4 Other Neuromorphic Systems *196*
6.4.1 Hyperdimensional Computing *196*
6.4.2 Dendritic Computing *196*
6.4.3 Reservoir Computing *202*
6.4.4 Oscillatory Neural Network *203*
6.4.5 Hopfield Neural Network and Simulated Annealing *203*
6.5 Summary and Outlook *204*
References *206*

7 Neuromorphic Perceptual Systems with Emerging Devices *217*
Ying Zhu, Changjin Wan, and Qing Wan
7.1 Background *217*
7.2 Sensation and Perception *218*
7.2.1 From Sensation to Perception *218*
7.2.2 Pursing Artificial Perception by Neuromorphic Devices *219*
7.3 Implementation of Artificial Perception *221*
7.3.1 Building Block of Artificial Perception: Artificial Sensory Neuron *221*
7.3.2 Intelligent Tasks Based on Artificial Perception *223*
7.3.2.1 Pattern Recognition Tasks *224*

7.3.2.2 Prosthetics and Robotics Applications *226*
7.3.3 A Roadmap Toward Neuromorphic Perceptual System *226*
7.4 Challenges and Perspectives *228*
7.4.1 Challenges *228*
7.4.2 Conclusions and Perspectives *229*
References *229*

Index *235*

Preface

Neuromorphic Devices for Brain-inspired Computing provides an overview of the development on neuromorphic electronics, which aimed at building a neuromorphic computing system from the bottom up. Neuromorphic computing, also called brain-like computing, is one promising computing paradigm in the post-Moore's Law era. The models, approaches, and hardware for neuromorphic computing are designed by replicating the neural network structures and information processing principles of the human brain. Just as the metal–oxide–semiconductor field-effect transistors (MOSFETs) to the integrated circuits, the fundamental neuromorphic device is the key element for building a neuromorphic computing system. Hence, it is urgent and meaningful to provide a comprehensive view on these neuromorphic devices and to look into the possibility of this new wave. Thus, we invited some experts from each neuromorphic electronics related sub-topic to fulfill the aforementioned target together. We have listed their names and affiliations in the list of contributors.

Chapter 1 introduces the device structure, materials, and switching behaviors of memristive devices, as well as the state-of-the-art memristive synapses, neurons, and neural networks.

Chapter 2 is focused on the spintronic device-based neuromorphic emulation and computing. Spintronic devices have unique properties such as low power, nearly infinite endurance, good scalability, and grate compatibility with the complementary metal–oxide–semiconductor (CMOS) technology and are recently recognized as promising candidates for neuromorphic devices.

Chapter 3 introduces multi-terminal devices, which are mainly transistors. Interesting dendritic integration functions and dendritic algorithms based on this type of devices are also discussed and briefly introduced.

Chapter 4 is devoted to neuromorphic devices based on chalcogenide materials that show the intriguing memory switch and threshold switch behaviors to emulate synaptic plasticity and neural firing functions.

Chapter 5 is devoted to neuromorphic devices based on organic materials. Organic neuromorphic devices are involved in both two-terminal devices and three-terminal devices, and they are notable for building bionic systems.

Chapter 6 is mainly about the neuromorphic computing implemented by emerging devices. The neuromorphic devices facilitate energy-efficient computing in deep neural networks (DNNs), spiking neural networks (SNNs), and other intelligent systems.

Chapter 7 describes another promising direction for neuromorphic electronics: the building of neuromorphic perceptual systems by emerging devices. This direction is aimed to endow intelligence to the bionic systems.

With more than one year of preparation, this book has finally been published. We all hope you will find our labors useful. Your comments and corrections will be highly appreciated.

Nanjing, China
May 2021

Qing Wan
Yi Shi

1

Two-Terminal Neuromorphic Memristors

Hui-Kai He, He-Ming Huang, and Rui Yang

Huazhong University of Science and Technology, School of Materials Science and Engineering, Wuhan 430074, P. R. China

1.1 Memristive Devices

A memristive device is a resistive device with an inherent memory; its theory was creatively conceived by Prof. Chua in 1971 [1] and was connected to the physical devices in 2008 by HP [2]. Since then, memristive devices have been extensively studied over the past decade due to their prominent advantages, such as simple structure, high operation speed, and low power consumption in applications of data storage, logic operation, and neuromorphic computation [3]. In this section, we will introduce traditional two-terminal memristive devices, mainly focusing on device structure and memristive materials.

1.1.1 Memristive Device Structure and Materials

1.1.1.1 Memristive Device Structure

Typically, a memristive device has a metal/insulator/metal (MIM) structure, composed of a switching layer sandwiched between two metal electrodes (possibly different), as shown in Figure 1.1a. Because of its simple structure, highly scalable cross-point and multilevel stacking memory structures have been proposed (Figure 1.1b), which is promising for the construction of huge neural networks and neuromorphic computing systems [3]. It is well known that electrodes play a crucial role in the resistive switching behavior of memristive devices. To date, in addition to metals (such as Ag [5], Cu [6], Pt [7], Au [8], Al [9], and W [10]), a variety of conductive materials have been explored as electrodes for memristors, including nitrides such as TiN [11], carbon materials such as graphene [12] and carbon nanotubes [13], conductive oxides such as ITO [14] and $SrRuO_3$ (SRO) [15], p- and n-type Si [16], and so on. Among these metals, Ag and Cu are the most popular ones due to their ability to dissolve in thin film electrolyte at low electric field and their high ionic mobility [17]. In addition to the electrodes, the switching layer where the resistive switching takes place is the key layer in memristive devices and has a great

Neuromorphic Devices for Brain-Inspired Computing: Artificial Intelligence, Perception, and Robotics, First Edition. Edited by Qing Wan and Yi Shi.

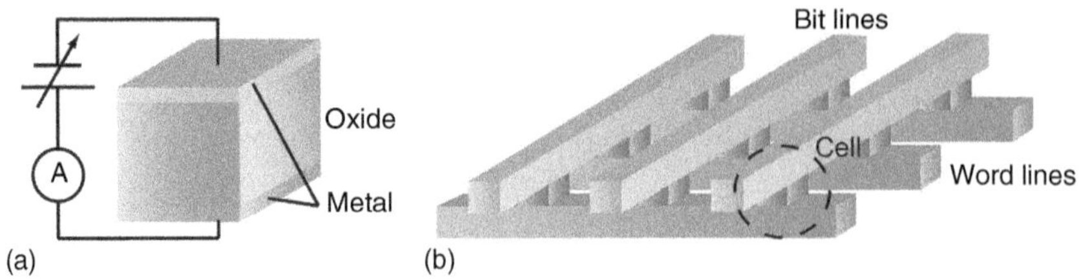

Figure 1.1 (a) Diagram of a memristive device with a capacitor-like structure in which a switching layer is sandwiched between two metal electrodes. (b) Diagram of a cross-point memory structure. Word and bit lines are used for selecting a memristive device and writing/reading data, respectively. Source: Sawa [4].

impact on the device performance. Typically, the switching layer is an insulator or a semiconductor. Also, it is normally in the form of thin film, which is compatible to large-scale integration in the semiconductor industry. Recently, other forms of the switching layer are also intensively investigated, including nanoparticles [18], nanowires [19], two-dimensional (2D) materials [20], three-dimensional nanoarrays, etc. Note that we mainly discuss memristors in the form of thin film in all below sections.

1.1.1.2 Memristive Materials

As mentioned above, the materials involved in memristors include switching materials and electrode materials. Here, we mainly focus on switching materials, which is also termed as memristive materials. Up to now, a great number of memristive materials have been explored for memristive devices used in neuromorphic computing. In this chapter, the memristive materials are subdivided into inorganic and organic materials. Generally speaking, inorganic materials have significant advantages over organic ones in switching stability and manufacturing technology, while organic ones stand out in terms of high-mechanical flexibility and low cost.

Inorganic materials for memristors can be loosely divided into binary oxides (e.g. TiO_x [21], TaO_x [22], HfO_x [23], WO_x [24], and ZnO [14]), perovskite oxides (e.g. $SrTiO_3$ [25] and $BiFeO_3$ [26]), and 2D materials (e.g. graphene [27], hexagonal boron nitride (h-BN) [28], and molybdenum disulfide (MoS_2) [20]). Among these inorganic materials, binary oxides have been intensively studied since they are the most abundant and show superior switching characteristics including ultrahigh ON/OFF ratio, sub-ns operation speed, and extreme endurance. In addition, their simple composition enables them to be easily fabricated by various film deposition technologies, mainly including magnetron sputtering [14, 24], atomic layer deposition (ALD) [29], thermal oxidation [30], and plasma oxidation [11]. Magnetron sputtering is a high-rate, high-efficient film deposition technology and is becoming increasingly popular owing its high-yield and low-cost production of uniform films over large areas. Recently, ALD has also attracted increasing attention for the deposition of binary oxides due to its ability to accurately control the thickness and uniformity of the films. Furthermore, binary oxides have good compatibility with conventional complementary metal oxide semiconductor (CMOS) process and good thermal stability. Thus, binary oxides have been the focus of both academic and industrial communities over the past decade. In particular, research interest in HfO_x and TaO_x has been extremely high in the past few years since they exhibit both sub-ns operation speed and extreme endurance of $>10^{10}$ cycles and may be the most promising memristive materials for practical applications in the near future.

In addition to binary oxide, perovskite oxides such as $SrTiO_3$, $SmNiO_3$, $BiFeO_3$, and $SrRuO_3$ have also been researched for memristors over the past few years. Among these perovskite oxides, $SrTiO_3$ receives the most attention for the implementation of memristive synapses because of its superior memristive properties and rich switching dynamics [25, 31, 32]. It has been found that perovskite oxides have advantages of excellent localized accumulation of oxygen ions and can be easily converted into a defective structure. However, it should be mentioned that they are generally obtained through pulsed laser deposition (PLD) under high temperature. Although this is an advanced deposition method that can obtain high-quality thin films with accurate stoichiometry, it is not widely used in the semiconductor industry due to its high-cost and the small uniform area of the deposited film, greatly hindering the development and application of perovskite oxides in memristors.

In recent years, 2D materials have become a new focus in memristors for the realization of artificial synapses and neurons due to their superior physical, chemical, and mechanical properties, including graphene, MoS_2, and h-BN. Graphene is one of the highly desirable materials for memristive bioinspired devices owing to its excellent properties of low cost, tunability, nontoxicity, flexibility, and biocompatibility [33, 34]. However, graphene is inherently a semimetallic material and needs to be oxidized or doped when it is used as a switching layer. In contrast to graphene, transition metal chalcogenides (TMDs), such as MoS_2 and tungsten selenide (WSe_2), are semiconductors with proper bandgaps from 1 to 2 eV [35]. Therefore, TMDs are considered as ideal substitutes for graphene. MoS_2, the common member of the TMDs family, has been intensively investigated and shows superior performance including ultrahigh ON/OFF ratio [36], ultralow operating voltage [37], and excellent thermal stability [38]. Reliable production of 2D materials with uniform properties is essential for translating their new electronic and optical properties into applications. Until now, various fabrication methods including mechanical exfoliation [39], liquid-phase exfoliation [40], and chemical vapor deposition (CVD)[41] have been employed to obtain atomically thin flakes of 2D materials. First discovered by Novoselov et al. in 2004, ultrathin 2D materials are peeled from their parent bulk crystals by mechanical exfoliation using adhesive tape. This method produces single-crystal flakes of high purity and cleanliness that are suitable for fundamental characterization. Liquid-phase exfoliation method is also a feasible way to prepare atomically thin 2D material. It can produce gram quantities of submicrometer-sized monolayers, but the resulting exfoliated material differs structurally and electronically from the bulk material [42]. To obtain large-area and uniform layers, CVD method is very promising. Such methods give reasonably good-quality material with typical flake sizes of hundreds of nanometers to a few centimeters, although the flake thickness is not conclusively shown to be monolayers.

Compared with inorganic materials, organic materials have the advantages of low cost, ease of fabrication, and, especially, high-mechanical flexibility. In addition, it is easy to modulate the electrical performance of organic materials by a designed molecular synthesis [43]. Accordingly, organic materials are attracting more and more attention that enable them to be promising for future flexible electronics, although most switching characteristics of organic materials are still not comparable with those of inorganic materials. It is well known that organic materials consist of large molecules with long chains of repeating monomer units. Hence, solution processes, for example, spin coating, are normally adopted to prepare organic films [44, 45]. Note that organic materials still suffer from some problems like poor thermal stability and bad compatibility with CMOS process [46].

1.1.2 Resistive Switching Behavior

The intrinsic physical phenomenon behind memristive devices is resistive switching (RS), which means the resistance can be reversely changed between low resistive state (LRS) and high resistive state (HRS) under external electric stimuli, resulting in a pinched hysteresis current–voltage (I–V) loop. If the resistance state of the memristive device changes from HRS to LRS, it is called an SET operation and is also considered as a "write" process. In contrast, if the resistance state is converted from LRS to HRS, it is called RESET operation and means an "erase" process (Figure 1.2).

Actually, the switching behavior can be classified into different types on the basis of I–V characteristics according to different criteria. Based on the polarity of the external electric field, the switching behaviors can be classified into two types: unipolar and bipolar resistive switching. The unipolar switching operates independently of the voltage polarity, which is also named nonpolar switching sometimes. Meanwhile, the bipolar switching shows a directional resistive switching, depending on the polarity of the applied voltage. On the basis of the switching dynamics, the switching behavior can be classified into digital and analog types. Generally, the digital resistive switching shows abrupt current jumping in the I–V curves, while the analog switching has continuous current curve during voltage sweeping. Digital resistive switching is preferable in the information storage because of its high ON/OFF ratio and fast switching speed [48, 49]. In contrast, the analog resistive switching is particularly concerned for artificial synaptic devices, since the gradual change of the resistance well resembles the potentiation/inhabitation of the synaptic weight in the adaptive learning process of a synapse [50]. The switching behaviors in Figure 1.2b,d are generally nonvolatile and widely used in the application of data storage. Notably, the analog switching shown in Figure 1.2c is usually nonvolatile but sometimes is volatile with current decay effect. The volatile analog switching with versatile time constants is desirable for simulating short-term and long-term synaptic plasticity. The unipolar digital switching shown in Figure 1.2a is volatile and is usually named as threshold switching, in which the device switches to LRS upon the application of a certain threshold voltage and then spontaneously decays back to HRS after the removal of the external voltage. This threshold switching is promising for implementing the threshold firing process of the neuron. Below, we mainly focus on the volatile and nonvolatile resistive switching.

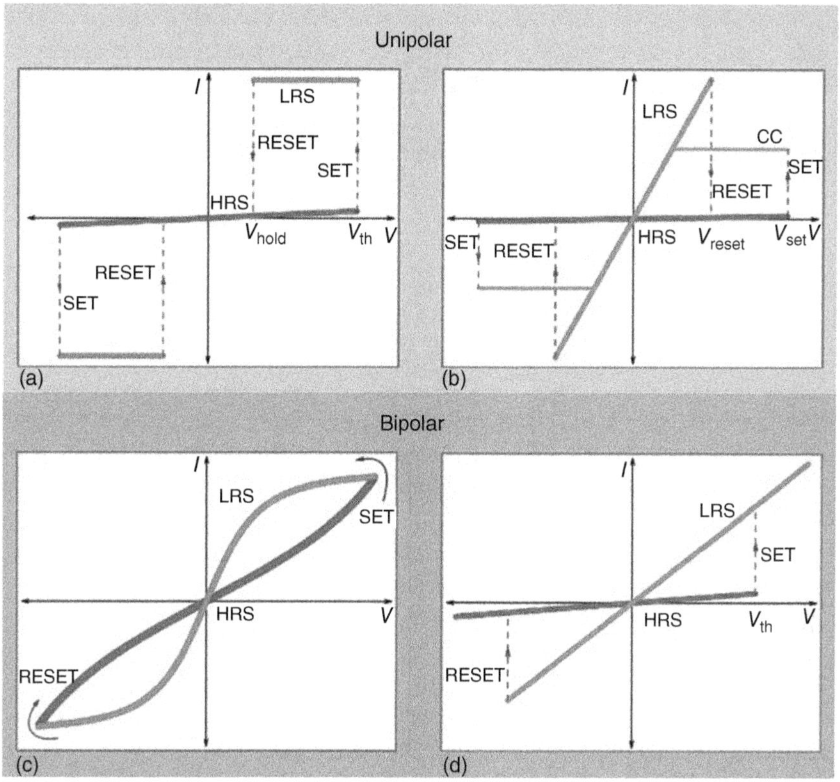

Figure 1.2 Typical I–V characteristics for different switching behaviors in memristive devices. (a) Unipolar threshold switching. (b) Unipolar digital switching with nonvolatility. (c) Bipolar analog switching with tunable volatility. (d) Bipolar digital switching with nonvolatility. Source: Yang et al. [47].

1.1.2.1 Volatile Resistive Switching

The current of the memristive device showing spontaneously decay after the removal of the external electric field is termed as volatile resistive switching, including volatile analog switching and threshold switching. Both of them can be utilized in the construction of artificial synapses and neurons.

To date, volatile analog switching has been observed in a variety of materials, such as WO_x [24], $Pr_{0.7}Ca_{0.3}MnO_3$ (PCMO) [51], and Nb-$SrTiO_3$ [11]. Moreover, it has been found that this switching is generally based on interface-type resistive switching, which will be discussed later. Note that the decay characteristics are essential in the emulation of both short- and long-term plasticity of biological synapses. Thus, we next give a brief introduction to the dynamic process of volatile analog switching using the Pd/WO_x/W memristor as an example [30]. After the stimulation of write pulses, current decay process of the device is carefully monitored by small read pulses, as shown in Figure 1.3a. The stimulation drives the current higher; however after the stimulation is removed, the current decreases with time. One can see that the decay appears to occur at two different time scales: right after stimulation, the current shows a very fast decay, and the decay becomes much slower after several seconds. Then, the decay can be well fitted by the sum of two stretched exponential functions with two time constants: a short-term effect with time constant ≈52.5 ms and a long-term time constant ≈92.5 seconds:

$$I = A_1 \times I_{\text{short}} + A_2 \times \text{I}_{\text{long}} = A_1 \text{I}_{0\text{s}} \exp\left[-\left(\frac{t}{\tau_{\text{s}}}\right)^{\beta_{\text{s}}}\right] + A_2 I_{01} \exp\left[-\left(\frac{t}{\tau_1}\right)^{\beta_1}\right]$$

where τ_s (τ_1), I_{0s} (I_{01}), and β_s (β_1) are the characteristic relaxation time, prefactor, and stretch index for the short-term (long-term term) process, respectively. Borrowing terms used in neuroscience, the first stage with time constant ≈52.5 ms is considered short-term, and the second stage with time constant ≈92.5 seconds is considered long-term. These two time constants differ by more than 3 orders of magnitude, which enable the memristor to emulate important rate- and timing-dependent behaviors at both short term and long term.

In addition to volatile analog switching, threshold switching is also volatile, and the current will decay over time once the applied voltage or pulse is removed. Recently, threshold switching was demonstrated by Wang et al. in two-terminal diffusive memristors based on Ag-doped SiO_xN_y [52]. Also, the decay process was attributed to the spontaneous rupture of the Ag filament driven by minimization of the interfacial energy between Ag and the dielectric after removing the external electric field. Figure 1.3b displays decay characteristics of the SiO_xN_y:Ag diffusive memristor showing variation of current (open circle) with applied voltage pulses (solid line). Under an applied pulse, the device exhibits threshold switching to an LRS after an incubation period (delay time). This delay time is related to the growth and clustering of silver nanoparticles to eventually form conduction channels. Following channel formation, the current jumps abruptly by several orders of magnitude and then slowly increases further under bias as the channel thickens. Once the voltage pulse is removed, the device relaxes back to its original HRS over

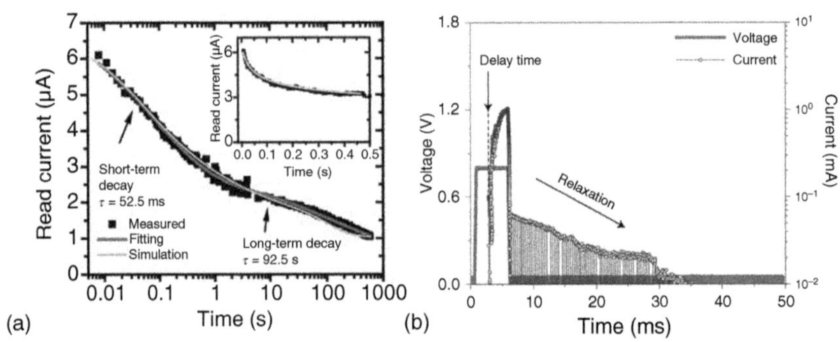

Figure 1.3 Decay characteristics of volatile analog switching and threshold switching devices. (a) Current decay of the Pd/WO_x/W device after the removal of the applied voltage pulses. The experimental data (black squares) can be fitted by the sum of two stretched exponential functions with distinct relaxation time constants. Source: Du et al. [30]. (b) Decay characteristics of the SiO_xN_y:Ag diffusive device showing variation of current (open circles) with applied voltage pulses. Source: Wang et al. [52].

a characteristic time (relaxation time). More importantly, the relaxation time is on the same order as the response of bio-synapses, that is, tens of milliseconds. Furthermore, it has been demonstrated that the relaxation time is related to the temperature, the voltage pulse parameters, operation history, Ag concentration, host lattice, device geometry, and humidity, which alone or in combination can be used to tune the desired dynamics for neuromorphic systems.

1.1.2.2 Nonvolatile Resistive Switching

Unlike volatile resistive switching, nonvolatile resistive switching, including bipolar digital switching and nonvolatile analog switching, means that the conductance is maintained after the removal of the external electric field. For bipolar digital switching, the device abruptly changes to LRS upon the application of a positive (negative) threshold voltage and then returns to HRS upon the application of a negative (positive) threshold voltage. Such switching is preferable in data storage due to its high ON/OFF ratio, fast switching speed, and long retention. Recently, thanks to high scalability and large dynamic range, this switching has also been used in memristors for the construction of artificial synapses [53, 54]. However, one disadvantage of this switching is the limited number of conductance states and low conductance update linearity, highly hindering its application and development in neuromorphic computing.

In contrast, nonvolatile analog switching, with multilevel conductance states and high conductance update linearity, has recently attracted increasing attention in the construction of neural network accelerators and neuromorphic systems. On the basis of the nonvolatile analog switching, vector–matrix multiplication (VMM) or weighted summation operation can be physically implemented in a cross-bar array of memristive devices based on Ohm and Kirchhoff laws [55, 56]. In this case, the VMM operation can be accelerated in a parallel, in-memory, and analog manner in the memristor crossbar, which could largely benefit data-centric and VMM-intensive algorithms, including a large majority of artificial neural network (ANN) algorithms and many other arithmetic calculations. This part will be introduced in detail in Section 1.5.

1.2 Resistive Switching Mechanisms

In the past few years, the switching mechanisms of memristors have been vigorously investigated to predict and manipulate the switching profiles to achieve superior switching performance [49, 57]. Especially for artificial synapses and neurons, the required characteristics are broadened to work sensitively to the input stimulus and generate controllable conductance change [58]. To closely mimic biological synapses and neurons, regulating the conductance change of the memristive device under electric stimuli becomes extremely important. Therefore, it is necessary for us to have a deep understanding of the switching mechanisms of memristors. Here, the common switching mechanisms are classified into two types including filamentary-type resistive switching and interface-type resistive switching according to the position where the resistive switching occurs.

1.2.1 Filamentary-Type Resistive Switching

In a memristive device base on the filamentary-type resistive switching, a forming process is usually required to obtain stable resistive switching behavior, during which conductive filaments (CFs) form and the memristive device reaches a LRS. Subsequently, local rupture and re-formation of the CFs occur during the reset and set processes, respectively, resulting in the alternation between HRS and LRS. Accordingly, the resistance value of LRS exhibits no or weak dependence on the device size in filamentary-type memristive devices [49]. Generally, the growth and rupture of CFs results from ion migration. Based on the polarities of the charges, there are two types of ions: cations and anions in nature, and they migrate in opposite directions under an external electric field. Thus, the filamentary-type resistive switching can be subdivided into cation migration-related filaments and anion migration-related filaments, and growth and rupture processes of the CFs based on each of these are discussed separately as follows.

1.2.1.1 Cation Migration-Related Filaments

The cation migration-based memristive devices typically have an electrochemically active electrode (AE), such as Ag or Cu; an electrochemically inert counter electrode (CE), such as Pt, Au, or W; and a thin film of a solid electrolyte, such as Ag^+ or Cu^+ ion conductor, sandwiched between both electrodes [59]. The CFs are formed via electrochemical dissolution and then redeposition of the active metal atoms. Therefore, such memristive devices are often called electrochemical metallization cells (ECM) and are also referred to programmable metallization cells (PMCs) or atomic switches in some literatures [60–62]. The first observation of a metal filament in ECM was achieved by Hirose in 1976. [63]. They demonstrated nonvolatile resistive switching behavior in the Ag/Ag–As_2S_3/Mo sandwich device. To clarify the switching mechanism, a planar Ag/Ag–As_2S_3/Au device was fabricated at the same time, and the growth of Ag filaments from the CE (Au electrode) to the AE (Ag electrode) was confirmed by optical microscopy.

The resistive switching process of an ECM device with Cu as the AE and Pt as the CE is schematically shown in Figure 1.4. The overall switching process consists of the following steps: (a) Upon applying a high enough positive voltage to Cu, metallic Cu is oxidized to Cu^+ ions in accordance with the reaction, $Cu \rightarrow Cu^+ + e^-$. (b) Cu^+ ions migrate along fast diffusion channels toward the CE driven by external electric field, and Cu^+ ions are reduced back to metallic Cu according to the reaction, $Cu^+ + e^- \rightarrow Cu$. (c) The Cu filament continues to grow, and two electrodes become connected. As a result, the ECM device switches from HRS to LRS. (d) Under negative voltage, the Cu filaments can be electrochemically dissolved with the help of Joule heating, thereby resetting the ECM device back to the HRS.

Although there is no doubt about the presence of metal filaments in cation migration-based memristors, the direct observation of metal filaments, especially their dynamic growth and rupture processes, has attracted great interest in academia. Originally, only ex situ observations of metal filaments in planar microscale ECM cells were reported. In recent years, thanks to advances in fabrication and characterization of nanomaterials, both ex situ and in situ observations

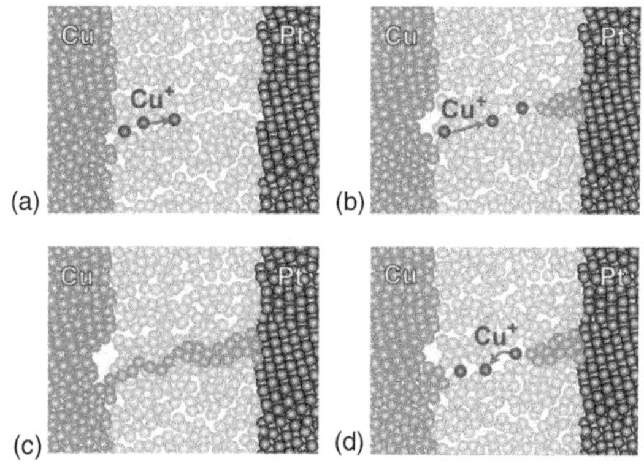

Figure 1.4 Schematic presentation of the resistive switching processes of an ECM device with Cu as the AE and Pt as the CE. (a) forming process, (b) SET process, (c) ON state, (d) RESET process. Source: Zidan et al. [64].

of metal filaments in vertical nanoscale ECM cells have been extensively explored. In 2012, Yang et al. reported the formation/rupture of nanoscale Ag filaments in Ag/SiO_2/Pt device by transmission electron microscopy (TEM) technique [65]. After application of a positive voltage bias to the Ag electrode, the as-fabricated device was switched to the LRS with an abrupt increase in current, as shown in Figure 1.5c. The electrical resistance change was found to accompany changes in the electrode and the switching material (Figure 1.5a), where Ag was injected into the insulating SiO_2 layer and the two electrodes were connected by the Ag

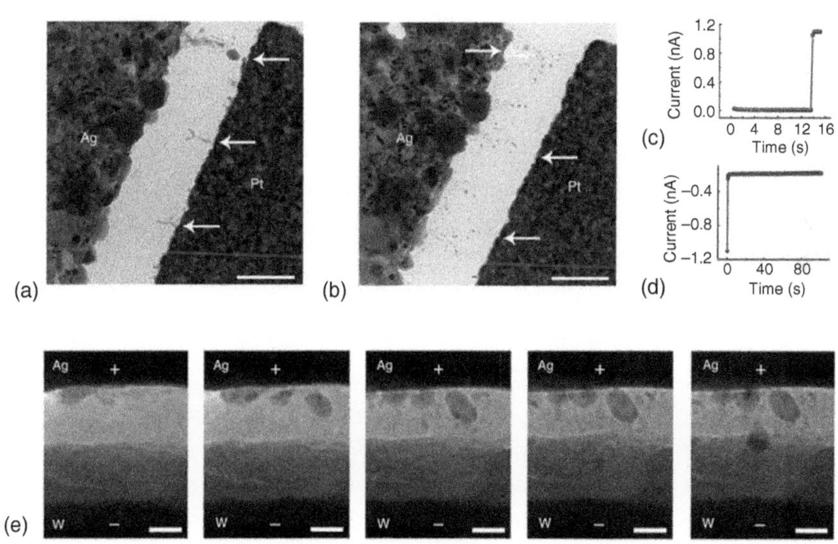

Figure 1.5 (a) TEM image of the Ag/SiO_2/Pt device after the forming process. Scale bar: 200 nm. (b) TEM image of the same device after erasing. Scale bar: 200 nm. (c) Corresponding *I*–*t* curve during the forming process that led to the image in (a). The applied voltage was 8 V. (d) Corresponding *I*–*t* curve during the erasing process that led to the image in (b). The applied voltage was −10 V. (e) Real-time structural evolutions of an Ag/a-Si/W-based device obtained through in situ TEM observation, showing the dynamic filament growth process that initiates from the reactive electrode. Source: Yang et al. [65].

filaments. Upon voltage bias of opposite polarity, the device switched back to the HRS (Figure 1.5d), corresponding to the rupture of the Ag filaments (Figure 1.5b). These observations unambiguously reveal the physical nature of the resistive switching process, where the physically displaced Ag atoms lead to the dramatic changes to the device's electrical properties. Besides direct observation of the metal filaments during resistive switching by ex situ TEM, in situ TEM technique provides important information concerning the microscopic dynamic ionic processes during filament growth. In Yang's work, the filament growth was further demonstrated in vertical Ag/a–Si/W ECM device via in situ TEM observation, as shown in Figure 1.5e. One can identify the real-time structural evolution of the ECM device since the filament growth is directly related to the electric measurements. The systematic in situ TEM analyses offers a complete picture of the different dynamic processes that occur in the ECM device, e.g. filament growth direction, position, and morphology [66].

Notably, the filament growth/dissolution dynamics can be affected by the ionic mobility and the redox reaction rate. Therefore, the filament dynamics can be classified into four categories (Figure 1.6). (i) In the case of high ion mobility and high redox rate, the ions can reach the inert electrode without agglomerating, thus avoiding nucleation within the insulating film so filament growth initiates from the inert electrode, and the large amount of ion supply due to high redox rate leads to the formation of cone-shaped filaments with its base at the inert electrode interface as shown in Figure 1.6a. Such case can be found in ECM devices with traditional solid electrolytes such as GeTe [67] that are known to be good ionic conductors. (ii) In the

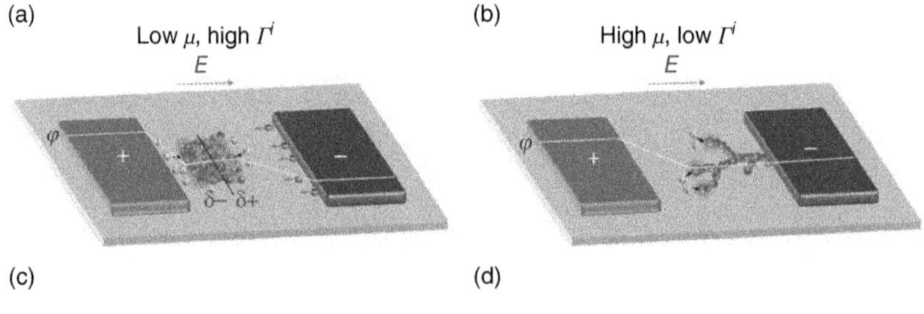

Figure 1.6 (a) Both ion mobility and redox rate are high, resulting in the filament growth from the inert electrode and an inverted cone shape. (b) Both ion mobility and redox rate are low, resulting in the filament growth from the active electrode with discrete nanoclusters and a forward cone shape. (c) Ion mobility is low, but redox rate is high, resulting in the filament nucleation inside the dielectric and reconnection with the source. (d) Ion mobility is high, but redox rate is low, resulting in the filament growth from the inert electrode and a branched structure. Source: Yang et al. [66].

case of low ion mobility and low redox rate (Figure 1.6b), the ions can pile and reach the critical nucleation conditions inside the dielectric, and further filament growth is fulfilled by cluster displacement via the repeated splitting–merging processes. An experimental example is the filament growth in amorphous Si, where the filament is initiated from the active electrode and grows toward the inert electrode as discrete nanoclusters [68]. (iii) In the case of low ion mobility and high redox rate, nucleation can occur inside the dielectric while large amounts of atoms can be deposited onto the cathode sides of the nuclei, leading to the gap filling shown in Figure 1.6c. (iv) In the case of high ion mobility and low redox rate (Figure 1.6d), nucleation only occurs at the inert electrode, but the limited ion supply means that the reduction predominately occurs at the edges with high field strengths, thus leading to the branched filament growth toward the active electrode.

Ion mobility and redox rate can be tuned by the careful selection of the electrode material and the switching materials, as well as the operating conditions since both can be strongly affected by the applied electric field and temperature. Hence, it is possible to adjust the filament dynamics by selecting proper electrode material, switching materials, and operating conditions. The behavior and characteristics of filament are important for artificial synapses and neurons because the filament dynamics is closely related to the synaptic plasticity.

1.2.1.2 Anion Migration-Related Filaments

Besides metal ions, the redox and migration processes of anions, mostly oxygen ions, are involved in the memristive devices based on transition metal oxides without electrochemical active electrodes. The migration of oxygen ions usually induces a redox reaction expressed by a valence change of the cation sublattice and leads to a stoichiometry change of the oxides. Therefore, anion migration-related filament mechanism is typically named as valence change mechanism (VCM). In a VCM device, resistive switching generally relies on the creation and annihilation of oxygen-deficient (or oxygen vacancies, $V_O^{\cdot\cdot}$) CFs based on oxygen migration. In contrast to ECM devices, VCM devices generally consist of inert electrodes and the switching materials. The switching layer is normally a transition metal oxide such as binary metal oxides and ternary perovskite oxide.

The switching process of a VCM device with inert electrodes and a transition metal oxide is schematically shown in Figure 1.7. $V_O^{\cdot\cdot}$ are randomly distributed at the initial state (Figure 1.7a). When the top electrode (TE) is positively biased, $V_O^{\cdot\cdot}$ migrate toward the bottom electrode (BE) and accumulate in the oxide/BE interface (Figure 1.7b). Afterward, $V_O^{\cdot\cdot}$ continue to accumulate under the positive voltage bias. Once the $V_O^{\cdot\cdot}$ CFs form and connect the TE and BE, the device switches into the LRS (Figure 1.7c). With the growth mode from the BE to the TE, the thinnest part of the formed CF should be located near the TE. When the TE is negatively biased, Joule heat is mainly concentrated on the thinnest part of the CF, and $V_O^{\cdot\cdot}$ in this region migrate toward the TE and then is replaced by the oxygen ions. As a result, the concentration of $V_O^{\cdot\cdot}$ in the thinnest part of the CF will be significantly decreased, leading to the rupture of the CF at that location (Figure 1.7d).

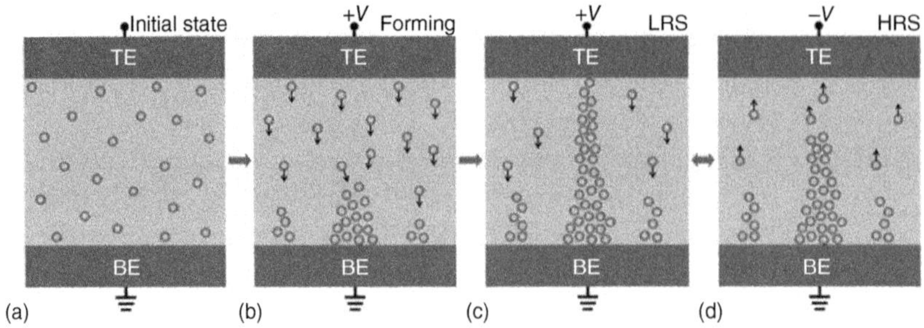

Figure 1.7 Schematic presentation of the switching process of a VCM device with inert electrodes and a transition metal oxide. (a) The initial state with randomly distributed $V_O^{\cdot\cdot}$. (b) The nucleation and subsequent growth from cathode to anode of $V_O^{\cdot\cdot}$ filament during forming process. (c) The LRS with a complete filament whose thinnest region is near the anode. (d) The HRS with a partially ruptured filament at its thinnest region. Source: Pan et al. [48].

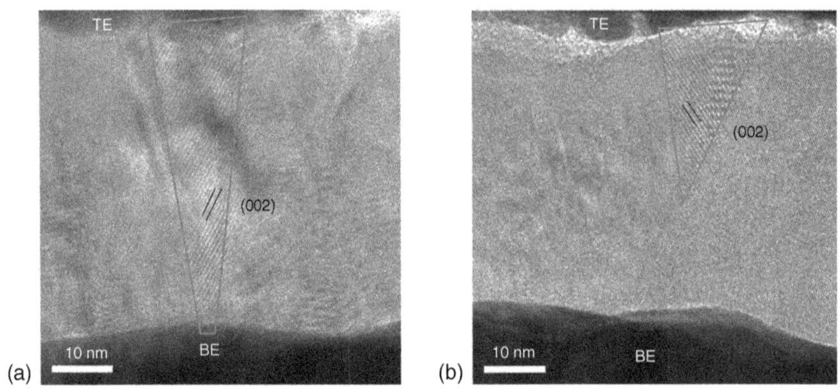

Figure 1.8 (a) High-resolution TEM image of a V_O nanofilament in Pt/TiO_2/Pt device after the set operation. (b) High-resolution TEM image of an incomplete filament in the same device after the reset operation. Source: Kwon et al. [68].

Similar to ECM devices, a host of material characterization techniques such as TEM and spectroscopic analysis have recently been employed to reveal the microscopic origin of resistive switching behaviors in VCM devices. Figure 1.8 shows high-resolution TEM (HRTEM) image of the same location in a Pt/TiO_2/Pt device after applying negative (set) and positive (reset) voltages, respectively [68]. Clear contrast differences in the HRTEM images can be observed, indicating changes in the composition of the film at different resistance states. In the set sample (Figure 1.8a), a connected V_O filament in the conical shape (marked by a gray line) was found between top electrode and bottom electrode. After the reset operation (Figure 1.8b), the connected filament ruptured and an incomplete filament was present near the top electrode, verifying the role of V_O migration during the resistive switching process of VCM devices.

It is generally more challenging to directly observe the V_O migration process and filament evolution in VCM devices since the filaments consist of native defects, i.e. $V_O^{\cdot\cdot}$ rather than foreign metallic species. Careful spectroscopic analysis is typically required to confirm the composition changes. Various state-of-the-art techniques such as X-ray absorption spectroscopy (XAS) [69, 70] and photoemission electron microscopy (PEEM) [71, 72] have recently been employed to comprehensively characterize the changes in film microstructure, composition, and chemical states accompanying the resistive switching process in VCM devices.

XAS Analysis Spatially resolved XAS using scanning transmission X-ray microscope (STXM) was performed to nondestructively investigate the chemical and structural changes during resistive switching. Figure 1.9a,b shows the presence of three distinct states of the TiO_2 within the junction area taken at an X-ray energy of 460.0 eV in the Ti L_3-edge in a $Pt/TiO_2/Pt$ device after electroforming and set/reset cycling [69]. Region is outside of the junction area and is most similar to the as-deposited TiO_2, which is known from X-ray diffraction (XRD) to be an amorphous phase. Within the junction (region ii), the spectrum strongly matches the known XAS for anatase, one of the crystalline polymorphs of TiO_2. The altered absorption spectrum in region iii matches that of reduced titanium oxide, in which the valence state of the Ti ions is reduced from +4 to +3. Furthermore, electron diffraction measurement revealed that the region iii is composed of the Ti_4O_7 phase with metallic conductivity, which proves the formation of localized conductive filament due to the migration of $V_O^{\cdot\cdot}$. Thermally driven radial migration of $V_O^{\cdot\cdot}$ was also demonstrated by in operando X-ray absorption spectromicroscopy in a Ta_2O_5-based device (Figure 1.9c,d) [70]. After 10^5

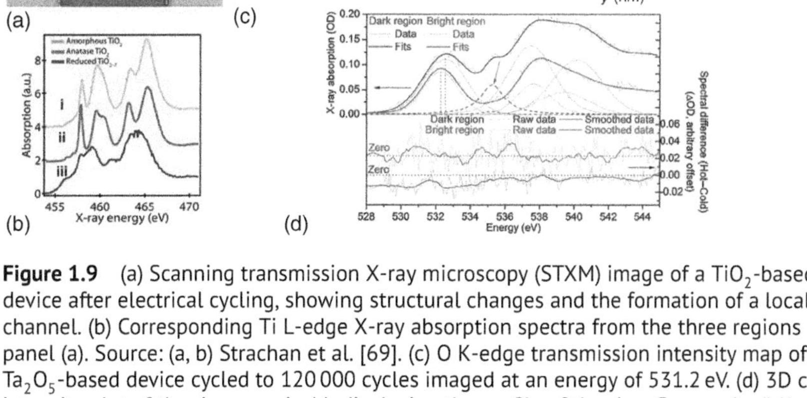

Figure 1.9 (a) Scanning transmission X-ray microscopy (STXM) image of a TiO_2-based device after electrical cycling, showing structural changes and the formation of a localized channel. (b) Corresponding Ti L-edge X-ray absorption spectra from the three regions in panel (a). Source: (a, b) Strachan et al. [69]. (c) O K-edge transmission intensity map of a Ta_2O_5-based device cycled to 120 000 cycles imaged at an energy of 531.2 eV. (d) 3D color intensity plot of the ring seen in (c), displaying the profile of the ring. Source: (c, d) Kumar et al. [70].

cycles, a ring-like feature with a bright inner core and a dark perimeter was observed as shown in Figure 1.9c. In addition to field driven drift of $V_O^{\cdot\cdot}$ in the vertical direction, temperature gradients due to Joule heating near the localized conduction channels produce thermophoretic forces that cause lateral migration of $V_O^{\cdot\cdot}$. As a result, lateral segregation of oxygen-deficient and oxygen-rich regions occurs, corresponding to the bright and dark regions in Figure 1.9c.

X-ray PEEM analysis In operando X-ray PEEM (XPEEM) analysis is a desirable technique to nondestructively reveal the changes in film microstructure, composition, and chemical states during the resistive switching process with high spatial resolution and interface or surface sensitivity. Recently, quantitative redox reactions were demonstrated via in operando XPEEM analysis in a $SrTiO_3$-based device with a graphene top electrode that offers excellent electrical conductivity and photoelectron transparency to allow electrical programming and spectromicroscopy measurements [71]. Noticeable changes were observed in the normalized intensity of O K-edge when the device alternated between the HRS and LRS, indicating the formation and dissolution of the $V_O^{\cdot\cdot}$ filament (Figure 1.10). Furthermore, quantitative information on the charge-carrier density differences between different resistance states is available since XPEEM is a highly surface-sensitive technique.

1.2.2 Interface-Type Resistive Switching

In contrast to the filamentary-type resistive switching, interface-type resistive switching generally originates from the variation of the Schottky barrier caused by migration of $V_O^{\cdot\cdot}$ [73]. In this case, the resistance in the LRS is inversely proportional to the electrode area, suggesting the entire electrode area is involved in the switching behavior. Furthermore, the interface-type switching normally does

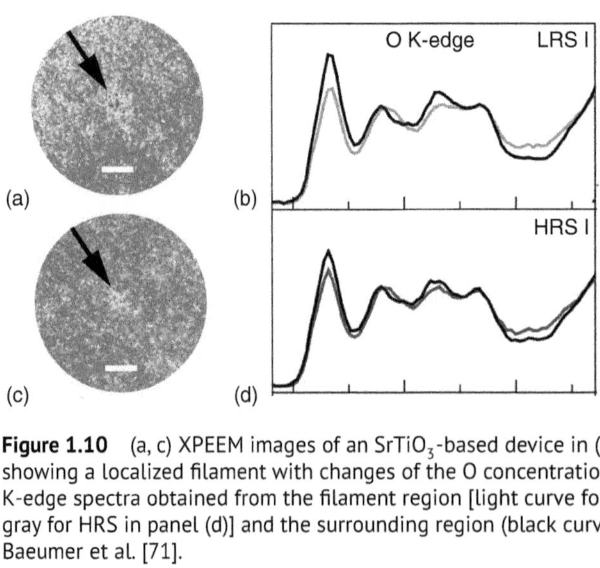

Figure 1.10 (a, c) XPEEM images of an $SrTiO_3$-based device in (a) LRS and (c) HRS, showing a localized filament with changes of the O concentration. Scale bars: 1 μm. (b, d) O K-edge spectra obtained from the filament region [light curve for LRS in panel (b) and dark gray for HRS in panel (d)] and the surrounding region (black curves). Source: Modified from Baeumer et al. [71].

not require a forming process and can easily obtain a multilevel storage because it changes the resistance value by a uniform interface effect. Overcoming the forming randomness of memristors, devices could achieve excellent reproducibility and cell-to-cell uniformity [74].

The interface-type resistive switching mechanism was first reported by Sawa in 2008 [4]. According to the article, the *I–V* curves of the Ti/PCMO, Au/Nb:STO, and SRO/Nb:STO interfaces show rectifying *I–V* behavior, thus revealing the presence of Schottky contacts. In addition to the rectification, *I–V* curves for these interfaces exhibit hysteretic behavior indicative of resistive switching. Furthermore, they found the migration of $V_O^{\cdot\cdot}$ at the interface plays a key role in the resistive switching. Up to now, different devices with interface-type switching have been explored, including Pt/WO_{3-x}/Pt [75], Pt/ZnO nanorod/FTO [76], Pt/Nb:STO/Al [77], ZrO_2:Y/PCMO [78], InGaZnO (IGZO)/a-IGZO[79], and so on.

The resistive switching process in the interface-type switching is schematically illustrated in Figure 1.11, using n-type semiconductor as an example [74, 80]. n-Type semiconductor forms a Schottky barrier when in contact with a metal with a large work function (such as Pt or SRO) and forms an Ohmic contact when associated with a low work function metal (such as Ti). In the conventional Schottky model, the capacitance is given by $C = \frac{\varepsilon_0 \varepsilon_S S}{W_d}$, where W_d is the width of the depletion layer, ε_0 is the relative dielectric constant of a vacuum, ε_s is the semiconductor relative dielectric constant, and S is the cell area. This equation indicates that LRS is achieved by tunneling through the Schottky barrier with a narrow W_d, while HRS is created by inhibited tunneling at the Schottky barrier with a wide W_d. With the application of a negative bias (Figure 1.11a), positively charged oxygen vacancies are attracted to the electrode narrowing the barrier, and the device switches to the LRS. A positive bias will result in the repulsion of the oxygen vacancies and in the restoration of the barrier width, and the device switches back to HRS (Figure 1.11b).

The current lack of full understanding of the underlying switching mechanisms remains a major challenge for interface-type memristive devices and a significant obstacle to their widespread applications. For this reason, in the past few years, a variety of materials characterization techniques have been employed to clarify the switching mechanisms in interface-type memristive devices, including the combination of TEM and electron energy-loss spectra (EELS) [81], spectromicroscopy studies [82], and so on.

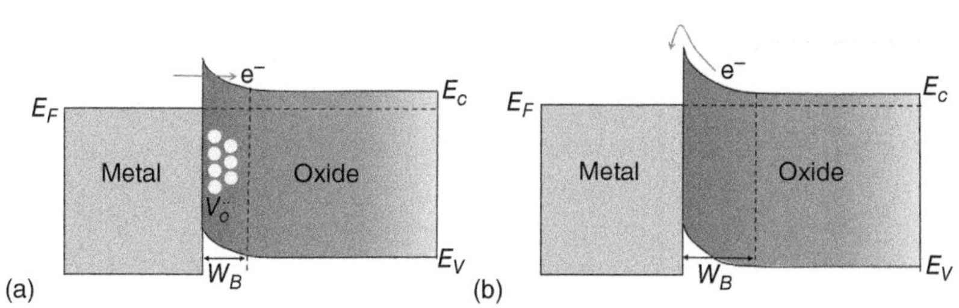

Figure 1.11 The interface-type resistive switching mechanism for a representative n-type semiconductor. (a) In the LRS, $V_O^{\cdot\cdot}$ accumulate at the interface, reducing the depletion width. (b) In the HRS, there are less $V_O^{\cdot\cdot}$ at the interface, increasing the depletion width.

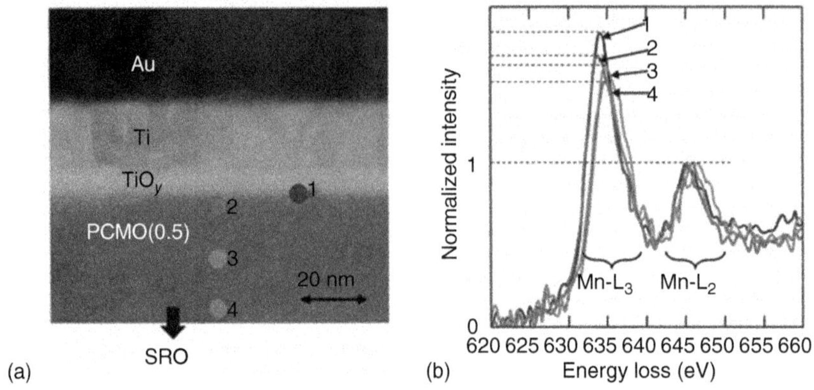

Figure 1.12 (a) Cross-sectional TEM image of Ti/PCMO junction. (b) Electron energy-loss spectra of Mn L-edge at several positions in Ti/PCMO junction indicated in (a). Source: (b) Asanuma et al. [81]. Copyright 2009, American Physical Society.

Combination of TEM and EELS As an example, the migration of $V_O^{\cdot\cdot}$ in the Ti/PCMO interface was demonstrated by the combination of cross-sectional TEM and the EELS spectra, as shown in Figure 1.12. An amorphous TiO_y layer between the Ti electrode and the PCMO layer was clearly observed in the cross-sectional TEM (Figure 1.12a). Furthermore, the EELS of the Mn L-edge obtained at different positions in the PCMO layer was carried out to confirm the diffusion of oxygen ions at the interface. As seen in Figure 1.12b, the peak intensity ratio of Mn-L_3 and Mn-L_2, I(L_3)/I(L_2), decreased with increasing distance from the boundary between the TiO_y and the PCMO layers, indicating the valence of the Mn ion in the vicinity of the interface was smaller than that far from the interface. The above results suggest that the number of oxygen vacancies in the vicinity of the interface was increased due to the migration of oxygen ions from the PCMO layer to the Ti electrode.

Spectromicroscopy Studies Figure 1.13 shows spectromicroscopy results using the absorption current and photoemission electrons of the Au/Al/PCMO/CMO/PCMO/Pt device in the LRS and HRS, respectively. There are three different regions in Figure 1.13a, b: Au/Al/PCMO, Au/PCMO, and PCMO. Au/PCMO and PCMO, respectively, are the most (light) and least (dark gray) conductive in both resistive states. A significant change in current occurred in the Au/Al/PCMO region, with high (light gray) and low (dark gray) current intensity in the LRS and HRS, respectively, indicating the Al top electrode plays a key role in the resistive switching. Furthermore, from Al 2p photoelectron peaks and spectromicroscopy spectra (Figure 1.13d–f), one can clearly observe the oxidized state of Al in the Au/Al/PCMO region changes a lot when the device switches from HRS to LRS. This indicates that interfacial oxidation of the Al electrode in Al/PCMO acts as a reservoir of oxygen during the resistive switching.

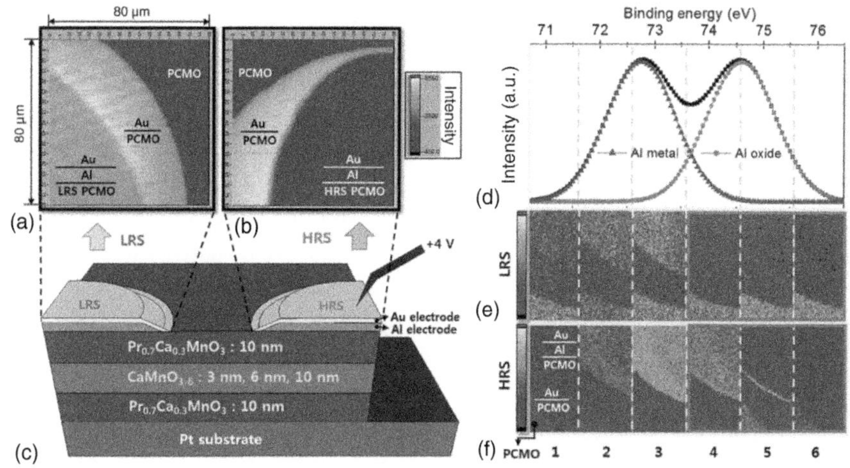

Figure 1.13 Spectromicroscopy results using the absorption current and photoemission electrons. Contrast images of absorbed electron intensity from the electrode areas of (a) LRS and (b) HRS. (c) The schematic diagram of the Au/Al/PCMO/CMO/PCMO/Pt (substrate) device. (d) Typical Al 2p photoelectron peaks of metallic (dark gray) and oxidized (light gray) states of aluminum as a function of binding energy. The *x*-axis energy intervals correspond to the window of integration for the corresponding lower panel of (e) and (f). (e) and (f) The panels show a color plot of the spatial distribution of the intensity of photoelectrons with energies within the respective energy interval [as indicated by the energy scale of (d)]. The six panels in (e) correspond to the images of an electrode in the LRS, and the panels in (f) to the respective images of an electrode in the HRS. The upper and lower panels 1–3 indicate the increasing (gray) intensity of the Al metal electrode. Panels 4–6 indicate the intensity of the Al oxide signal. Clearly, the signal is very weak in the upper panels corresponding to the LRS. In contrast, the lower panels clearly indicate oxidation of Al in the HRS. In particular, panel 5 of (f) shows a strong and thin signal coming from the rim of the Al electrode. Source: Lee et al. [82].

1.3 Memristive Bioinspired Devices

The human brain is far superior to the digital computer in cognitive and classification tasks, especially in solving probabilistic and unstructured problems requiring low power dissipation. This is because the architecture of the human brain is completely different from that of the digital computer [83, 84]. In a standard digital computer, the data memory and processor are physically separated by a channel, and constant data movement is required while working. In the human brain, huge and complex neural networks composed of gigantic amounts of neurons massively interconnected by an even larger number of synapses are responsible for computing and memory. Unlike a digital computer, information storage and processing happen at the same time in the brain [85].

Taking inspiration from the architecture and principle of the human brain, neuromorphic computing that aims to construct energy efficient and fault tolerant computing systems may have great potential in the future artificial intelligence (AI), big data, Internet of things (IoT), etc. [86]. In particular, devices that can reproduce the functionalities of biological synapses and neurons are building blocks in the construction of such neuromorphic systems. Fortunately, the emergence of memristive devices, with inherent dynamics resembling biological synapses and neurons, provides a feasible way to construct neuromorphic computing systems [87, 88]. For instance, one compact memristive device is enough to reproduce some functions of a biological synapse. Moreover, the vector–matrix multiplication, which is believed to be the most data-intensive work in neural networks, can be naturally performed in a cross-bar array of memristive devices on the physical level based on Ohm and Kirchhoff laws. These features endow memristive devices ideally suitable for implementing highly efficient neuromorphic computing systems.

1.3.1 Memristive Synapses

It has been found that the human brain is composed of $\sim 10^{11}$ neurons and $\sim 10^{15}$ synapses. All human perception, emotion, learning, forgetting, and memory are performed by this complex neural network with building blocks of neurons and synapses. A neuron works to generate and transmit action potentials in the neural networks. A synapse is a specialized junction where two neurons contact and communicate with each other, as shown in Figure 1.14a, and plays a crucial role in the process and storage of information. The presynaptic side of the synapse is an axon terminal of the presynaptic neuron, while the postsynaptic side is the dendrites. An action potential, or nerve impulse, arrives at the axon terminal and triggers the presynaptic neuron to release neurotransmitters; these neurotransmitters bind to the receptors on the postsynaptic membrane and causes a transient postsynaptic membrane depolarization and then generate the excitatory or inhibitory postsynaptic potential (EPSP or IPSP), depending on the transmitting efficiency of the synapse. This transmitting efficiency of a synapse is not fixed but can vary as a result of the activities of pre- and postsynaptic neurons, which is termed as synaptic plasticity. In other words, the synaptic weight, that is, contact strength, is flexible and can be modified according to the activities of the relative neurons [90].

In biological system, synaptic plasticity performs different time courses ranging from milliseconds to hours, days, and even the whole lifetime. Therefore, with regard to the time course of the plasticity, synaptic plasticity can be roughly divided into short-term plasticity (STP) and long-term plasticity (LTP), as shown in Figure 1.14b [91]. It is found that STP allows the synapses to perform critical functions, such as fast response and information filtering. Furthermore, the synaptic plasticity exhibits the activity-dependent characteristic; weak tetanic induces only STP, whereas strong trains of stimulus induce LTP. Different from STP, LTP is thought to be crucial during synaptic development and for regulating neural circuits in the adult brain.

Given that one nanoscale memristive device can replicate diverse synaptic functions, it shares a similar structure and working principle to the biological synapse. These enable the memristive device to be a promising candidate as synaptic element to build neuromorphic computing systems. To date, encouraging progress has been achieved in the implementation of both STP and LTP using

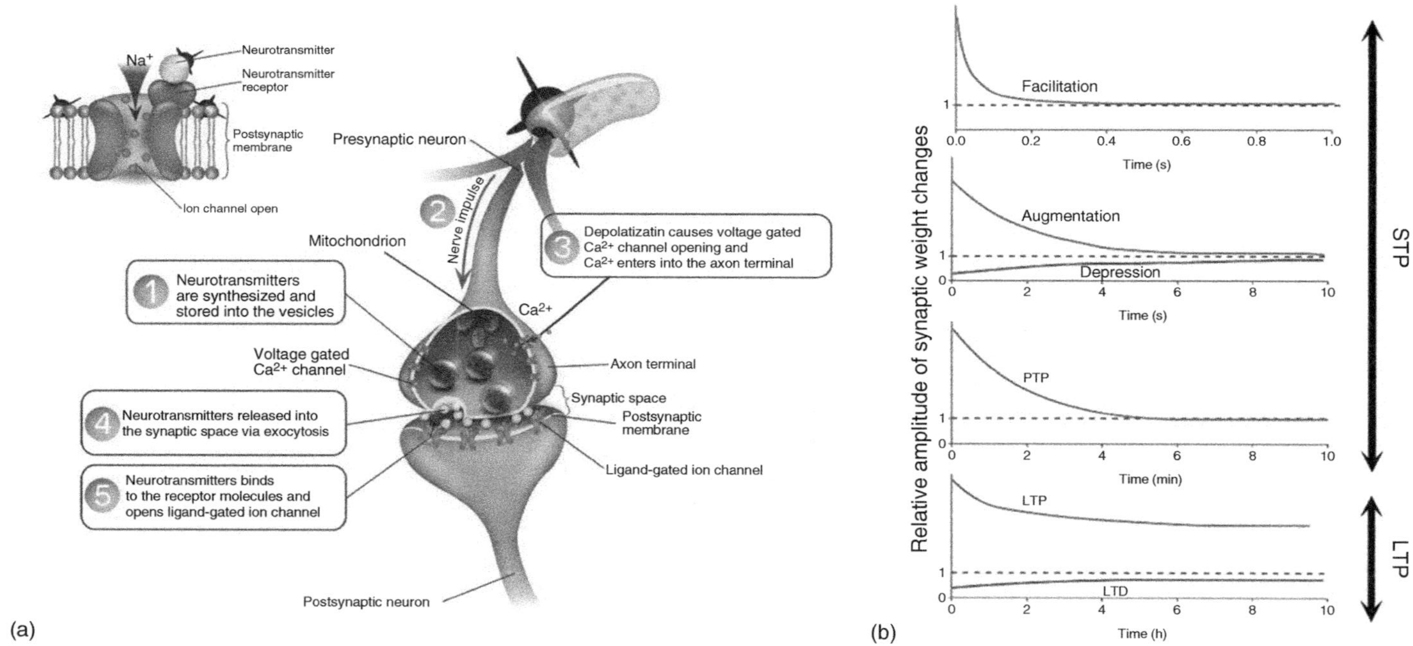

Figure 1.14 Biological synapses and its plasticity. (a) A biological synapse is formed at the axon terminal with the dendrites of other neurons. When a nerve impulse arrives in the presynaptic axon terminal, neurotransmitter molecules are released from synaptic vesicles into the synaptic cleft. Neurotransmitters are then bound to specific receptor proteins, causing the generation of electrical or chemical signals in the postsynaptic cell. Source: John et al. [89]. (b) Different forms of activity-induced synaptic plasticity in biological synapses: synaptic facilitation/depression, augmentation, post-tetanic potentiation (PTP), and long-term potentiation/depression. The first three are the short-term plasticity (STP), and the last one is the long-term plasticity (LTP). Source: Nicholls et al. [94].

memristive devices. Therefore, below we discuss short-term memristive synapses and long-term memristive synapses reported recently.

1.3.1.1 Short-Term Memristive Synapses

Numerous forms of synaptic plasticity lasting on the order of milliseconds to several minutes have been termed as STP, including synaptic facilitation/depression, augmentation, and post-tetanic potentiation (PTP), as shown in Figure 1.14b [91]. Synaptic facilitation (or depression) occurs when a brief train of stimuli is applied to a presynaptic nerve, in which the amplitude of EPSP progressively increases (decreases). In contrast, PTP happens when a relatively long high-frequency train of stimuli is applied to the presynaptic neuron and an intermediate phase between synaptic facilitation and PTP is an augmentation, which is introduced by stimulus trains with moderate duration. Although facilitation, PTP, and augmentation are all referred to as STP, they spin different temporal domains. Facilitation decays about a few hundred milliseconds, whereas the decay of augmentation and depression lasts over a period of several seconds. PTP can persist for tens of minutes. It is found that these STP play important roles in short-term adaptations to sensory inputs and short-term memory and are helpful in filtering out some unnecessary information.

In biological systems, paired-pulse facilitation (PPF) and paired-pulse depression (PPD) are the most important learning rules of STP and play a key role in decoding the temporal information for auditory or visual signals. In PPF, synaptic current evoked by the second spike is increased when the second spike closely follows a previous spike, as shown in Figure 1.15a [92, 94]. Furthermore, the magnitude of the enhancement is determined by the time interval between the two spikes, and a smaller interval gives rise to a larger synaptic current enhancement. In contrast, PPD describes the opposite phenomena, where the synaptic current evoked by the second spike is decreased when the second spike closely follows a previous spike [93], and the magnitude of the decrease is determined by the interval between the two spikes, which is similar to PPF (Figure 1.15b). It is found that PPF (PPD) is highly related to an increased (decreased) residue in presynaptic Ca^{2+} concentration left over from the first stimulus, which results in an increase (decrease) in the number of neurotransmitters released at the synapse when the second stimulus is applied [95].

As the most important STP, PPF has been implemented in many memristive devices with volatile resistive switching behavior [96–99]. Zhu et al. demonstrated PPF in the three-terminal in-plane lateral memristive devices [96], as shown in Figure 1.16. The indium–zinc–oxide (IZO) channel was self-assembled on phosphorus (P)-doped nanogranular SiO_2 proton conductive films at room temperature. The channel conductance can be modulated in a volatile way due to the proton-related electrical double-layer effect. In this device, the presynaptic spikes are applied on the in-plane gate, and the conductance of the channel is regarded as the postsynaptic response, that is, excitatory postsynaptic current (EPSC) or inhibited postsynaptic current (IPSC) (see Figure 1.16b). Figure 1.16c shows the EPSC evolution over time by applying two consecutive presynaptic spikes on the in-plane gate. It is clearly observed that the peak value (A2) of the EPSC triggered

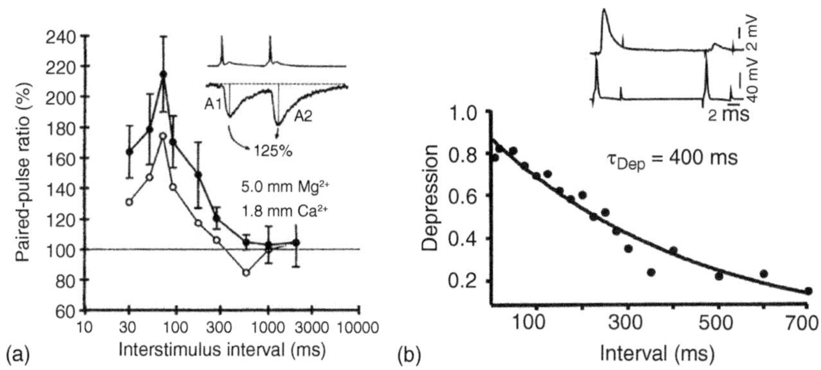

Figure 1.15 Short-term synaptic plasticity. (a) Paired-pulse facilitation (PPF). Source: Debanne et al. [92]. (b) Paired-pulse depression (PPD). Source: Waldeck et al. [93].

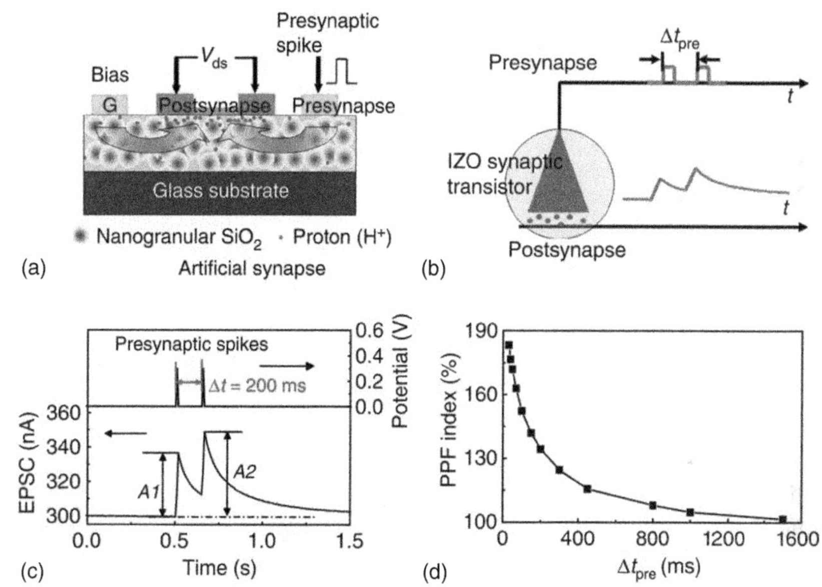

Figure 1.16 PPF implementation in the three terminal in-plane lateral memristive device. (a) Schematic diagram of the device structure. (b) Schematic image of the implementation approaches. (c) A pair of presynaptic spikes applied and the experimental response of the device. *A*1 and *A*2 represent the amplitudes of the first and second EPSCs, respectively. (d) PPF index, defined as the ratio of *A*2/*A*1, plotted as a function of interspike interval Δt_{pre}. Source: Zhu et al. [96].

by the second presynaptic spike is larger than that by the first presynaptic spike (A1). The PPF index, defined as the ratio of A2/A1, gradually decreases with the increase of the interspike interval, as shown in Figure 1.16d. That is because the transient accumulation of protons after the first spike will gradually fade as the interspike interval increases. The dependences of the PPF index on the interval follow a double exponential function in biological synapses and some memristive emulators.

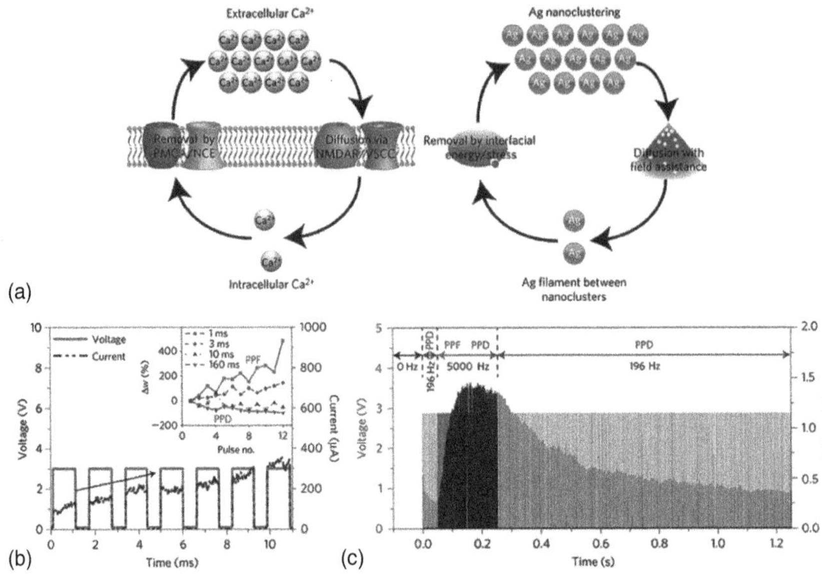

Figure 1.17 Biorealistic memristive synapses realized in diffusive memristive devices. (a) Schematic illustration of the analogy between Ca^{2+} and Ag dynamics in the diffusive SiO_xN_y:Ag memristive device. (b) Experimental demonstration of short-term synaptic PPF and PPD behavior by applying multiple subsequent voltage pulses (3 V, 1 ms) with different time intervals between 1 and 160 ms. (c) Experimental demonstration of PPD following PPF in the diffusive SiO_xN_y:Ag memristive device. The inflection effect from PPF to PPD under stimuli with the same frequency (5000 Hz) is a result of the gradual depletion of silver at one electrode and accumulation at the other. Source: Wang et al. [52].

Recently, Wang et al. first realized both PPF and PPD in two-terminal diffusive memristive devices with threshold switching behavior based on Ag-doped SiO_xN_y [52], as displayed in Figure 1.17. In their device, threshold switching is related to the formation of Ag filament through Ag cluster migration driven by electric field, and then the filament spontaneously ruptures driven by minimization of the interfacial energy between Ag and the dielectric after removing the external electric field. It is proposed that this kind of diffusive memristive devices performs significant similarities to the biological synapse in terms of ion diffusive dynamics, dynamic balance, and regulating roles, as schematically shown in Figure 1.17a. The experimental demonstration of PPF and PPD is shown in Figure 1.17b,c. High-frequency pulses (short t_{zero}) trigger facilitation (PPF), while low-frequency pulses trigger depression (PPD). More interestingly, the inflection from facilitation to depression in biological synapses, an effect solely induced by an increased number of stimulation pulses at the same frequency, that is, PPD following PPF induced by a series stimuli with same frequency, can also been biologically mimicked (see the current evolution under stimuli with 5000 Hz shown in Figure 1.17c). This inflection effect from PPF to PPD under stimuli with the same frequency is attributed to the gradual depletion of silver at one electrode and accumulation at the other. Similar PPD following PPF phenomena have been recently demonstrated in a biomolecular memristive device by Najem and his coworkers [100].

As mentioned before, there are several forms of STP with different time courses, including facilitation (tens of milliseconds), synaptic augmentation (seconds), and PTP (tens of seconds to minutes). It means that STP in biological synapses spans broad timescales from milliseconds to minutes. However, the most reported STP behaviors in memristive devices only exhibit certain specific timescales limited by their intrinsic micro-physicochemical processes during resistive switching. Recently, Zhang et al. emulated dynamic synaptic plasticity over broad timescales from milliseconds to days in the Ag/MgO/Pt ECM device [101]. Their Ag/MgO/Pt device exhibits volatile switching behavior, and its data retention of LRS can be tuned from millisecond to several 10 seconds through controlling the amplitude of applied external pulses. Therefore, many forms of STP have been mimicked including facilitation, augmentation, and PTP. It is proposed that the switching dynamics is attributed to the formation and spontaneous rupture of silver nanofilaments. By adjusting the thickness of filaments via applying the stimulus with different amplitudes, different timescales of STP can be realized. In addition to tuning external programming stimulus, mediating STP can also be achieved by controlling the orientation of silica mesopores in the device of Ag/Ag_2S/silica mesopore/ITO. [98]. The dielectric silica layer with vertical mesopores can facilitate the rupture of a conductive pathway, which results in short relaxation time for LRS (0.5 seconds) compared with parallel mesopores (3 seconds) and dense amorphous silica (>10 seconds). Thus, the time constants of PPF emulated in these devices present different timescales.

1.3.1.2 Long-Term Memristive Synapses

Unlike STP, LTP including long-term potentiation (LTPot) and long-term depression (LTD) refers to synaptic plasticity that lasts for hours or even days. LTPot was first observed in the hippocampus, where high-frequency repetitive activation of excitatory synapses caused a potentiation of synaptic connection that could last for hours or even days. Note that LTPot lasts longer than STP, but it still decays over time, as shown in Figure 1.18a. Moreover, the decay rate of LTPot depends on the duration and intensity of the stimulus [102, 104]. Figure 1.18a shows the LTPot response of the CA1 pyramidal neurons to three different stimulus conditions. The LTPot triggered by a single 100 Hz stimulus train decays back to zero over the next two hours; the potentiation decays less rapidly after four stimulus trains and becomes much slower after eight stimuli trains – with 80% of the potentiation persisting after two hours. Similar to LTPot, LTD also performs different time course of decay depending on the stimulus conditions. Both LTPot and LTD involve an increase in postsynaptic calcium concentration. A relatively large increase results in LTPot; a smaller increase leads to LTD [105]. Recently, LTP has garnered many interests because it provides an important key to understanding some of the cellular and molecular mechanisms of memory formation.

For LTP, the synaptic plasticity is dependent on the activities of pre- and postsynaptic neurons according to STDP [103] or Bienenstock–Cooper–Munro (BCM) learning rule [90, 106, 107]. The classical pair-STDP rule is a temporally specific extension of the Hebbian associative plasticity, which describes the synaptic weight depending on

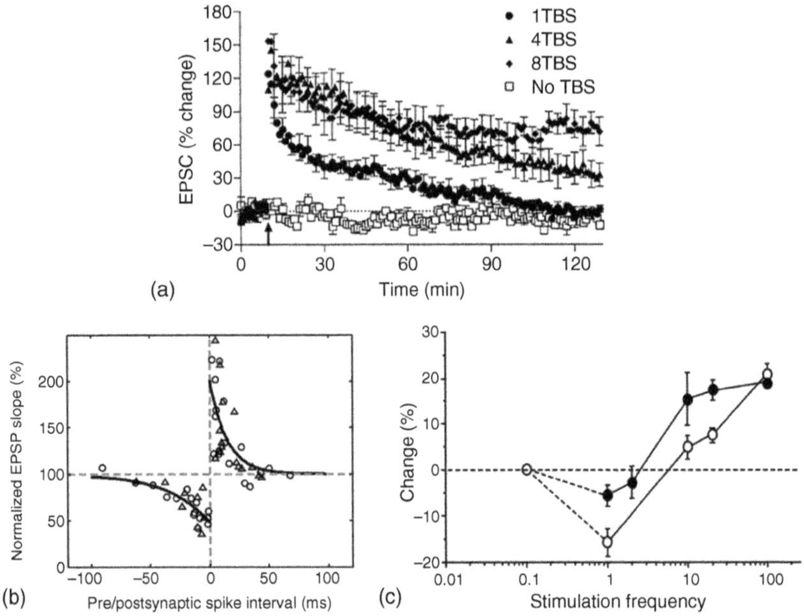

Figure 1.18 LTP in biology. (a) Long-term potentiation (LTPot) response of the CA1 pyramidal neurons to three different stimulus conditions. Source: Raymond and Redman [102]. (b) Spiking time-dependent plasticity (STDP). Source: Froemke and Dan [103]. (c) Bienenstock–Cooper–Munro (BCM) learning rule. Source: Kirkwood et al. [90].

the interval between pre- and postsynaptic spikes, as shown in Figure 1.18b. Specifically, if a presynaptic spike precedes a postsynaptic spike with a short interval, the synaptic weight enhances, whereas the synaptic weight decreases if the postsynaptic spike precedes a presynaptic spike. Moreover, there is no change of synaptic weight if the time interval between pre- and post-spike is too large. Actually, the abovementioned pair-STDP is Hebbian STDP. There are other three types of STDP, including the anti-Hebbian STDP, the symmetrical STDP, and the visual STDP learning rule observed in different types of synapses [108, 109].

In addition to STDP, BCM learning rule is another fundamental activity-dependent learning rules. According to BCM learning rule[107], the synaptic weight modulation performs frequency-dependent and sliding threshold characteristics, as schematically shown in Figure 1.18c [90]. In particular, high-frequency stimulation leads to potentiation, and low-frequency stimulation leads to depression. Furthermore, there exists a threshold frequency at which the synaptic weight can be maintained. More importantly, this threshold frequency is not fixed, and it will shift according to the priming neuron activity. For example, after a period of excited synaptic activity, the threshold will slide to a higher frequency, which can effectively prevent serious excitotoxicity stimulation. Similarly, the threshold will slide to lower frequency after a period of decreased activity, which is helpful for improving the synaptic modulation efficiency.

In this part, we will focus on the implementation of STDP and BCM learning rules in the memristive devices. It is well known that STDP is one of the most important LTP and the prevalent weight updating rules used in the spiking neuron networks [110]. There are two approaches to realize STDP. One is phenomenological implementation through overlapping the manually and carefully designed programming pulses to encode the spike timing information in the amplitude or duration of the effective applied pulses on the devices. The other one is biorealistic implementation through non-overlapping programming pulses using second-order memristive devices or combining a volatile memristive device with a nonvolatile memristive device. In addition, BCM learning rule describes the frequency-dependent synaptic plasticity with sliding threshold characteristics. However, previously demonstrated BCM rules in memristive devices mostly perform frequency-dependent characteristics without a sliding threshold feature, which is actually spike-rate-dependent plasticity (SRDP). Similar to STDP, biorealistic implementation of BCM rules requires second-order memristive devices with internal short-term dynamics.

Recently, phenomenological implementation of STDP learning rules has been intensively demonstrated using memristive devices in different material systems [79, 111–115]. However, this kind of STDP implementation is not biorealistic since the presynaptic and postsynaptic spikes pose the same profile and STDP is achieved via simple overlapping spikes. In fact, synaptic plasticity in biology is in turn modulated by other secondary state variables such as the postsynaptic calcium ion (Ca^+) concentration, rather than directly by the spikes. It is found that the postsynaptic Ca^+ concentration rises right after the excitation stimulus and then spontaneously decays over time after stimuli, which provides an internal timing mechanism to encode the relative timing information of the spikes [116]. Lu and his coworkers developed second-order memristive devices to biorealistically implement STDP learning rules through non-overlapping approach [30, 117]. In the so-called second-order memristive devices, there are two state variables. The one with long-term dynamics directly determines the device conductance, while the second one with short-term dynamics has an impact on the conductance by affecting the first state variable. This second state variable plays a role similar to postsynaptic Ca^+ concentration in regulating synaptic plasticity; therefore, biorealistic implementation of STDP without the need of pre/postsynaptic pulses overlapping can be achieved in second-order memristive devices. The local temperature [117] or the mobility of oxygen vacancies [30] can be treated as the second state variables, since they perform short-term dynamics in memristive devices.

Furthermore, STDP rule is implemented in the second-order memristive device of $Pd/Ta_2O_{5-x}/TaO_y/Pd$ [117]. In this device, the size of the conduction filaments is the first state variable with long-term dynamics, while the local temperature of the device is the second-state variable with short-term dynamics. There are two parts in the pre- and postsynaptic spikes: a programming element with high-voltage short duration and a heating element with low-voltage long duration. The device local temperature elevates, but conductance does not change upon applying the heating element alone. Then, the local temperature spontaneously decays after removing the pulses, but this elevated local temperature facilitates the filament growth during the programming element. When the presynaptic spike arrives earlier than the postsynaptic spike (i.e. in the case of $\Delta t > 0$), the postsynaptic spike will be affected by the temporal heating effect from the heating element in the presynaptic spike. The smaller Δt induces the much higher local temperature; as a result, much bigger size of filament is obtained in the device. The sign of conductance change is dominated by the final spike applied on the device, whereas the size of the conductance change relies on the pre/postsynaptic spike interval. In the opposite case ($\Delta t < 0$), the conductance decrease can be obtained by similar arguments. This local temperature plays a role similar to postsynaptic Ca^+ concentration in encoding the timing information of pre/postsynaptic spikes. It should be noted that the obtained conductance change in STDP rules stems from the modulation of the first state variable (i.e. the size of conduction filaments) and thus is long term, even though the timing information is encoded in the local temperature with short-term dynamics. The nonvolatile memristive devices without short-term dynamics can also be adopted in biorealistic STDP implementation by combining with a volatile element in series [52].

Biorealistic implementation of BCM learning rules with sliding threshold frequency also requires second-order memristive devices, since the sliding threshold frequency calls for variables with short-term dynamics. Lu and his coworkers demonstrated BCM rules in the Pd/WO_x/W second-order memristive devices [30]. The current of LRS state spontaneously decays after the removal of the external electric field, as shown in Figure 1.19a–c. The current decay process is also known as forgetting effect, retention loss, or memory loss characteristics. Taking advantage of this forgetting effect, one can realize biorealistical implementation of BCM learning rule with a sliding threshold frequency. The conductance of the device at any given instant depends not only on the stimulus but also on the state of the device, which is related to the forgetting process. As given in Figure 1.19b, the same stimulus (10 Hz) can trigger either current decrease (in step 2, following 200 Hz stimulus) or current increase (in step 4, following 1 Hz stimulus) depending on the previous activity. Thus, the modulation efficiency of a stimulus with a given frequency is highly related to the previous stimulus frequency, which is in line with the sliding threshold frequency in biological BCM rules. The realized BCM rules are given in Figure 1.19c. The detailed measurement approach is given as follows: First, one of three previous activities (ten pulses at either 10, 20, or 50 Hz) are applied on the device with the same state. Second, a stimulus with different frequency (the abscissa) is applied on the device, and the resultant net current change (the ordinate) is recorded. It is clearly shown that high-frequency stimulation leads to the potentiation and low-frequency stimulation leads to the depression. Furthermore, the threshold frequency slides as a function of the previous stimulus. This result is in good agreement with the biological BCM rules. This biorealistic BCM learning rule has also been demonstrated in cone-shaped ZnO memristive device with forgetting effect [119]. Very recently, Xiong et al. [118] realized the tunable forgetting rate by engineering the electrode/oxide interface by controlling the electrode composition in STO memristive devices, as shown in Figure 1.19d,e. The BCM rules with tunable sliding frequency thresholds are biorealistically demonstrated (see Figure 1.19f).

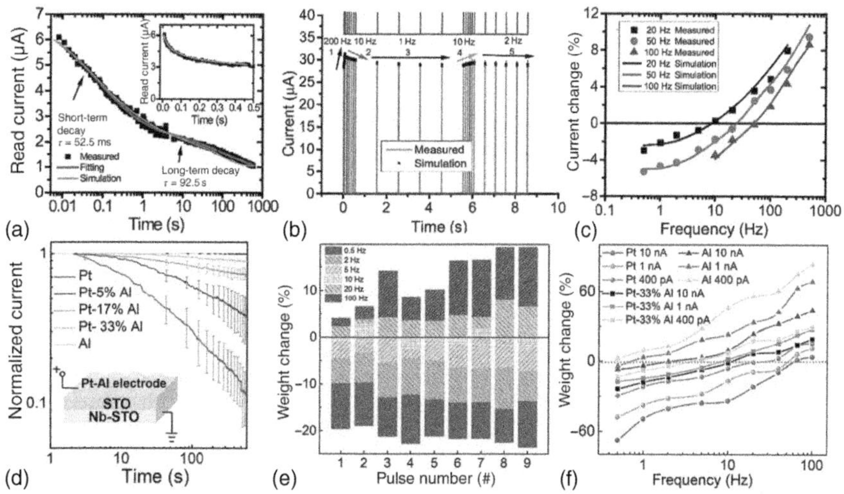

Figure 1.19 Biorealistic BCM learning rules obtained in second-order memristive devices. (a) Memristive device forgetting effect observed in the device of Pd/WO_x/W. (b) The memristive device response to consecutive programming pulse trains (1 V, 1 ms) at different frequencies. The same stimulus (10 Hz) can trigger either current decrease (in step 2, following 200 Hz stimulus) or current increase (in step 4, following 1 Hz stimulus) depending on the previous activity. (c) The experimental BCM rules obtained by controlling the priming stimuli frequency. Source: (a–c) Du et al. [30]. (d) Tunable forgetting effect observed in $SrTiO_3$ (STO) memristive device with different electrodes (Pt, Pt-5% Al, Pt-17% Al, Pt-33% Al, Al). (e) The current change of the Pt-33% Al/STO/Nb-STO device by applying nine pulses (1 V, 10 μs) with different frequencies. (f) Frequency threshold of the BCM rules can be tunable through modulating the Al content in the electrodes (0%, 33%, 100%) or controlling the device states (400 pA, 1 nA, 10 nA). The frequency threshold shifts to low frequencies when the Al content in the electrodes is higher and the conductance level is lower. Source: (d–f) Xiong et al. [118].

1.3.2 Memristive Neurons

Besides synapses, another fundamental element in biological neural networks is neuron. A biological neuron consists of several parts: the soma, the dendrites, the axon, and the neuronal membrane [120], as schematically shown in Figure 1.20a. The information encoding function of neurons is mainly realized by the change in their membrane potential. The resting equilibrium membrane potential is determined by the concentration gradients of ions across the membrane. The ion channels on the membrane are much more permeable to K^+ than to Na^+, so the equilibrium potential of K^+ is about −70 mV when the neuron is at rest. The action potential is generated by the flow of ions through voltage-gated channels, as shown in Figure 1.20b,c. That is, the depolarization of the membrane caused by external stimulations will change the configuration of the Na^+ channels, resulting in the higher permeability to Na^+. As a result, there will be Na^+ influx through voltage-gated Na^+ channels, which forms the rising edge of the action potential and reverses the polarity of the plasma membrane. The voltage-gated Na^+ channels will then be closed, and the K^+ channels are activated, hence leading to an outward flux

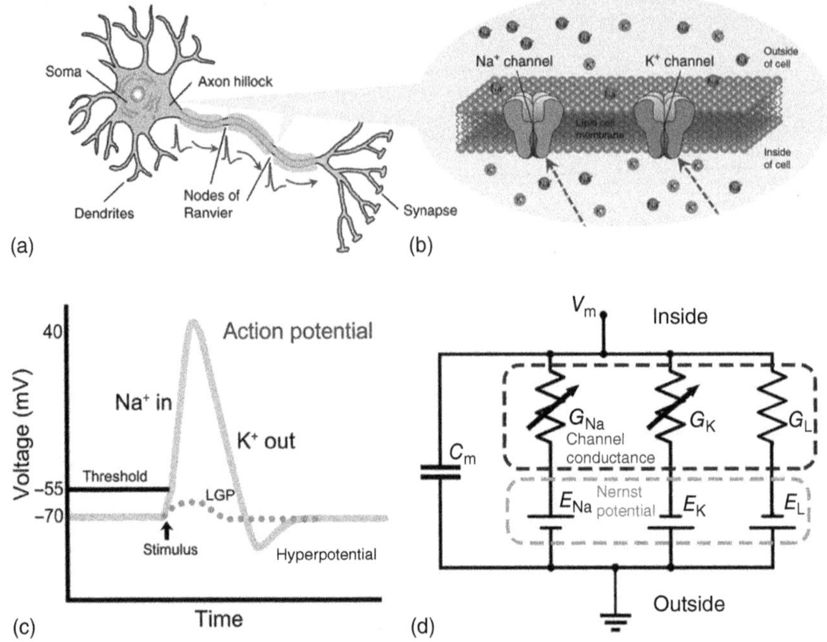

Figure 1.20 Biological neurons. (a) Schematic of a biological neuron, including soma, dendrites, axon, and nucleus. (b) Schematic of neuronal membrane. Source: (a, b) Yi et al. [120]. (c) Action potential in a biological neuron. (d) Hodgkin–Huxley neuron model. Source: (c, d) Yang et al. [47].

of K^+ ions. This forms the dropping edge of the action potential and returns the membrane potential to the resting state.

The above neuronal dynamics can be described by the integrate-and-fire (IF) model, leaky-integrate-and-fire (LIF) model, and Hodgkin–Huxley (HH) model [121]. IF and LIF models are mainly based on the insight that pulsatile spike is a threshold process. In the IF neuron model, the spikes of the presynaptic neuron, scaled by the strength of the corresponding synapses connecting with the dendrites, are applied to the neuron and then cause EPSCs or IPSCs due to the membrane depolarization. These EPSCs will be temporally and spatially integrated and induce the neuron membrane boost that is known as the local graded potential (LGP). An action potential is triggered when LGP reaches the threshold. If the LGP is below the threshold, IF neuron will retain that voltage boost continuously until it fires again. This characteristic is not in line with observed neuronal behavior. In fact, LGP shows short-term dynamics and results in a sub-threshold membrane boost that soon leaks out in a short time in biological neurons. Thus, a leaky term is added to the membrane potential to form an LIF neuron model. In an LIF neuron, the sub-threshold membrane boost will leak out in a period of time, which endows the neuron to implement time-dependent memory. The IF and LIF models give a bioplausible description of neural behavior but provide only a weak link to the underlying biophysical causes of electrical activity, and these two models are rather limited in predicting the precise time course of voltage during and after a spike.

In contrast, the HH model is the detailed biophysical neuron model that describes the dynamics of ion channels on the neuron cell when it is spiking and widely accepted in computational neuroscience. A typical HH neuron model based on the parallel thought of a simple circuit is shown in Figure 1.20d. The HH model treats each component of a neuron as an electrical element. The neuron membrane is represented as a capacitance (C_m); the conductance of ion channels is presented by resistors with programmable resistance. The leak channels are presented by a linear resistor (G_L). Nernst potential generated by the difference in ion concentration is represented by a battery (E_{ions}). Compared with IF and LIF neurons, HH neuron is very complex because it is a nonlinear system and cannot be solved analytically, which limits its application in artificial neuron networks. However, the development of memristive devices opens a new avenue for hardware artificial neurons.

In biological systems, scalable neurons are indispensable building blocks, in addition to compact synapses, for constructing large-scale neuromorphic computing systems. Thus, the development of scalable artificial neurons is an active area of research in the past few years, and some stirring progress have been made with the emergence of memristive devices [122–124]. Note that the design of memristive neurons is much different from that of artificial synapses, since they play quite different roles in neural networks. Rich synaptic functions can be generally emulated in a single memristive device via its conductance changes representing synaptic weight modulation. In contrast, one memristive alone normally is not sufficient as a substitute of a neuron; a hybrid circuit consisting of CMOS transistors and memristive devices is mostly needed to emulate a neuron [125]. Here, the goal is to mention that the introduction of memristive devices significantly simplifies the neuron circuits compared with those based on CMOS transistors only.

As mentioned above, there are many neuron models including IF neuron, LIF neuron, HH neurons, and some extended and simplified versions of these models [121, 126, 127]. Actually, these neuron modes can be divided into two classes: bioplausible neurons consisting of IF and LIF neurons and biophysical neurons mainly referring to HH neurons [121]. IF and LIF neurons are focused on the threshold process generating a spike, while HH neurons are based on the biophysical dynamic details of ion channels in the neuron membrane. Bioplausible neurons are simple but prevalent in neuromorphic computing, while biophysical neurons describe some salient features of biological neurons that bioplausible ones fail to capture, such as hyperpolarization. It is found the hyperpolarization can prevent the spiking rate from going extremely high value, which is critical in the neuromorphic computing systems coding based on spike frequency and timing [128]. Some representative reports about bioplausible and biophysical neurons based on memristive devices are given in Table 1.1.

Table 1.1 Memristive neurons.

Memristive neuron	Auxiliary circuit	Device	Device size	Operation (voltage/ current)	Switching type	Mechanism	References
Bioplausible neuron							
IF	Comparators	$Ge_2Sb_2Te_5$	Φ 20 nm	6 V/700 μA	Nonvolatile	PCM	[124]
LIF	Capacitor	Pt/SiO_xN_y:Ag/Pt	—	1.4 V/40 μA	Threshold	ECM	[129]
	Capacitor	$Ag/SiO_2/Pt$	$5 \times 5\ \mu m^2$	2 V/100 μA	Threshold	ECM	[130]
	MOSFET	Pt/SiO_x:Ag/Pt/ $TaO_xTa_2O_5/Pt$	$10 \times 20\ \mu m^2$	2 V/300 μA	Pseudo-memcapacitor	ECM and VCM	[131]
	—	$GaTa_4Se_8$	$40\ \mu m^2$	40 V/20 mA	Threshold	MIT	[132]
	—	$Pt/FeO_x/Ag$	$2 \times 2\ \mu m^2$	4 V/10 mA	Threshold	ECM	[133]
Biophysical neuron							
HH	Voltage source and capacitor	$Pt/NbO_2/Pt$	$110 \times 110\ nm^2$	1.6 V/300 μA	Threshold	MIT	[123]
	—	$Pt/VO_2/Pt$	$50 \times 50\ nm^2$	1.2 V/3 mA	Threshold	MIT	[134]
	Comparator and resistor	W/WO_3/PEDOT: PSS/Pt	$50 \times 50\ \mu m^2$	2.7 V/1 μA	Bipolar/volatile	Proton inserting/ extracting	[10]

IF neuron, the integrate-and-fire neuron; LIF, leaky-integrate-and-fire neuron; HH, Hodgkin–Huxley neuron; PCM, phase change memory cell; VCM, valence change mechanism; ECM, electrochemical metallization cell; MIT, metal–insulator transition.

1.3.2.1 Bioplausible Memristive Neurons

In 2016, Tuma et al. [124] successfully developed an IF neuron in a hybrid circuit consisting of a phase-change ($Ge_2Sb_2Te_5$, GST for short) device and an auxiliary circuit including comparators. The phase-change device is used to represent the membrane potential and realize the integration function with innate stochastic dynamics that arise directly from the physics of melt-quench-induced reconfiguration of the amorphous phases in GST. The fire function is realized by triggering a spike generator when LGP reaches the threshold values set by the comparator. Owing to the nonvolatile nature of resistive switching in GST, the leaky function is missed in this stochastic neuron. In 2018, Yang et al. [25] designed an LIF neuron in which an ideally volatile memristive device (Memristor 2) was adopted to represent membrane potential with short-term dynamics and thus realize leaky function. The circuit of this LIF neuron is shown in Figure 1.21. Memristor 1 presents nonvolatile resistive switching behavior with or without forgetting effect, which is used to represent the synapse receiving input spikes. These input signals scaled by the relative synaptic strengths then temporally and spatially integrate as V_{add} via the adder OP_1, which is the integration function of the neuron. Then, V_{add} is applied on Memristor 2 and evokes the resistance decrease of Memristor 2, which results in a boost of the divided voltage of RC (V_{out}). V_{out} plays a role like LGP does in biological neurons. A spike is

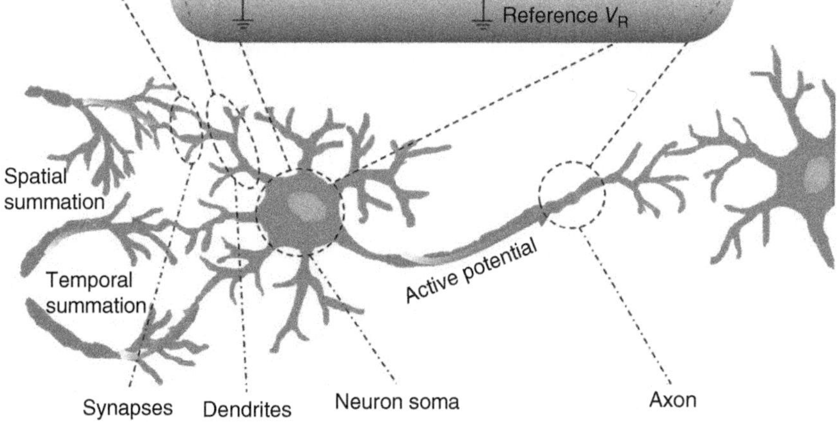

Figure 1.21 The LIF neuron implementation in the threshold-less memristive devices combined with comparators. In this LIF circuit, Memristor 1 represents the synapses to receive inputs from upstream neurons. Memristor 2 is a volatile second-order memristive device, which is used to mimic the dynamics of the local graded potential. Source: Yang et al. [25].

generated if V_{out} is larger than the reference value of comparator OP_2 (V_R), realizing the fire function. If V_{out} is lower than V_R, V_{out} will gradually leak out, owing to the volatile feature of Memristor 2, which is the leaky function of this neuron. Note that the resistive switching process in Memristor 2 is analog and without obvious threshold value. Thus, a comparator is required to realize the threshold spiking process. Furthermore, the spike generation is implemented with the help of a pulse generator (P), which makes it difficult for such neurons to achieve high-density integration.

In contrast, if the memristive device performs intrinsic threshold switching process, the LIF function of neurons can be realized using a much simpler auxiliary circuit without comparator and external pulse generator [135]. In this case, the threshold switching memristive device handles threshold spiking process. A capacitor is typically connected to the threshold switching memristive device in parallel to implement integration effect via the charging process. Wang et al. [129] recently developed an LIF neuron with a simple circuit composed of a diffusive memristive device and a capacitor in parallel, as shown in Figure 1.22a. In this LIF neuron, the capacitor acts as the membrane and realizes the leaky and integration effects of neurons, while the diffusive memristive device functions as the ion channel and implements the fire function. The output signal in this neuron is the current flow across the diffusive memristive device. To obtain a voltage spike, a resistor can be connected in series with a threshold switching device, as shown in Figure 1.22, and the output signal is represented by the divided voltage of this series resistor. When input pulses are applied on this LIF neuron, the capacitor charges with a time constant of R_sC and lifts up the voltage across the diffusive memristive device (CL loop in Figure 1.22b). The memristive device turns on by forming conducting filaments when the voltage reaches its threshold, and soon, the capacitor performs discharging through the memristive device (DL loop in Figure 1.22b) and induces the plunge of the voltage across the memristive device, resulting in it recovering to high resistance. Accompanied by this capacitor charging and discharging process, a voltage spike is generated. Zhang and his coworkers [130] developed this type of LIF neuron using an ECM cell of $Ag/SiO_2/Au$, and it was found that the spike frequency could be tuned by controlling the strength of input signals. In comparison with the abovementioned artificial neuron composed of comparators, this hybrid LIF

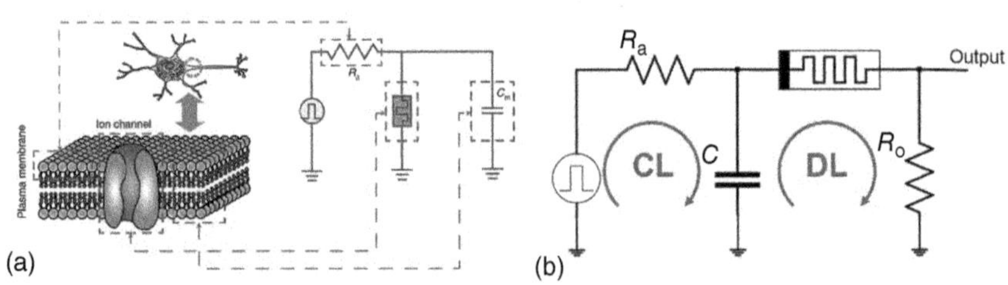

Figure 1.22 The implementation of an LIF neuron in the threshold memristive devices with parallel capacitors. (a) Schematic illustration of an artificial LIF neuron composed of a threshold memristive device and a parallel capacitor, in which the memristive device functions as the ion channel and the capacitor acts as the membrane. Source: Wang et al. [129]. (b) Schematic illustration of the LIF neuron circuit with voltage output.

neuron circuit is much simpler, since it typically comprises a threshold switching device and a capacitor in parallel. Thus, the reported artificial LIF neurons mostly belong to this type of LIF neurons so far. It is noteworthy that some LIF neurons with tunable integration time have been realized using memcapacitors [136, 137] or pseudo-memcapacitors (a diffusive memristive device with a series capacitor) [131] to replace normal capacitors.

Note that two-terminal memristive devices share similar structure with normal capacitors, that is, MIM sandwich structure. Thus, it is not surprising to find that some memristive devices work as capacitors at a given time, in particular at HRS state, and a capacitor is included in the equivalent circuits of some memristive devices. Therefore, it is not necessary to add a capacitor to implement the integration function of LIF neurons, at least for some memristive devices, meaning that an LIF neuron can be realized with a single component. In 2017, Pablo and his coworkers [132] reported a LIF neuron based on a single MIT memristive device, as shown in Figure 1.23. The sample shown in this figure is an MIT device based on $GaTa_4Se_8$

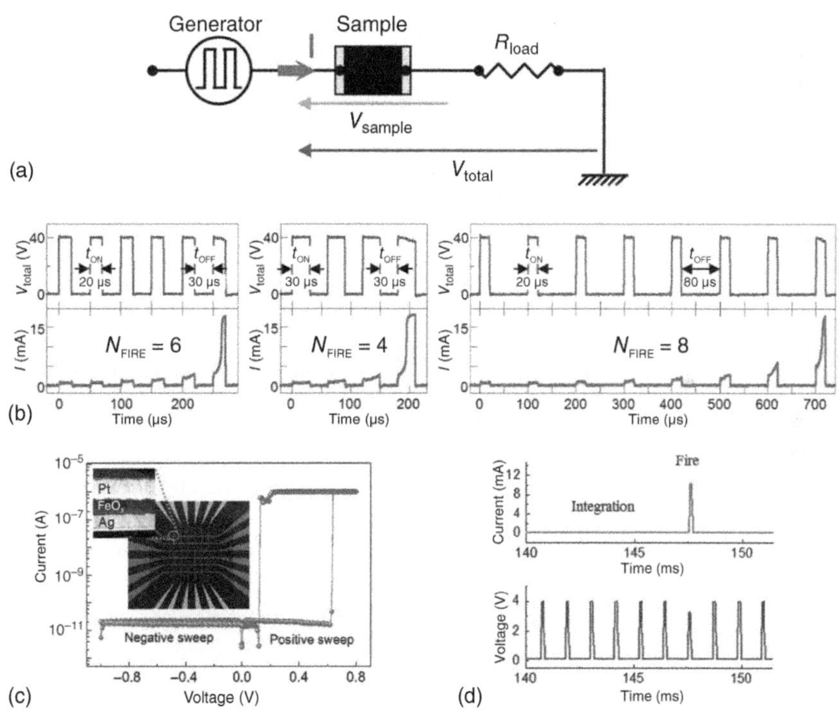

Figure 1.23 LIF neuron based on a single component of threshold memristive device. (a) Experimental setup in which the current flowing through the memristive device represents the output spike. (b) Experimental integrate-and-fire functionality realized in the single component of $GaTa_4Se_8$ MIT cell. Source: (a, b) Stoliar et al. [132]. (c) *I–V* characteristics of the $Pt/FeO_x/Ag$ threshold memristive device. Inset: The cross-section view of the memristive device under high-resolution transmission electron microscopy (HRTEM) inspection. (d) The demonstration of integrate-and-fire behavior under continuous voltage pulse train based on the single $Pt/FeO_x/Ag$ threshold memristive device. Source: (c, d) Zhang et al. [133].

with threshold switching behaviors. The current flowing through the MIT cell is the output signal. The working process of this LIF neuron is similar to those based on hybrid circuits with capacitors. Furthermore, the spiking frequency can also be tuned by controlling the input pulse parameters (see Figure 1.23b). In 2018, Zhang et al. [133] realized a highly compact LIF neuron with low power consumption using a single threshold switching device of Pt/FeO_x/Ag in a similar way, as shown in Figure 1.23c,d. Detailed experimental results indicate that the formation and automatic retrieval of silver filaments are the physical mechanism of threshold switching behavior. From the above discussions, we can conclude that threshold switching with volatile nature is very useful for implementing LIF neurons. Thus, this observed in many oxides has been exploited to realize bioplausible neurons with a compact structure [138]. In addition to memristive devices, some magnetic devices [139–141] were also exploited to realize some bioplausible functions of neurons, which is beyond the scope of this article.

1.3.2.2 Biophysical Memristive Neurons

Unlike bioplausible neurons that can be realized in a single memristive device, the biophysical neuron, that is, the HH neuron, involving complex dynamics of Na^+ and K^+ ion channels, requires much more complex circuits, and the progress on realization of artificial HH neurons is quite limited. In 2012, Pickett and his coworkers [123] developed an HH axon using two elements, each comprising an NbO_2 cell and a capacitor in parallel to function as sodium and potassium ion channels in the neuron membrane. These two elements are coupled with each other through a load resistor (R_{L2}), as shown in Figure 1.24a. The adopted NbO_2 memristive device performs threshold switching (see Figure 1.24b). The spike generated by this axon well resembles the biological spikes with hyperpolarization, as displayed in Figure 1.24c. The extended version with the stochastic effect of this HH axon is simulated by Saeed in 2017, [142]. Moreover, in 2018, Yi et al. [134] experimentally demonstrated the stochastic HH axon built with VO_2 MIT memristive devices in a similar circuit and realized most of the known neuron dynamics. Note that this circuit is typically termed as HH axon rather than HH neuron due to the fact that it mainly deals with the way of generating a spike highly resembling biological axons. Very recently, Huang and his coworkers [10] designed an artificial quasi-HH neuron with LIF functions based on a volatile memristive device with distinct nano-battery effect. The memristive device adopted here performs volatile analog switching behaviors, as shown in Figure 1.25. Unlike normal oxide memristive devices based on oxygen ion migration, the analog switching obtained in this device stems from proton migration with much higher mobility than oxygen ions. The protons come from the poly(3,4-ethylenedioxythiophene):polystyrene sulfonate (PEDOT:PSS) bottom electrode. The high ionic conductivity of protons fosters the enhancement of the battery effect in the device of W/WO_3/PEDOT:PSS. The hybrid neuron circuit is shown in Figure 1.25c. There are two memristive devices, M_1 and M_2, in the dedicated electrical circuit, in which M_1 acts as neuron membrane and realizes the LGP dynamics while M_2 functions as neuron axon and realizes firing an HH neuron-like spike. Since this memristive device has no threshold, the comparator is utilized to implement LIF functions in the first part of this circuit. If LGP reaches the reference

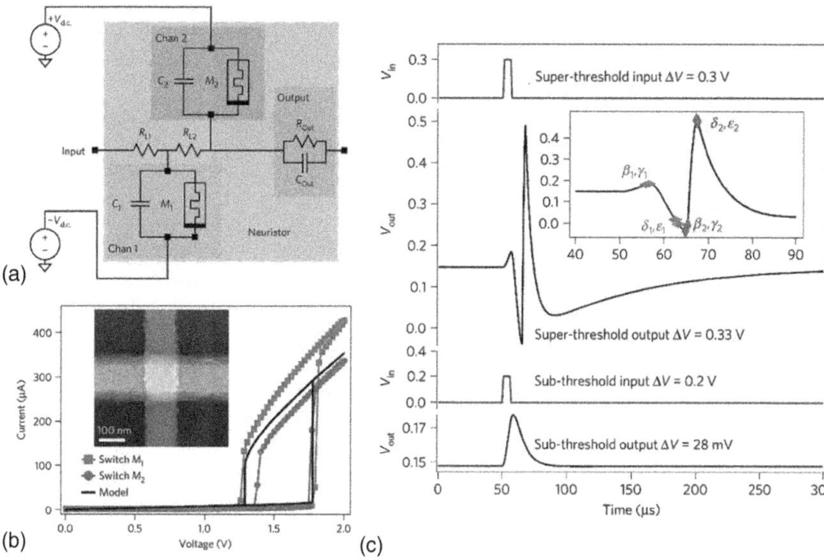

Figure 1.24 Memristive HH axon. (a) Circuit diagram of the HH axon. The channels consisting of Mott memristive devices (M_1 and M_2), each with a characteristic parallel capacitance (C_1 and C_2) and coupled through a load resistor of R_{L2}. (b) Threshold switching observed in the Mott NbO_2 memristive devices. (c) Simulated super-threshold 0.3 V input pulse and its corresponding spike output highly resembling the biological spikes. Source: Pickett et al. [123].

value of comparator, a falling edge is generated by the comparator and then triggers timer to generate a pulse with fixed width. This pulse carrying the timing information of input signals is applied on a simple circuit consisting of M_2 and a resistor R3 in series. The output spike is the divided voltage on R3, which highly resembles the neuron-like spike (see Figure 1.25d). In particular, the hyperpotential (part d) arises from the battery potential of M_2 memristive device after removing external electric potential. The output spike frequency is tunable through control of the input signals, as given in Figure 1.25e, and it is found that the interaction of protons and electrons in M_2 resembles the coordination of K^+ and Na^+ ion channels in the neuron membrane. To the best of our knowledge, this is the first physical demonstration of artificial neurons with functions described in both biophysical and bioplausible neuron models.

In these years, memristor has been intensively investigated for the construction of artificial synapses and neurons because of its simple structure and rich dynamics. In this chapter, we first briefly introduce the state-of-art memristive synapses including short-term memristive synapses and long-term memristive synapses. Subsequently, recently reported memristive neurons, such as bioplausible neurons and biophysical neurons, are discussed. Here, bioplausible neurons consist of IF neurons and LIF neurons, and biophysical neurons refer to HH neurons.

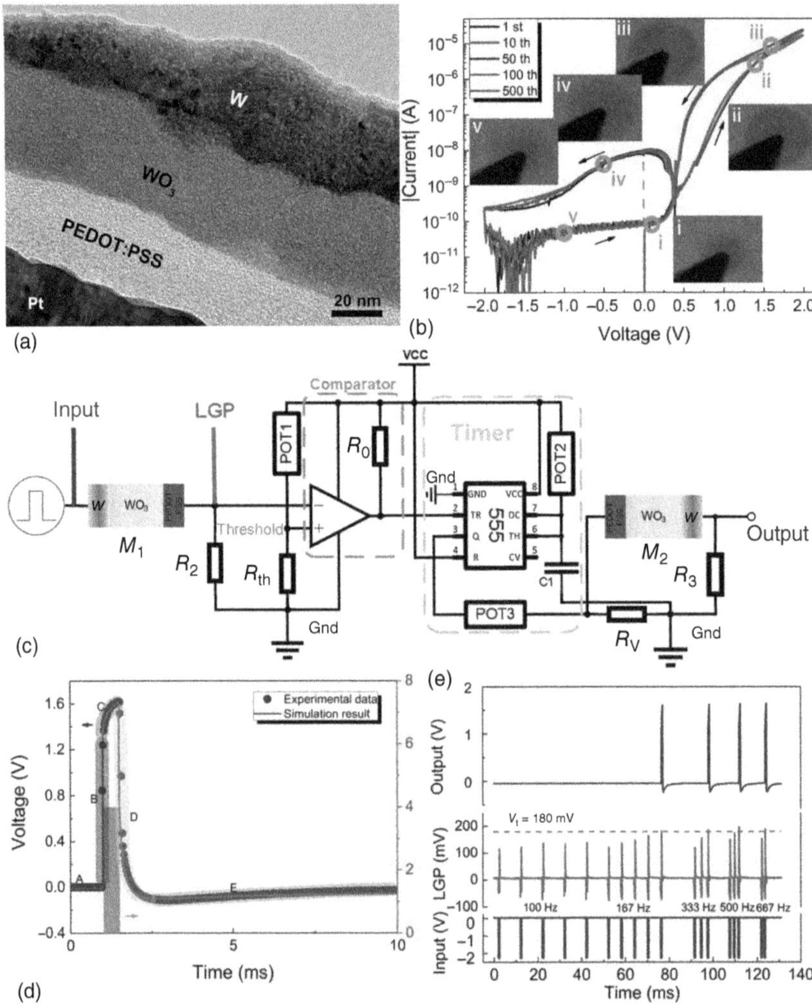

Figure 1.25 HH neuron with LIF functions. (a) The cross-sectional TEM image of the W/WO_3/PEDOT:PSS/Pt device. (b) $I-V$ curves of the analog switching. (c) Electrical circuit of the proposed HH neuron with two W/WO_3/PEDOT:PSS/Pt memristive devices. The first memristive device handles the tasks of integration and leaky, while the second one is used to realize the fire function. (d) Experimental and simulation results of the firing spike, which is in good agreement with biological spikes. (e) Temporal integration and bioinspired fire realized with the circuit. Source: (b–e) Huang et al. [10].

1.4 Memristive Neural Networks

1.4.1 Memristive ANN Computing

Two training approaches have been proposed to realize memristive ANN computing, offline training, and online training. In the case of offline training, the training process is done by software with the memristive device model. Then, the obtained synaptic weight is mapped into the memristive device crossbar. The salient features of quickly parallel operation in memristive device crossbar will accelerate the working process of networks. In contrast, in online training approach, the training process is directly done on the memristive device crossbar, which enables parallel operation in both training and working processes.

In 2017, Sheridan et al. [143] implemented sparse coding algorithms in a 32×32 crossbar array of WO_x analog memristive devices, as shown in Figure 1.26a. This memristive network was offline trained by software; the obtained synaptic weight was then mapped to synaptic crossbar array via controlling the input pulse duration. The reconstructed image composed of the individual patches based on sparse-coding algorithm reproduces the original image well; see Figure 1.26b. Actually, this offline training approach is compatible with almost all algorithms used in ANNs; thus neural networks with different goals can be built in this way. In 2018, Hu and his coworkers [144] demonstrated a 128×64 one-transistor-one-memristive device (1T1M) array and achieved an 89.9% recognition accuracy for 10 000 Modified National Institute of Standards and Technology (MNIST) handwritten digit test set in a single-layer neural network, as shown in Figure 1.26c,d. Very recently, long short-term memory networks have been developed based on the same 128×64 1T1M array, with which predicting airline passenger numbers and identifying an individual human-based gait were successfully realized [151]. They then combined the convolutional neural networks (CNN) with LSTM networks to demonstrate the ConvLSTM, which classifying the MNIST-sequence videos [152].

Unlike offline training, online training approach means most of the training computation works are directly done on the memristive device crossbar. Moreover, the online learning self-adaptively mitigated the impact of the hardware imperfections [152]. In 2015, Prezioso et al. [145] developed a single-layer perceptron using 10×6 fragment of the TiO_{2-x} memristive device crossbar array; see Figure 1.26e. This network was online trained using the Manhattan update rule, i.e. a coarse–grain variety of delta rule, to perform the classification of 3×3-pixel black/white image into three classes (Figure 1.26f). It is found that Manhattan update rule is more feasible for the online training in memristive device crossbars, as given in Figure 1.26g. In 2017, Yao and his coworkers [146] experimentally demonstrated a 128×8 1T1M array and realized gray-scale face classification through parallel online training based on Manhattan update rule; see Figure 1.26h. In addition to Manhattan update rule, Sanger's rule can also be used to update synaptic weight to realize principal component analysis in memristive networks with 9×2 TaO_x array [147] (see Figure 1.26i). Compared with single-layer networks, multilayer networks are generally much more powerful in defects tolerance, as shown in Figure 1.26j. That is because the hidden neurons are helpful for minimizing the impact of device defects. Bayat et al. [153] experimentally demonstrated a multilayer memristive networks composed of two 20×20 transistor-free crossbar arrays and realized classification of letters. Li and his coworkers [148] demonstrated a multilayer network composed of 128×54 array for the first layer and 108×10 array for the second layer. They achieved much higher accuracy (91.7%) than the single layer networks (89.9%) [144] on the MNIST

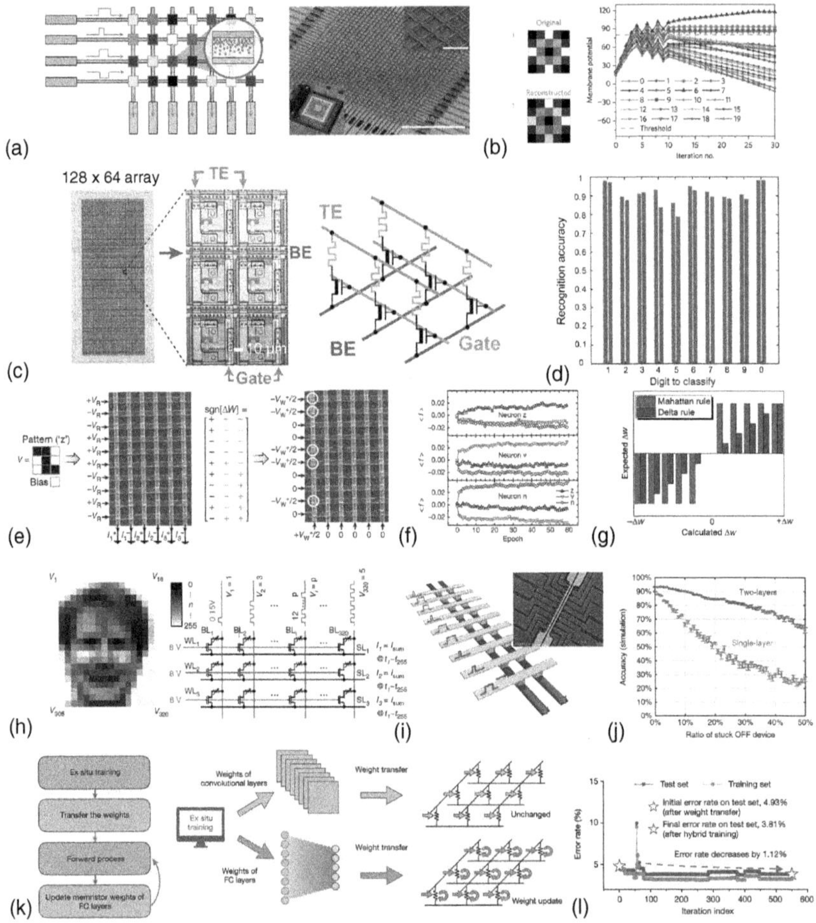

Figure 1.26 Memristive ANN computing. (a) Left: Schematic of a memristive device crossbar-based computing system, showing the input neurons (the left), the memristive device crossbar array, and the output neurons (the bottom). A memristive device is formed at each cross-point and can be programmed to different conductance states. Right: Scanning electron micrograph (SEM) image of a fabricated 32×32 WO_x memristive device array. (b) The original image and the reconstructed image based on sparse coding in the memristive device networks. The membrane potentials of the neurons as a function of iteration number during locally competitive algorithm (LCA) analysis. Source: (a, b) Sheridan et al. [143]. (c) The fabricated 128×64 1T1M memristive device array. (d) Total recognition accuracy for each digit for 10 000 images from the MNIST database based on the128×64 1T1M memristive device array. Source: (c, d) Hu et al. [144]. (e) The fabricated 10×6 TiO_{2-x} memristive device array without transistors. (f) The evolution of output signals in this 10×6 TiO_{2-x} memristive device array, averaged over all patterns of a specific class. Source: (e, f) Prezioso et al. [145]. (g) The Manhattan rule and delta rule show different Δw in the same calculated Δw. The sign is the only parameter considered in the Manhattan rule, while the amplitude is also included in the delta rule. (h) The schematic of parallel read operation and how a pattern is mapped to the input in the 128×8 1T1M array for face classification [146]. (i) Schematic diagram and SEM image of a 2×9 memristive device array for principal component analysis. Source: Choi et al. [147]. (j) The multilayer network helps with defect tolerance. If one device is stuck, the associated hidden neuron will adjust the connections accordingly Source: Li et al. [148]. (k) Schematic diagram of hybrid trained method. Source: Yao et al. [149]. (l) Error-rate traces over 550 hybrid training iteration cycles. Source: Based on Cai et al. [150].

handwritten recognition. Wang et al. [154] realized a reinforcement learning on a three-layer network and performed two control problems of the cart-pole and mountain car simulations.

Recently, a new hybrid training method is put forward, including the offline trained weights of the convolutional kernel and the online trained weights of the full connected layer. Yao and his coworkers [149] firstly implemented a memristor-based CNN with the fully hardware. They achieve a high accuracy (>96%) for MNIST handwritten digits by utilizing hybrid training (as shown in Figure 1.26k) to accommodate nonideal device characteristics and accelerate the process by parallel computing. In addition, the energy efficiency is much greater (more than 2 orders of magnitude) than that of state-of-the-art graphics processing. Cai et al. [150] integrated a passive crossbar array with traditional CMOS circuits (Figure 1.26l), including a full set of mixed-signal interface blocks and a computing processer based on digit computing. The integrated system provided the hardware solutions for neural networks with high speed and efficient power consumption.

1.4.2 Memristive SNN Computing

In comparison with memristive ANN networks, the progress on memristive SNN networks is quite limited due to the lack of mature algorithms and comparable simple circuits for SNN. Even so, some encouraging breakthroughs have been made in memristive SNN computing. In 2015, Nishitani and his coworkers [155] proposed a supervised learning model with simplified error backpropagation in a memristive SNN, in which a three-terminal ferroelectric memristive device functions as synapses to store analog synaptic weight. In this spiking backpropagation method, the input signal and output signal are all spiking timing. Also, the synaptic weight update process is divided into two steps. In the first step, only the output layer is updated, whereas in the second step, both output layer and hidden layer were updated. When the output spikes approach the desired timing, the residual error rapidly drops, meaning that the supervised learning process succeeded. Li and his coworkers [156] applied the volatile device to emulate the forgetting phenomenon of the biological brain. In the training procedure, the potentiation and forgetting of the synaptic weights are realized based on the electric pulses and the time intervals, respectively. Recently, Wang and his coworkers [131] developed a prototypical integrated capacitive switching neural network and realized associative learning and signal classification, as shown in Figure 1.27a–c. In their prototypical networks, neurons are represented by a pseudo-memcapacitor composed of a volatile diffusive memristive device and a series capacitor (Figure 1.27a), and the synapses share the similar structure with neurons but using a nonvolatile memristive device to replace the volatile diffusive memristive devices.

In addition to supervised learning, unsupervised learning has also been experimentally implemented in memristive SNN. In 2016, Serb et al. [157] experimentally realized weight-dependent STDP in the TiO_{2-x} memristive devices; see Figure 1.27d. Then, they demonstrated unsupervised learning in a prototype probabilistic neural network (Figure 1.27e), in which four input synapses are implemented by hardware

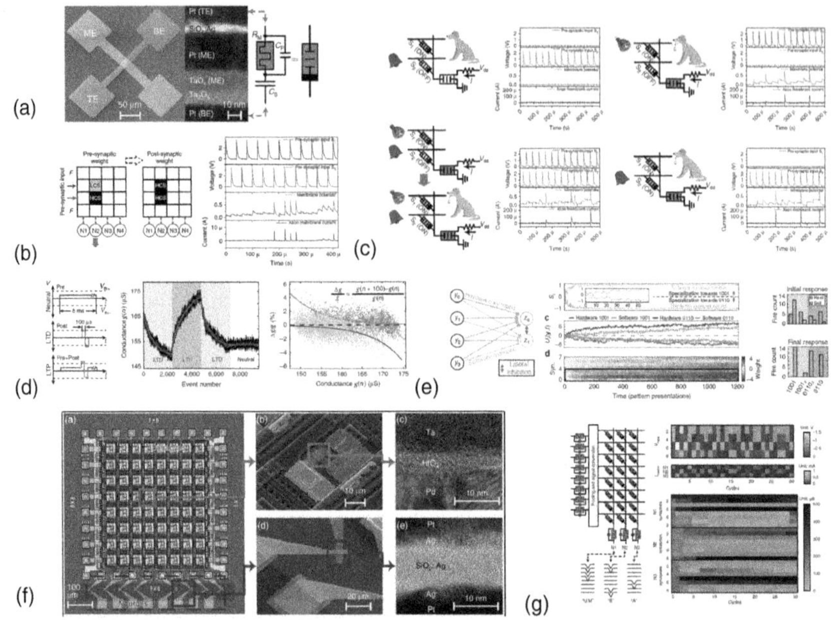

Figure 1.27 Memristive SNN computing. (a) Scanning electron micrograph of the plan view of the integrated dynamic pseudo-memcapacitor composed of a volatile diffusive memristive device and a series capacitor and a transmission electron micrograph of the cross section. (b) Schematic illustration and experimental observation of the synapse programming using the Hebbian-like mechanism. Simultaneous presynaptic signals (dark gray and light lines in ii) were applied to both the low capacitance state (LCS) synapse and the high capacitance state (HCS) synapse. (c) Capacitive network for associative learning based on the Hebbian-like mechanism. Two presynaptic signals model the sight of food and the sound of a bell, respectively. The postsynaptic neuron models the salivation of a dog. Source: (a–c) Wang et al. [131]. (d) Weight-dependent STDP in the TiO_{2-x} memristive devices. (e Unsupervised learning in a winner-take-all network with a mixture of software and memristive device synapses. Source: (d, e) Serb et al. [157]. (f) All memristive neural network, consisting of an 8×8 1T1R memristive synapse crossbar interfacing with eight diffusive memristive device artificial neurons. (g) Unsupervised training of an all-memristive neural network based on STDP learning rules. Source: (f, g) Wang et al. [129].

TiO_{2-x} memristive devices while the other four synapses are implemented by software. The final classification results are optimized when enough unlabeled inputs are feed in the networks through the introduction of probabilistic in the trigger conditions of neurons. Moreover, successful learning can be achieved, even if there are corrupted inputs. Wang et al. [129] recently built a fully memristive neural network, in which the LIF neurons composed of a diffusive memristive device and a parallel capacitor were directly connected with the memristive 1T1M arrays, as shown in Figure 1.27f. This fully memristive network is capable of performing unsupervised learning using STDP learning rule for pattern classification. Software pooling and signal conversion are used to convert the Rectified Linear Units (ReLUs) layer output to timing signals. Then these signals acting as inputs are fed into the fully connected layer using STDP rules; see Figure 1.27g. This is an important step toward the hardware implementation of all memristive spiking neural networks sharing extremely similar neuron and synaptic dynamics to their biological counterparts. As for trained methods for unsupervised SNN, Guo and his coworkers [158] utilized the STDP as the learning rule and proposed to dilute spike events in encoding the input data, endowing the networks with the ability of working with the large conductance variations of the synaptic devices.

In the last, we briefly introduce the state-of-art memristive neural networks in the final section, including offline/online training in memristive crossbar for ANN and supervised/unsupervised learning achieved in memristive SNN.

1.5 Summary and Outlook

In this chapter, we have given an overview on two-terminal neuromorphic memristors used in brain-inspired computing. The device structure, materials, resistive switching behaviors, and mechanisms are presented considering unique requirements for neuromorphic computing. Furthermore, state-of-art demonstrations of memristive synapses, neurons, and corresponding neural networks including ANN and SNN have been discussed in detail. Despite encouraging progresses have been achieved in last few decades, memristor-based brain-inspired computing is still in its infancy, and there are many challenges to be resolved at both device level and system level.

At the device level, an ideal synaptic element combining high yield, superior stability, and small cycle-to-cycle and device-to-device variations is still unavailable, which is obstacle for maximizing the computing accuracy of offline training memristive systems. This is because the switching mechanisms of most memristive synapses are based on metal or oxygen ions migration, leading to the stochastic formation/rupture of conductive filaments. This inevitably induces low yield and large variations. Therefore, considerable efforts should be devoted to explore new switching mechanisms with high controllability and reliability. Furthermore, high linearity and symmetry of the conductance modulation are key parameters in maximizing the accuracy in online training networks. However, most reported memristive synapses show large nonlinearity and asymmetry. To address this issue, further device engineering is needed, and co-optimization studies between devices, algorithms, and circuits might also be necessary. For memristive neurons, the exploration of some dynamic physicochemical processes in memristive devices to faithfully reproduce the neuron dynamics is required for the implementation of all memristive bioinspired computing.

At the system level, the sneak current, which flows through unselected cells, can severely restricts the size of synapse array. Although the 1T1M configuration can effectively address the sneak path problem, the additional three-terminal transistors will remarkably increase the circuit complexity and affect the integration density of crossbar arrays. Fortunately, memristive synapses with nonlinear or self-rectifying switching behavior are preferable for the construction of transistor-free crossbar arrays. In addition, the learning algorithms of memristive ANN and SNN are still under development. Therefore, it is necessary to develop new network topology and learning algorithms that are prevalent for memristive ANN and SNN, in particular those that can take advantage of the unique properties of memristors.

Last but not the least, close collaboration between materials scientists, computer architects, and neuroscientists is crucial to demonstrate large-scale memristive bioinspired computing systems for practical applications.

References

1 Chua, L.O. (1971). *IEEE Trans. Circuit Theory* 18: 507.
2 Strukov, D.B., Snider, G.S., Stewart, D.R., and Williams, R.S. (2008). *Nature* 453: 80.
3 Jeong, D.S., Thomas, R., Katiyar, R.S. et al. (2012). *Rep. Prog. Phys.* 75: 076502.
4 Sawa, A. (2008). *Mater. Today* 11: 28.
5 Gao, S., Zeng, F., Chen, C. et al. (2013). *Nanotechnology* 24: 335201.
6 Gao, S., Song, C., Chen, C. et al. (2012). *J. Phys. Chem. C* 116: 17955.
7 Chen, C., Yang, Y.C., Zeng, F., and Pan, F. (2010). *Appl. Phys. Lett.* 97: 083502.
8 Peng, C.N., Wang, C.W., Chan, T.C. et al. (2012). *Nanoscale Res. Lett.* 7: 559.
9 Ma, L.P., Liu, J., and Yang, Y. (2002). *Appl. Phys. Lett.* 80: 2997.
10 Huang, H.-M., Yang, R., Tan, Z.-H. et al. (2018). *Adv. Mater.*: 1803849.
11 Lee, M.-J., Lee, D., Cho, S.-H. et al. (2013). *Nat. Commun.* 4: 2629.
12 Son, J.Y., Shin, Y.-H., Kim, H., and Jang, H.M. (2010). *ACS Nano* 4: 2655.
13 Tsai, C.-L., Xiong, F., Pop, E., and Shim, M. (2013). *ACS Nano* 7: 5360.
14 Hu, D.-C., Yang, R., Jiang, L., and Guo, X. (2018). *ACS Appl. Mater. Interfaces* 10: 6463.
15 Moors, M., Adepalli, K.K., Lu, Q.Y. et al. (2016). *ACS Nano* 10: 1481.
16 Tang, G.S., Zeng, F., Chen, C. et al. (2013). *J. Appl. Phys.* 113: 244502.
17 Yang, Y., Gao, P., Li, L. et al. (2014). *Nat. Commun.* 5: 4232.
18 Wang, D.T., Dai, Y.W., Xu, J. et al. (2016). *IEEE Electron Device Lett.* 37: 878.
19 Hong, D.S., Chen, Y.S., Sun, J.R., and Shen, B.G. (2016). *Adv. Electron. Mater.* 2: 1500359.
20 Jiang, J., Guo, J.J., Wan, X. et al. (2017). *Small* 13: 1700933.
21 Yin, X.B., Tian, K., Tan, Z.H. et al. (2015). *J. Electrochem. Soc.* 162: 271.
22 Lee, M.-J., Lee, C.B., Lee, D. et al. (2011). *Nat. Mater.* 10: 625.
23 Kumar, S., Wang, Z., Huang, X. et al. (2016). *ACS Nano* 10: 11205.
24 Tan, Z.H., Yang, R., Terabe, K. et al. (2016). *Adv. Mater.* 28: 377.
25 Yang, R., Huang, H.M., Hong, Q.H. et al. (2018). *Adv. Funct. Mater.* 28: 1704455.
26 Shi, T., Yang, R., and Guo, X. (2016). *Solid State Ionics* 296: 114.
27 Jeong, H.Y., Kim, J.Y., Kim, J.W. et al. (2010). *Nano Lett.* 10: 4381.
28 Shi, Y., Liang, X., Yuan, B. et al. (2018). *Nat. Electron.* 1: 458.
29 Wu, Y., Lee, B., and Philip Wong, H.-S. (2010). *IEEE Electron Dev. Lett.* 31: 1449.
30 Du, C., Ma, W., Chang, T. et al. (2015). *Adv. Funct. Mater.* 25: 4290.

31 Yin, X.-B., Tan, Z.-H., and Guo, X. (2015). *Phys. Chem. Chem. Phys.* 17: 134.
32 Tan, Z.-H., Yin, X.-B., Yang, R. et al. (2017). *Sci. Rep.* 7: 713.
33 Tian, H., Mi, W.T., Zhao, H.M. et al. (2017). *Nanoscale* 9: 9275.
34 Kostarelos, K. and Novoselov, K.S. (2014). *Nat. Nanotechnol.* 9: 744.
35 Wang, Q.H., Zadeh, K.K., Kis, A. et al. (2012). *Nat. Nanotechnol.* 7: 699.
36 Bessonov, A.A., Kirikova, M.N., Petukhov, D.I. et al. (2015). *Nat. Mater.* 14: 199.
37 Cheng, P., Sun, K., and Hu, Y.H. (2016). *Nano Lett.* 16: 572.
38 Wang, M., Cai, S., Pan, C. et al. (2018). *Nat. Electron.* 1: 130.
39 Novoselov, K.S., Jiang, D., Schedin, F. et al. (2005). *Proc. Natl. Acad. Sci.* 102: 10451.
40 Coleman, J.N., Lotya, M., O'Neill, A. et al. (2011). *Science* 331: 568.
41 Li, X.S., Cai, W.W., An, J.H. et al. (2009). *Science* 324: 1312.
42 Eda, G., Yamaguch, H., Voiry, D. et al. (2011). *Nano Lett.* 11: 5111.
43 Chen, Y., Liu, G., Wang, C. et al. (2014). *Mater. Horiz.* 1: 489.
44 Wang, Z.S., Zeng, F., Yang, J. et al. (2010). *Appl. Phys. Lett.* 97: 253301.
45 Yang, J., Zeng, F., Wang, Z.S. et al. (2011). *J. Appl. Phys.* 110: 114518.
46 Son, D.I., Kim, T.W., Shim, J.H. et al. (2010). *Nano Lett.* 10: 2441.
47 Yang, R., Huang, H.M., and Guo, X. (2019). *Adv. Electron. Mater.* 5: 1900287.
48 Pan, F., Gao, S., Chen, C. et al. (2014). *Mater. Sci. Eng. R-Rep* 83: 1.
49 Waser, R., Dittmann, R., Staikov, G., and Szot, K. (2009). *Adv. Mater.* 21: 2632.
50 Wang, J. and Zhuge, F. (2019). *Adv. Mater. Technol.* 4: 1800544.
51 Sawa, A., Fujii, T., Kawasaki, M., and Tokura, Y. (2004). *Appl. Phys. Lett.* 85: 4073.
52 Wang, Z., Joshi, S., Savel'ev, S.E. et al. (2017). *Nat. Mater.* 16: 101.
53 Jo, S.H., Chang, T., Ebong, I. et al. (2010). *Nano Lett.* 10: 1297.
54 Shi, Y., Nguyen, L., Oh, S. et al. (2018). *Nat. Commun.* 9: 5312.
55 Zidan, M.A., Strachan, J.P., and Lu, W.D. (2018). *Nat. Electron.* 1: 22.
56 Yu, S.M. (2018). *Proc. IEEE* 106: 260.
57 Yang, J.J., Miao, F., Pickett, M.D. et al. (2009). *Nanotechnology* 20: 215201.
58 Kim, S.G., Han, J.S., Kim, H. et al. (2018). *Adv. Mater. Technol.* 3: 1800457.
59 Valov, I., Waser, R., Jameson, J.R., and Kozicki, M.N. (2011). *Nanotechnology* 22: 254003.
60 Lu, W., Jeong, D.S., Kozicki, M., and Waser, R. (2012). *Mater. Res. Soc. Bull.* 37: 124.
61 Hasegawa, T., Terabe, K., Tsuruoka, T., and Aono, M. (2012). *Adv. Mater.* 24: 252.
62 Yang, Y. and Lu, W. (2013). *Nanoscale* 5: 10076.
63 Hirose, Y. and Hirose, H. (1976). *J. Appl. Phys.* 47: 2767.
64 Zidan, M.A., Chen, A., Indiveri, G., and Lu, W.D. (2017). *J. Electroceram.* 39: 4.
65 Yang, Y., Gao, P., Gaba, S. et al. (2012). *Nat. Commun.* 3: 732.
66 Yang, Y., Gao, P., Li, L. et al. (2014). *Nat. Commun.* 5: 2382.
67 Choi, S.-J., Park, G.-S., Kim, K.-H. et al. (2011). *Adv. Mater.* 23: 3272.
68 Kwon, D.-H., Kim, K.M., Jang, J.H. et al. (2010). *Nat. Nanotechnol.* 5: 148.
69 Strachan, J.P., Pickett, M.D., Yang, J.J. et al. (2010). *Adv. Mater.* 22: 3573.
70 Kumar, S., Graves, C.E., Strachan, J.P. et al. (2016). *Adv. Mater.* 28: 2772.

71 Baeumer, C., Schmitz, C., Ramadan, A.H.H. et al. (2015). *Nat. Commun.* 6: 8610.

72 Baeumer, C., Raab, N., Menke, T. et al. (2016). *Nanoscale* 8: 13967.

73 Bagdzevicius, S., Maas, K., Boudard, M., and Burriel, M. (2017). *J. Electroceram.* 39: 157.

74 Hansen, M., Ziegler, M., Kolberg, L. et al. (2015). *Sci. Rep.* 5: 13753.

75 Yang, R., Terabe, K., Liu, G. et al. (2012). *ACS Nano* 6: 9515.

76 Zhou, W., Yang, R., He, H.-K. et al. (2018). *Appl. Phys. Lett.* 113: 061107.

77 Mikheev, E., Hoskins, B.D., Strukov, D.B., and Stemmer, S. (2014). *Nat. Commun.* 5: 3990.

78 Arndt, B., Borgatti, F., Offi, F. et al. (2017). *Adv. Funct. Mater.* 27: 1702282.

79 Wang, Z.Q., Xu, H.Y., Li, X.H. et al. (2012). *Adv. Funct. Mater.* 22: 2759.

80 Ha, S.D. and Ramanathan, S. (2011). *J. Appl. Phys.* 110: 071101.

81 Asanuma, S., Akoh, H., Yamada, H., and Sawa, A. (2009). *Phys. Rev. B.* 80: 235113.

82 Lee, H.S., Park, H.H., and Rozenberg, M.J. (2015). *Nanoscale* 7: 6444.

83 Versace, M. and Chandler, B. (2010). *IEEE Spectrum* 47: 30.

84 Strukov, D.B. (2011). *Nature* 476: 403.

85 Thomas, A. (2013). *J. Phys. D Appl. Phys.* 46: 093001.

86 Mead, C.A. (1990). *Proc. IEEE* 78: 1629.

87 Ma, W., Zidan, M.A., and Lu, W.D. (2018). *Sci. China Inform. Sci.* 61: 060422.

88 Thomas, A. (2018). *Nat. Electron.* 1: 261.

89 John, R.A., Liu, F., Chien, N.A. et al. (2018). *Adv. Mater.* 30: 1800220.

90 Kirkwood, A., Rioult, M.G., and Bear, M.F. (1996). *Nature* 381: 526.

91 Nicholls, J.G., Martin, A.R., Fuchs, P.A. et al. (2012). *From Neuron to Brain.* Sunderland, MA: Sinauer Associates.

92 Debanne, D., Guerineau, N.C., Gähwiler, B., and Thompson, S.M. (1996). *J. Physiol.* 491: 163.

93 Waldeck, R.F., Pereda, A., and Faber, D.S. (2000). *J. Neurosci.* 20: 5312.

94 Atluri and Regehr, W.G. (1996). *J. Neurosci.* 16: 5661.

95 Kirischuk, S., Clements, J.D., and Grantyn, R. (2002). *J. Physiol.* 543: 99.

96 Zhu, L.Q., Wan, C.J., Guo, L.Q. et al. (2014). *Nat. Commun.* 5: 3158.

97 Xu, W., Cho, H., Kim, Y.H. et al. (2016). *Adv. Mater.* 28: 5916.

98 Li, B., Liu, Y., Wan, C. et al. (2018). *Adv. Mater.* 30: 1706395.

99 Milano, G., Luebben, M., Ma, Z. et al. (2018). *Nat. Commun.* 9: 5151.

100 Najem, J.S., Taylor, G.J., Weiss, R.J. et al. (2018). *ACS Nano* 12: 4702.

101 Zhang, Y., Zhong, S., Song, L. et al. (2018). *Appl. Phys. Lett.* 113: 203102.

102 Raymond, C.R. and Redman, S.J. (2006). *J. Physiol.* 570: 97.

103 Froemke, R.C. and Dan, Y. (2002). *Nature* 416: 433.

104 Raymond, C.R. (2007). *Trends Neurosci.* 30: 167.

105 Bear, M.F. and Malenka, R.C. (1994). *Curr. Opin. Neurobiol.* 4: 389.

106 Bienenstock, E.L., Cooper, L.N., and Munro, P.W. (1982). *J. Neurosci.* 2: 32.

107 Benuskova, L., Diamond, M.E., and Ebner, F.F. (1994). *Proc. Natl. Acad. Sci. U. S. A.* 91: 4791.

108 Serrano-Gotarredona, T., Masquelier, T., Prodromakis, T. et al. (2013). *Front. Neurosci.* 7: 2.

109 Abbott, L.F. and Nelson, S.B. (2000). *Nat. Neurosci.* 3: 1178.

110 Prezioso, M., Merrikh Bayat, F., Hoskins, B. et al. (2016). *Sci. Rep.* 6: 21331.

111 Kuzum, D., Jeyasingh, R.G.D., Lee, B., and Wong, H.S.P. (2012). *Nano Lett.* 12: 2179.

112 Alibart, F., Pleutin, S., Bichler, O. et al. (2012). *Adv. Funct. Mater.* 22: 609.

113 Li, Y., Zhong, Y.P., Zhang, J.J. et al. (2014). *Sci. Rep.* 4: 4906.

114 Kim, S., Choi, S., Lee, J., and Lu, W.D. (2014). *ACS Nano* 8: 10262.

115 He, W., Huang, K.J., Ning, N. et al. (2014). *Sci. Rep.* 4: 4755.

116 Rubin, J.E., Gerkin, R.C., Bi, G.Q., and Chow, C.C. (2005). *J. Neurophysiol.* 93: 2600.

117 Kim, S., Du, C., Sheridan, P. et al. (2015). *Nano Lett.* 15: 2203.

118 Xiong, J., Yang, R., Shaibo, J. et al. (2019). *Adv. Funct. Mater.*: 1807316.

119 Sokolov, A.S., Jeon, Y.-R., Kim, S. et al. (2019). *NPG Asia Mater.* 11: 5.

120 Yi, W., Tsang, K.K., Lam, S.K. et al. (2018). *E. A. J. N. c. Flores, Nat. Commun.* 9: 4661.

121 Gerstner, W. and Naud, R. (2009). *Science* 326: 379.

122 Bao, L., Kang, J., Fang, Y.C. et al. (2018). *Sci. Rep.* 8: 13727.

123 Pickett, M.D., Medeiros-Ribeiro, G., and Williams, R.S. (2012). *Nat. Mater.* 12: 114.

124 Tuma, T., Pantazi, A., Le Gallo, M. et al. (2016). *Nat. Nanotechnol.* 11: 693.

125 Ziegler, M., Wenger, C., Chicca, E., and Kohlstedt, H. (2018). *J. Appl. Phys.* 124: 152003.

126 Amirsoleimani, A., Ahmadi, M., and Ahmadi, A. (2017). Presented at the 2017 International Joint Conference on Neural Networks, Anchorage, AK, May.

127 Bao, H., Liu, W., and Hu, A. (2019). *Nonlinear Dyn.* 95: 43.

128 Nicholls, J.G., Martin, A.R., Wallace, B.G., and Fuchs, P.A. (2001). *From Neuron to Brain*. Sunderland, MA: Sinauer Associates.

129 Wang, Z., Joshi, S., Savel'ev, S. et al. (2018). *Nat. Electron.* 1: 137.

130 Zhang, X., Wang, W., Liu, Q. et al. (2018). *IEEE Electron Device Lett.* 39: 308.

131 Wang, Z., Rao, M., Han, J.-W. et al. (2018). *Nat. Commun.* 9: 3208.

132 Stoliar, P., Tranchant, J., Corraze, B. et al. (2017). *Adv. Funct. Mater.* 27: 1604740.

133 Zhang, Y., He, W., Wu, Y. et al. (2018). *Small* 14: 1802188.

134 Yi, W., Tsang, K.K., Lam, S.K. et al. (2018). *Nat. Commun.* 9: 4661.

135 Lim, H., Kornijcuk, V., Seok, J.Y. et al. (2015). *Sci. Rep.* 5: 9776.

136 Ignatov, M., Ziegler, M., Hansen, M. et al. (2015). *Front. Neurosci.* 9: 376.

137 Feali, M.S., Ahmadi, A., and Hayati, M. (2018). *Neurocomputing* 309: 157.

138 Mehonic, A. and Kenyon, A.J. (2016). *Front. Neurosci.* 10: 57.

139 Agrawal, A. and Roy, K. (2019). *IEEE Trans. Magn.* 55: 1400107.

140 Chen, X., Kang, W., Zhu, D. et al. (2018). *Nanoscale* 10: 6139.

141 Mizrahi, A., Hirtzlin, T., Fukushima, A. et al. (2018). *Nat. Commun.* 9: 1533.

142 Feali, M.S. and Ahmadi, A. (2017). *Neural Processing Lett.* 45: 1.

143 Sheridan, P.M., Cai, F., Du, C. et al. (2017). *Nat. Nanotechnol.* 12: 784.

144 Hu, M., Graves, C.E., Li, C. et al. (2018). *Adv. Mater.* 30: 1705914.

145 Prezioso, M., Merrikh-Bayat, F., Hoskins, B.D. et al. (2015). *Nature* 521: 61.

146 Yao, P., Wu, H., Gao, B. et al. (2017). *Nat. Commun.* 8: 15199.

147 Choi, S., Shin, J.H., Lee, J. et al. (2017). *Nano Lett.* 17: 3113.

148 Li, C., Belkin, D., Li, Y. et al. (2018). *Nat. Commun.* 9: 2385.

149 Yao, P., Wu, H., Gao, B. et al. (2020). *Nature* 577: 641.

150 Cai, F., Correll, J.M., Lee, S.H. et al. (2019). *Nat. Electr.* 2: 290.

151 Li, C., Wang, Z., Rao, M. et al. (2019). *Nat. Mach. Intell.* 1: 49.

152 Wang, Z., Li, C., Lin, P. et al. (2019). *Nat. Mach. Intell.* 1: 434.

153 Bayat, F.M., Prezioso, M., Chakrabarti, B. et al. (2018). *Nat. Commun.* 9: 2331.

154 Wang, Z., Li, C., Song, W. et al. (2019). *Nat. Electr.* 2: 115.

155 Nishitani, Y., Kaneko, Y., and Ueda, M. (2015). *IEEE Trans. Neural Netw. Learn. Syst.* 26: 2999.

156 Li, J., Xu, H., Sun, S.-Y. et al. (2020). *Neurocomputing* 408: 21.

157 Serb, A., Bill, J., Khiat, A. et al. (2016). *Nat. Commun.* 7: 12611.

158 Guo, Y., Wu, H., Gao, B., and Qian, H. (2019). *Frontiers in Neuroscience*: 13.

2

Spintronic Neuromorphic Devices

Deming Zhang[1], Sai Li[1,2], Xinran Wang[1], Ao Du[1], Wang Kang[1], Lang Zeng[1], and Weisheng Zhao[1]

[1] *Beihang University, Fert Beijing Institute, MIIT Key Laboratory of Spintronics, School of Integrated Circuit Science and Engineering, XueYuan Road No.37, Beijing 100191, P. R. China*
[2] *Beihang University, Shenyuan Honors College, XueYuan Road No. 37, Beijing, 100191, P. R. China*

2.1 Introduction

Spintronic devices are regarded as next-generation microelectronic devices due to their low energy dissipation, non-volatility, high speed, and potential for pure spin current transport. They are used to mimic brain-inspired artificial synapses and neurons. A number of different proposals for spintronic neuromorphic implementations exist. Basically, all of these spintronic neuromorphic devices can be classified into two categories:

- Spintronic neuromorphic devices based on static property: The most important and widely used property of spintronic devices is its magneto resistive effect, which also is the 2007 Nobel Prize in Physics awarded to Albert Fert and Peter Grünberg. This effect is the basis for storage usage of spintronic devices. From another viewpoint, the tunning of the magnetoresistive effect by the magnetic state of spintronic devices can be regarded as updating the synaptic weights in neuromorphic computation. In addition, with the help of the peripheral circuit, the behaviors of neurons can be obtained. Most of the neuromorphic computation proposals based on magnetic tunneling junction, spintronic memristor, and some of the neuromorphic computation proposals based on skyrmions can be subjected into this category.
- Spintronic neuromorphic devices based on dynamic property: Besides static magnetoresistive effect, the spintronic devices have ample dynamic properties which can be used to design novel neuromorphic computation implementation. Such dynamical properties include the stochastic switch of magnetic tunneling junction, the breathing mode of skyrmion size oscillation, the temporal memory effect of spin torque oscillators (STOs), and the mutual synchronization of STO. Reservoir computing (RC) and recurrent computing can be implemented with such

Neuromorphic Devices for Brain-Inspired Computing: Artificial Intelligence, Perception, and Robotics, First Edition. Edited by Qing Wan and Yi Shi.

dynamic property. Since the dynamic property of spintronic devices is complicated enough, the neuromorphic computation task such as spoken digit recognition can be performed by even one single spintronics device.

2.2 Magnetic Tunnel Junction for Neuromorphic Computing

2.2.1 Device Structure and the Write/Read Operation

2.2.1.1 Binary Magnetic Tunnel Junction

As the key element of spintronics, magnetic tunnel junction (MTJ) [1, 2] is composed of two ferromagnetic (FM) layers (e.g. CoFeB) [3] separated by one ultra-thin oxide barrier layer (e.g. MgO) as shown in Figure 2.1a. Based on the stable magnetic direction in the ferromagnetic layer, the MTJ can be classified into two types as shown in Figure 2.1b: (i) the MTJ with the in-plane magnetic anisotropy (IMA-MTJ) and (ii) the MTJ with the perpendicular magnetic anisotropy (PMA-MTJ). Generally, the magnetization of one FM layer, called pinned layer (PL), is fixed in one direction, whereas the magnetization of the other FM, called free layer (FL), is reversible. Depending on the relative magnetization orientation of these two FM layers, i.e. parallel (P) or anti-parallel (AP), one MTJ device can exhibit two stable resistance states, i.e. R_P (low resistance state) or R_{AP} (high resistance state), which are used to represent binary logic information "0" and "1," respectively. In addition, the resistance difference can be characterized by the tunnel magnetoresistance ratio, which can be expressed as follows [4]:

$$\text{TMR} = \frac{R_{AP} - R_P}{R_P} \times 100\% \tag{2.1}$$

The read operation of the MTJ can be achieved by applying a relatively small read current across it to sense its resistance state. Generally, it can be implemented by using a sense amplifier. As for its write operation, the mainstream approach is to use the spin transfer torque (STT) effect owing to its low critical current, high speed, and simple realization [1]. As illustrated in Figure 2.1c, taking the PMA-MTJ for

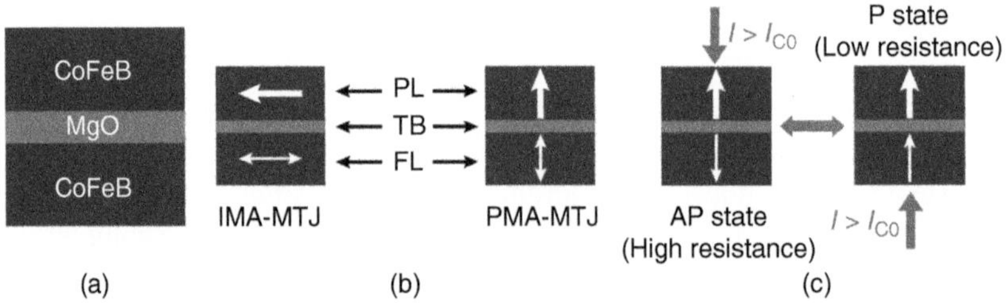

Figure 2.1 (a) Structure schematic of the MTJ composed of CoFeB/MgO/CoFeB thin films; (b) Two types of the MTJ, i.e. IMA-MTJ and PMA-MTJ; (c) Spin Transfer Torque (STT) switching mechanism, taking the PMA-MTJ for example.

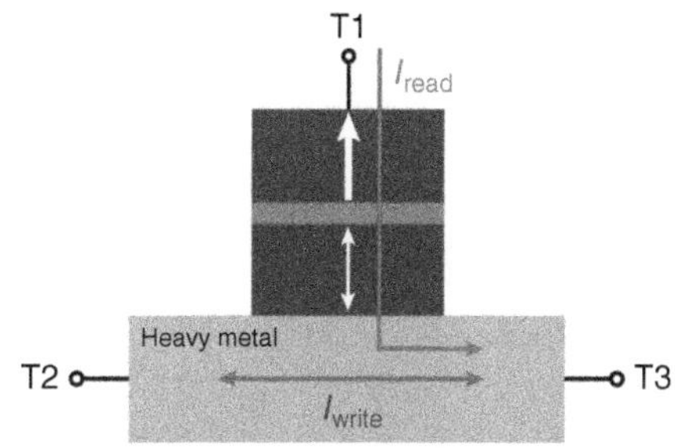

Figure 2.2 Structure schematic of the three-terminal SOT-MTJ device.

example, the MTJ device can be written by just a bidirectional current. In particular, the PMA-MTJ device can be switched from the AP (P) state to P (AP) state when the applied current that flows from the FL (PL) to the PL (FL) exceeds the critical current I_{C0}, which can be described as follows [5]:

$$I_{C0} = \alpha \frac{\gamma e}{\mu_B g} \mu_0 M_S H_k V \tag{2.2}$$

where α is the Gilbert damping factor, γ is the gyromagnetic ratio, e is the electron charge, μ_B is the Bohr magneton, g is a parameter corresponding to the spin polarization, and the magnetization orientation angel between the FL and the PL, μ_0 is the permeability in the free space, M_S is the saturation magnetization of the free layer, H_k is the PMA field, and V is the volume of the FL.

However, the STT-MTJ still suffers from some serious challenges before large scale applications [6–9]: (i) a large incubation delay; (ii) a large write current to achieve high write success rate; and (iii) poor write endurance owing to a large write current across it [10]. To address these issues, an emerging write mechanism, i.e. the spin-orbit torque (SOT) effect [11, 12], is proposed. As shown in Figure 2.2, the spin-orbit torque-magnetic tunnel junction (SOT-MTJ) is a three-terminal device. Different from the STT-MTJ, the SOT-MTJ employs an additional heavy metal stacked on the FL of the MTJ [13], which is used as the write current path. Compared to the STT-MTJ, the SOT-MTJ has some advantages: [14–16] (i) it can achieve lower write energy dissipation and smaller write delay; (ii) its read current (I_{read}) and write current (I_{write}) paths are separated, allowing to the independent optimization for its read and write performance; and (iii) the I_{write} does not flow across the MTJ, thus avoiding the breakdown of the thin oxide barrier layer and achieving higher endurance.

2.2.1.2 Multi-level Spintronic Memristor

Figure 2.3a describes the device structure of the spintronic memristor [17] with the STT write mechanism, which constitutes of a magnetic strip, an ultra-thin tunneling barrier layer, reference layer and two fixed magnetic domains (d1 and d2). As seen, the magnetic domain d1 and d2 are connected laterally to the magnetic strip at its both ends and they have fixed anti-parallel magnetization orientation. The magnetic strip includes two magnetic domains with opposite magnetization orientation separated by one domain wall (DW) [19]. The magnetic strip can be regarded as a "free" layer since its magnetization state can be manipulated by the domain wall motion

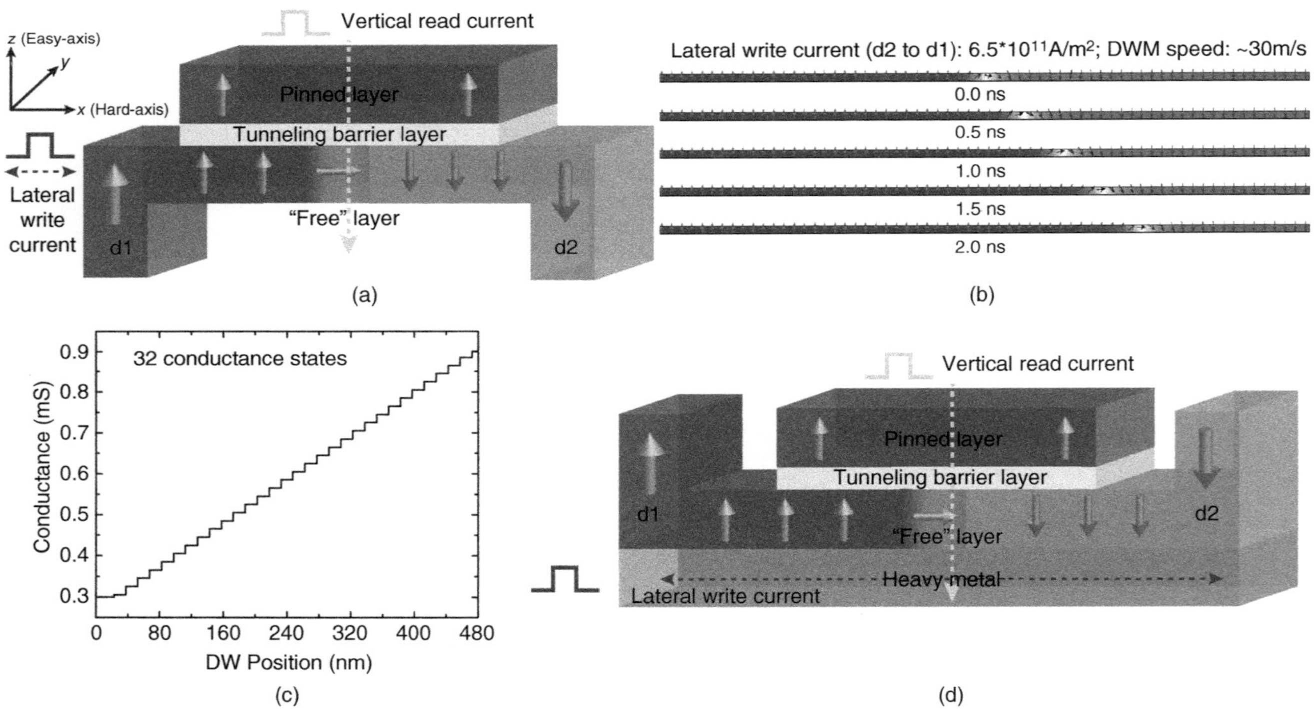

Figure 2.3 (a) The device structure of the spintronic memristor with the STT write mechanism [17]; (b) Magnetization dynamics of the "Free" layer with micro-magnetic simulations when the applied lateral write current density is 6.5×10^{11} A/m^2; (c) The conductance of the spintronic memristor versus the position of the DW; (d) The device structure of the spintronic memristor with the SOT write mechanism [18]. Source: (a) Zhang et al. [17], (d) Sengupta et al. [18].

(DWM) [20]. Thus, the spintronic memristor can be regarded as a MTJ device with three terminals.

For this spintronic memristor, the magnetization state of the magnetic strip, i.e. the position of the DW, is used to represent the stored information. For its write operation, the position of the DW can be manipulated by applying a bidirectional lateral write current due to the STT effect as shown in Figure 2.3b. As for its read operation, the conductance of the spintronic memristor that depends on the position of the DW as shown in Figure 2.3c can be sensed by applying a small vertical read current owing to the TMR effect [21].

Figure 2.3d shows the device structure of the spintronic memristor with the SOT write mechanism, where a heavy metal is added as the write current path in comparison to the spintronic memristor with the STT write mechanism [22]. For its write operation, the position of the DW can be programmed by applying a bidirectional write current (I_{write}) flowing through the heavy metal according to the SOT effect. For its read operation, a read current (I_{read}) flowing from the READ terminal to GND terminal can be applied to sense the conductance of the spintronic memristor.

The spintronic memristor with the PMA can achieve a smaller critical write current density in the range of 10^{11} A/m^2 [23] than that with IMA. It is worth noting that the critical write current density with the SOT effect is less than that with the STT effect [24]. Moreover, the length of the DW in the magnetic strip with PMA is generally smaller than that with IMA [25]. In summary, the advanced SOT writing mechanism and the magnetic strip with PMA are better choices for the spintronic memristor.

2.2.2 Working as the Synaptic Device

As one of the basic computational elements in artificial neural networks (ANNs), the synapse has two important features: (i) it can store analogue synaptic weight (memory) and (ii) it can update its synaptic weight depending on the information of the pre-neuron and post-neuron connected to it, i.e. the learning or training. In the field of synaptic electronics, the synaptic weight is often characterized by the conductance of synaptic device. It means that the synaptic device should have both memory and tunable conductivity characteristics. In the following, it will be introduced that how the MTJ device behaves as the synaptic device.

2.2.2.1 Stochastic Binary Synaptic Device

For memory application, one big challenge for the STT-MTJ is its intrinsic stochastic switching induced by the unavoidable thermal fluctuations of the initial magnetization direction of the FL of the MTJ, which has been revealed in the STT switching experiments [26, 27]. However, the work in [28–30] proposed that such stochastic switching feature of the STT-MTJs can be exploited for mimicking the functionality of the synapse. In these works, the stochastic STT-MTJs are employed as synapses and they were used to construct a spiking neural network (SNN) in the form of crossbar topology, in which each input neuron is connected to each output neuron by one single STT-MTJ, thus achieving high area-efficiency. The matrix–vector operation in the feedforward phase of the SNN can be implemented by the parallel read operation of the crossbar architecture. For the learning of the SNN, although the conductance of the STT-MTJ cannot be tuned gradually, its switching probability under a voltage pulse can be tuned between 0% and 100% by the switching (programming) pulse width, which allows to a simplified spike timing dependent plasticity (STDP)[28] rule. System-level simulation results show that such SNN with the STT-MTJ as the synapses can detect cars in a video [28]. Moreover, Monte Carlo simulation results show that it has high robustness to device variations. Additionally, among the three current regimes for the stochastic switching of the STT-MTJ (i.e. the low current regime, the intermediate current regime, and the high current regime), the intermediate current regime can achieve less energy consumption and higher robustness to device variations [28].

It is worth noting that during the feedforward phase, the sneak path issue of the crossbar of the STT-MTJs can be mitigated owing to its parallel read operation, but it affects the learning phase. To address this issue, the selector device, such as the transistor, is generally used, resulting in the loss of its compactness.

2.2.2.2 Analog-Like Synaptic Device

Intrinsically, the STT-MTJ is still a binary device, and can only encode two synaptic weights [31]. However, the synaptic weights in the ANNs or biological brain, are usually analogue rather than binary. To address this issue, one consequent scheme is to employ multiple STT-MTJs to jointly behave as one single synapse, which is envisioned to obtain an analog-like weight spectrum [32]. To achieve this goal, there are three approaches. One approach is to compound multiple (N) STT-MTJs connected in parallel operating in a stochastic regime to jointly behave as one single synapse [33], termed as CMS, which can be implemented with the crossbar topology of the STT-MTJs as shown in Figure 2.4a. For its plasticity, it can be implemented by using a simplified stochastic STDP learn rule as shown in Figure 2.4b, thanks to the intrinsic stochastic switching behavior of STT-MTJ. As seen in Figure 2.4c, with such learning rule, the CMS with N STT-MTJs can achieve $N+1$ discrete conductance levels due to the summing up of parallel conductance, ranging from the minimum compound synaptic weight $W_{\text{min}} = N \times G_{\text{AP}}$ to the maximum weight $W_{\text{min}} = N \times G_{\text{P}}$ with a step of $\Delta w = G_{\text{P}} - G_{\text{AP}}$, where G_{AP} and G_{P} are the conductance of STT-MTJ in the AP and P state, respectively. That is, such scheme allows to build an analog-like conductance behavior for a binary device at the cost of the binary device number required to build a biological synapse. Moreover, the impact of the CMS on the integration density for physical fabrication can be significantly suppressed by the 3D schematic topology of the CMS array. Taking the CMS composed of 2 STT-MTJs (cylinder), for example, as shown in Figure 2.4d, each STT-MTJ is sandwiched between the vertical electrodes and the horizontal electrodes, and these two vertical electrodes and two horizontal electrodes were connected to the post-neurons and the pre-neurons, respectively.

The second approach is to vertically stack multiple (N) STT-MTJs to jointly behave as a synaptic device [34], termed as CSS, as shown in Figure 2.5a. According to the theoretical work [35], the proposed CSS can exhibit multiple states with well design and manipulation of materials/thickness of the capping layer and thickness of MgO

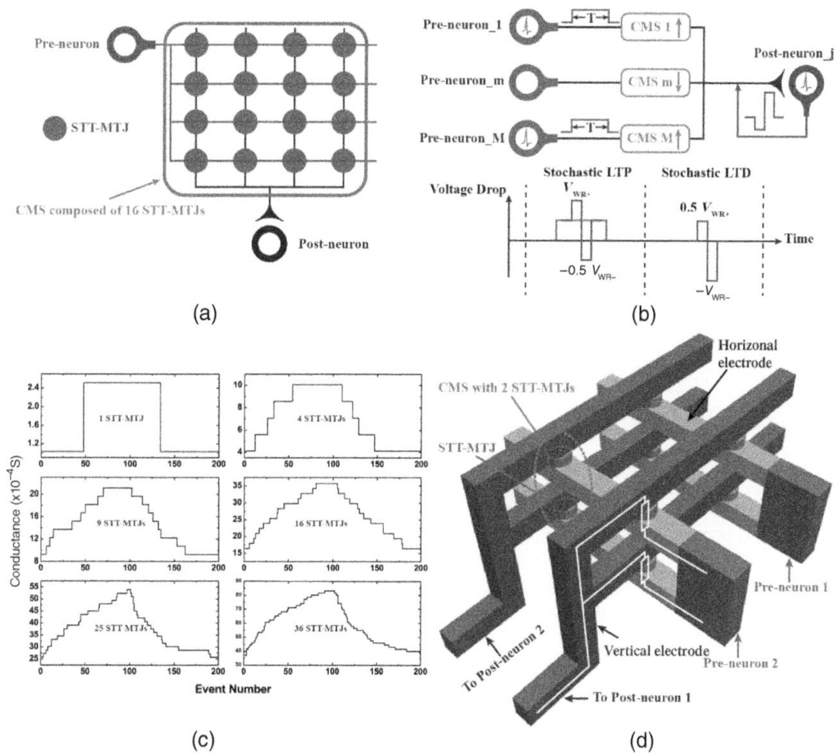

Figure 2.4 (a) Schematic topology of the proposed CMS composed of multiple ($N = 16$) STT-MTJs connected in parallel; (b) 3D schematic topology of the proposed CMS with 2 STT-MTJs; (c) Pulsing scheme for the simplified stochastic STDP learning rule; (d) The conductance evolution of the proposed CMS composed of $N = 1, 4, 9, 16, 25, 36$ p-MTJs under 100 stochastic long-term potentiation (SLTP) operations with $P_{ON} = 0.02$ followed by 100 stochastic long-term depression (SLTD) operations with $P_{OFF} = 0.02$. Source: Zhang et al. [33].

tunneling barrier of each STT-MTJ. With the increasing number of MTJs in one CSS, its conductance states increase more rapidly, thus achieving an analog-like synaptic weight spectrum. In ideal case, N vertically stacked STT-MTJs can achieve 2^N states. For example, when one CSS is composed of three vertically stacked STT-MTJs, it can exhibit eight distinct discrete resistance states even if device variations are taken into account as shown in Figure 2.5b. However, such synaptic device suffers from more and more severe fabrication variations with increasing N.

The third approach is to use one single spintronic memristor as an analogue synaptic device [36–40] since it can achieve simultaneously the learning and memory features required for the synaptic device for neuromorphic computing. Moreover, its high endurance enables a large number of learning cycles. However, one of the biggest challenges for the spintronic memristor is its scalability, i.e. keeping its analog feature with the decrease of its device dimension.

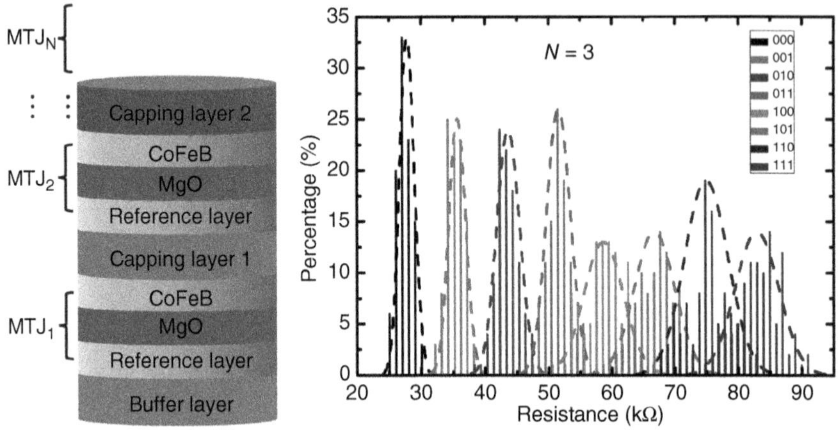

Figure 2.5 (a) Schematic of the proposed CSS device, where the capping layer materials/thickness and MgO tunneling barrier thickness of each STT-MTJ can be manipulated separately to obtain desired multilevel states. In ideal case, N vertically stacked MTJs can exhibit 2^N states; (b) the proposed CSS with $N = 3$ STT-MTJs can exhibit 8 distinct discrete resistance states. Source: Zhang et al. [34].

2.2.3 Working as the Neural Device

2.2.3.1 Spiking Neural Device

In neural networks, the neuron is another computation element. So far, many different abstract neural computing models that emulate the information processing functionality of the biological neurons have been developed. Among them, the leaky-integrate-fire (LIF) model for the spiking neuron is of the highest degree of bio-fidelity, which is described by [41]

$$C_{\text{mem}} \frac{\mathrm{d}V_{\text{mem}}}{\mathrm{d}t} = -\frac{V_{\text{mem}}}{R_{\text{mem}}} + \sum_j \delta(t - t_{f,j}) \times W_j \tag{2.3}$$

where V_{mem} is the membrane potential, C_{mem} is the membrane capacitance, R_{rest} is the membrane resistance, W_j is the synaptic weight for the j-th input neuron, and $\delta(t - t_{f,j})$ is the spiking event that occurs at the time $t_{f,j}$. As seen in Figure 2.6a, the membrane potential of one neuron integrates the input spikes and leaks when there is no input spike, and when the membrane potential exceeds its threshold voltage, it will fire.

Generally, it is area-expensive for complementary metal oxide semiconductor (CMOS)-based hardware implementation of such spiking neuron with the LIF model. To address this issue, there are two approaches. The spiking neuron can be emulated by using one single STT-MTJ device with IMA free layer. The magnetization dynamics of the FL of the STT-MTJ is similar to the LIF model, which can be illustrated by the Landau–Lifshitz–Gilbert (LLG) equation with the STT term [42, 43]:

$$\frac{1 + \alpha^2}{\gamma} \frac{\mathrm{d}\boldsymbol{m}}{\mathrm{d}t} = -(\boldsymbol{m} \times H_{\text{eff}}) - \alpha(\boldsymbol{m} \times \boldsymbol{m} \times H_{\text{eff}}) + \frac{1}{qN_{\text{s}}}(\boldsymbol{m} \times I_{\text{s}} \times \boldsymbol{m}) \tag{2.4}$$

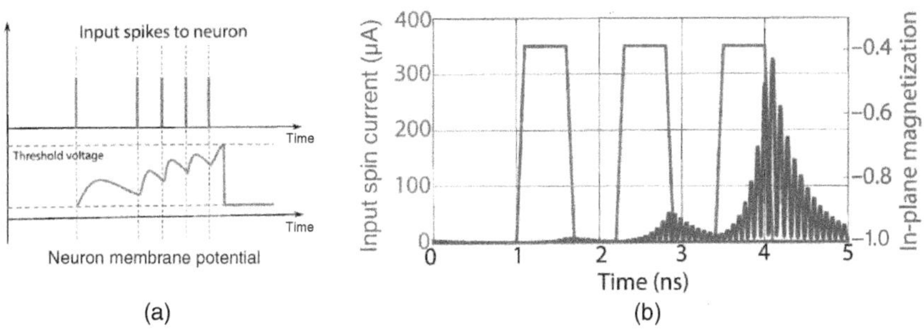

(a) (b)

Figure 2.6 (a) The membrane potential of a spiking neuron with the LIF model integrates input spikes and leaks when there is no input spike. It spikes when the membrane potential exceeds its threshold. (b) The magnetization dynamics of the IMA STT-MTJ device due to the applied three input pulses. When the three input pulses are applied, the magnetization of the FL begins integrating, and starts leaking once the applied input pulses are removed. Source: Sengupta et al. [42].

where $\boldsymbol{m}$ is the unit vector of the magnetization direction of the FL, γ is the gyromagnetic ratio for electron, α is Gilbert's damping ratio, H_{eff} is the effective magnetic field, q is the elementary charge, N_s is the number of spins in the FL, and I_s is the input spin current induced by charge current flowing across the PL due to the STT effect [44–46]. The first two terms in the right side of the LLG equation can be regarded as the "Leaky" term for the magnetization dynamics of the FL, while the last term can be seen as the "Integrate" term for the magnetization dynamics of the FL with respect to the applied input pulses. As seen in Figure 2.6b, when the input pulses are applied, the magnetization of the FL of the IMA STT-MTJ starts integrating and begins leaking once the applied input pulses are removed. The firing of the spiking neuron is equivalent to the magnetization reversal of the FL of the STT-MTJ. Once the STT-MTJ based spiking neuron fires, it will be required to be reset back to its initial resistance state. As a result, the operation of the STT-MTJ based spiking neuron involves three phases, i.e. the "write" phase, the "read" phase, and the "reset" phase. During the "write" phase, the STT-MTJ spiking neuron receives the sum of the input synaptic weighted current. During the "read" phase, it is employed to determine whether the STT-MTJ based spiking neuron fires. During the "reset" phase, the fired STT-MTJ based spiking neuron will be reset back to its initial resistance state, which is similar to the "refractory" period observed in biological neurons [47].

Furthermore, with consideration of the thermal noise, the magnetization switching of the STT-MTJ is intrinsically stochastic and its switching probability with respect to the sum of the input synaptic weighted current is a non-linear relation as shown in Figure 2.7, which is similar to the functionality of the stochastic spiking neuron observed in the cortex [48–51]. It is worth noting that the switching probability of the STT-MTJ based spiking neuron corresponding to the sum of the input synaptic weighted current can be tuned by varying the energy barrier of the FL and the duration of the summarized input synaptic weighted current pulse as shown in Figure 2.7.

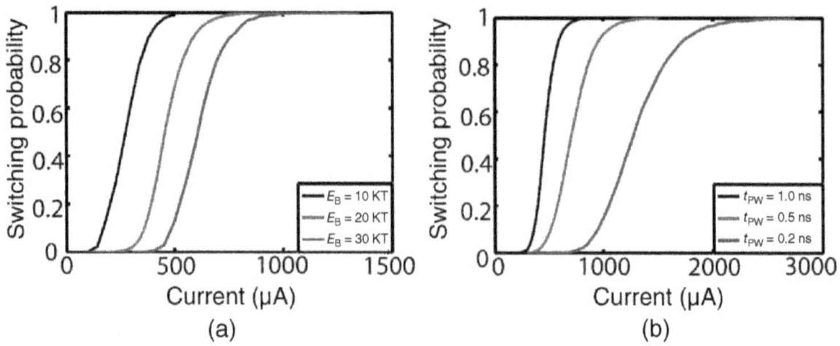

Figure 2.7 The switching probability of the IMA STT-MTJ can be tuned by (a) the energy barrier of the FL and (b) the duration of the summarized input synaptic weighted current pulse. Source: Sengupta et al. [42].

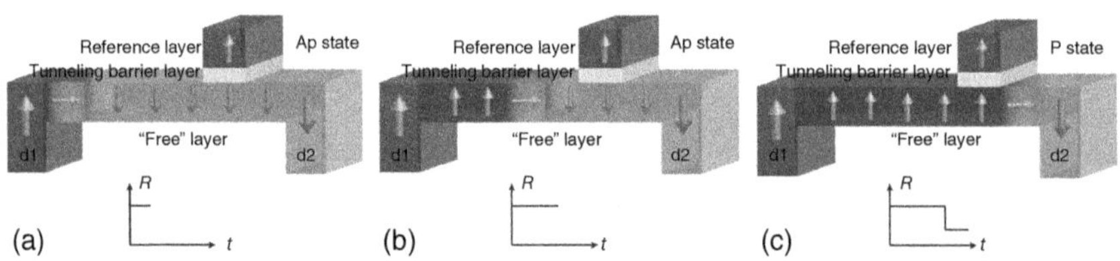

Figure 2.8 The spiking neuron can be emulated by a DW based spintronic memristor, which involves three operation phases: (a) reset phase, during which the DW is set to the left of the magnetic strip, and the MTJ on the right side is in the AP state, i.e. high resistance state; (b) integration phase, during which it receives the sum of the synaptic weighted currents (I_{sum}) and the DW moves toward right; (c) firing phase, during which it fires or not depending on the resistance state of the MTJ. If the DW is moved to the right side of the magnetic strip during the integration phase, the MTJ is switched into the P state (low resistance state), i.e. spiking neuron fires. Otherwise, it does not fire. After that, it will go back to the reset phase again. Source: Grollier et al. [29].

Additionally, the SOT-MTJ can also be employed to emulate the functionality of the stochastic spiking neuron with the similar method [52–55].

The other one is to use a spintronic memristor to mimic the spiking neuron with integrate-fire (IF) model as shown in Figure 2.8. It has three operation phases, i.e. the reset phase, the integration phase, and firing phase. During the reset phase as seen in Figure 2.8a, the DW is positioned to the left of the magnetic strip far away from the MTJ, and the MTJ is in the P state, i.e. low resistance state. During the integration phase, the spintronic memristor receives the sum of the synaptic weighted currents (I_{sum}), and the DW will move toward the right as shown in Figure 2.8b. The larger the magnitude and pulse width of the I_{sum} is, the further the DW goes to right. During the firing phase, the spintronic memristor based spiking neuron fires or not depending on the resistance state of the MTJ on the right side. If the DW moves to the right of the magnetic strip, the MTJ on the right side will be switched into the AP state (high resistance state) as shown in Figure 2.8c, i.e. spiking neuron fires. Otherwise, it does not fire. After firing, it will be reset as shown in Figure 2.8a. In this way, such spintronic memristor can mimic the functionality of the spiking neuron.

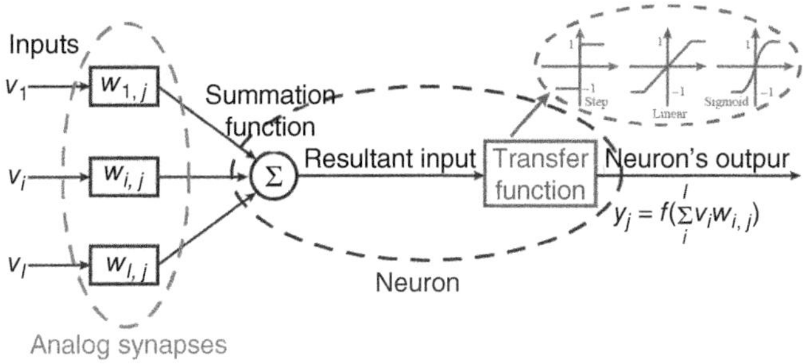

Figure 2.9 Schematic of the ANN, where the neuron is generally characterized by a transfer function.

2.2.3.2 Artificial Neural Device

In ANNs, each neuron is characterized by a transfer function (activation function). Depending on its received synaptic weighted input signals, the neuron can produce an output signal, which will be transmitted via the axon as the input to its fan-out neurons. Generally, the transfer function can be divided into two classes: step transfer function and non-step (e.g. linear and sigmoid) transfer function as seen in Figure 2.9.

For the step transfer function, it can be emulated by a single MTJ device [56], corresponding to the switching between its two stable resistance state by applying a resultant synaptic weight current that is larger than its threshold current. For example, a single SOT-MTJ device with a two-step switching scheme is proposed to function as a "step" neuron as shown in Figure 2.10. For the first step as shown in Figure 2.10b, a charge current (I_{clock}) across the heavy metal (HM) between the terminal B and C is applied, which can induce a SOT to make the magnetization direction of the FL aligned along its hard-axis (i.e. ±y-axis). Then in the second step, this neural device receives the resultant synaptic weight current between its terminal

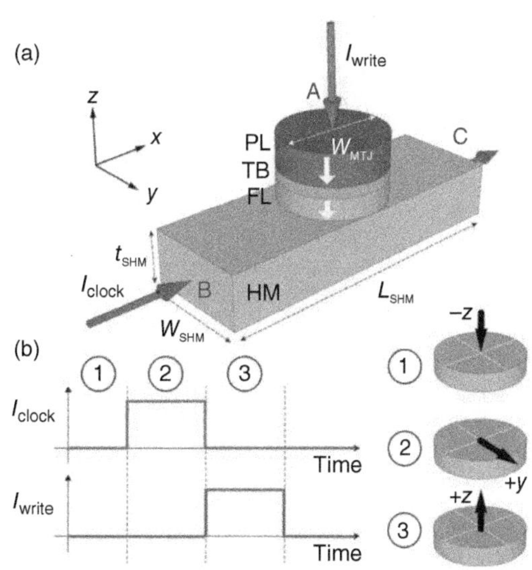

Figure 2.10 (a) SOT-MTJ as the "step" neural device with (b) two-step switching scheme. Source: Sengupta et al. [56].

A and C as the write current (I_{write}), which flows across the MTJ and can exert a STT to align the magnetization direction of the FL along its easy-axis (i.e. $\pm z$-axis). Depending on the polarity of the I_{write}, the SOT-MTJ can be switched into two different stable resistance state. If the I_{write} is positive, its induced STT will make the magnetization of the FL anti-parallel to that of the PL, i.e. the SOT-MTJ is in AP (high resistance) state. On the other hand, the magnetization of the FL will be switched parallel to that of the PL, i.e. the SOT-MTJ is in P (low resistance) state. These two different resistance states of the SOT-MTJ can be used to represent the binary outputs of a "step" neuron. Functionally, the proposed "step" neural device is similar to a spiking neuron that fires a pulse when the received resultant synaptic weight current is larger than a certain threshold.

As for the non-step transfer function, it is more attractive for complex pattern recognition tasks since more information can be encoded in the neuron's output. For this reason, the spintronic devices are also exploited to implement the functionality of the "non-step" neuron.

Noting that the above proposed CSS in [34] can achieve multiple resistance state, it can be used to implement a multi-step transfer function. Ideally, the CSS with N stacked MTJs can achieve a $(N+1)$-step transfer function as shown in Figure 2.11, which is different from the 2^N discrete states when it works as an artificial synaptic device.

Figure 2.12a shows the schematic of the proposed CSS based neural circuit [34], termed as CSN circuit. It involves three operation phases, i.e. encode, read, and reset phase. During the encode phase in Figure 2.12b, the resultant synaptic weighted current I_{synapse}, can be encoded into the CSN resistance state. The higher the magnitude of the I_{synapse} is, the larger resistance state the CSN can be switched into. During the read phase as shown in Figure 2.12c, the CSN can generate an output voltage signal V_{out} to the next layer neurons depending on its resistance state. Note that the output voltage signal V_{out} can be tuned by varying I_{read}. After the information is transferred to the next layer, the CSN circuit is reset for next operation as shown in Figure 2.12d.

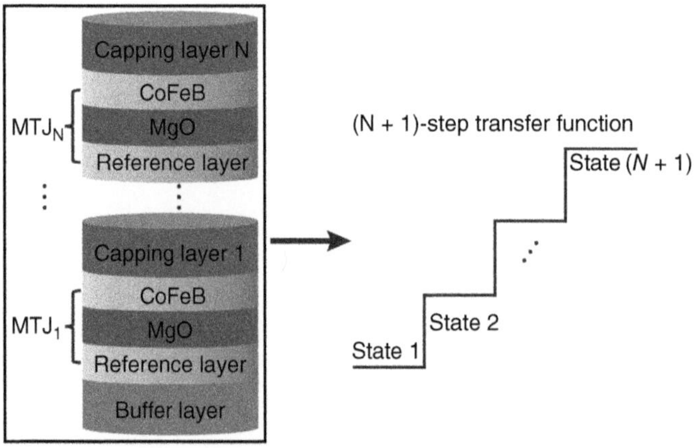

Figure 2.11 Proposed CSN, which can implement a multi-step transfer function. Source: Zhang et al. [34].

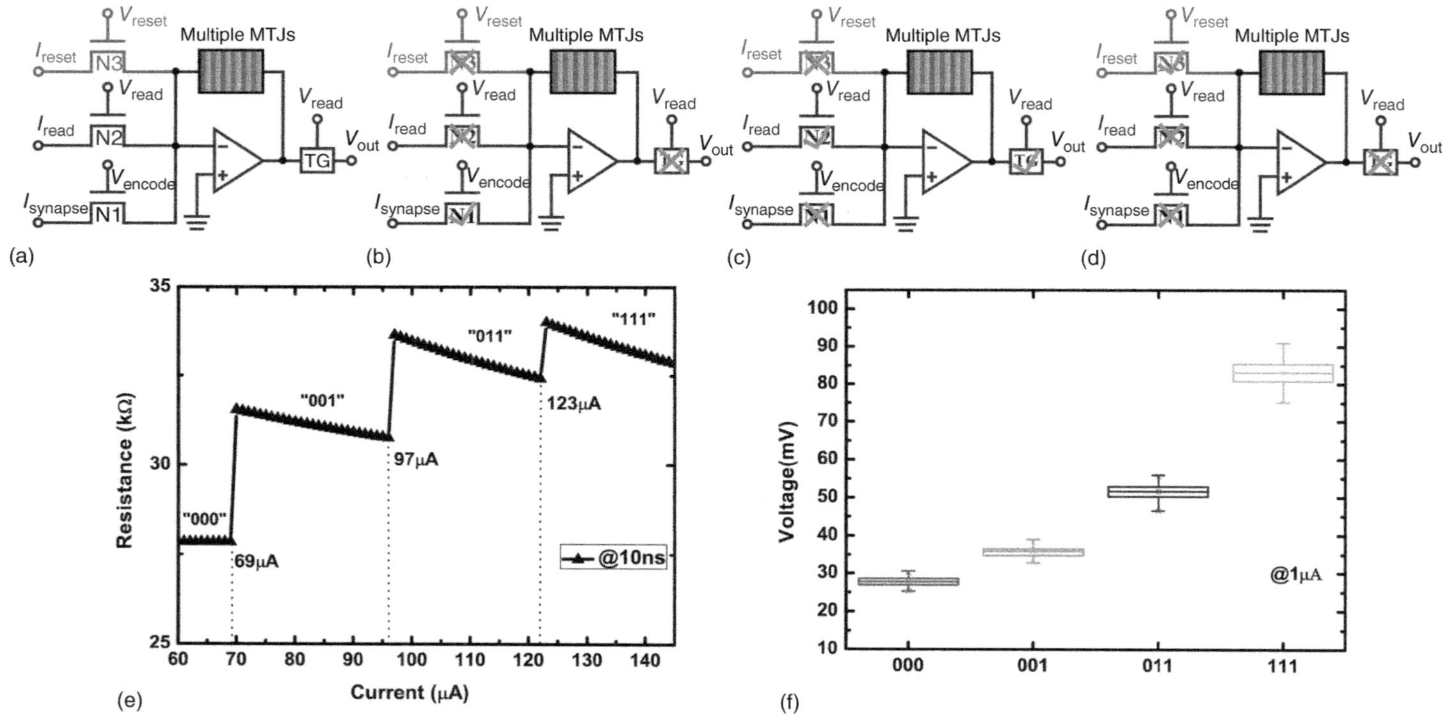

Figure 2.12 (a) The proposed CSN circuit, which involves three operation phases, i.e. (b) encode phase, (c) read phase, (d) reset phase; (e) The simulated threshold switching current for different CSN resistance states and its resistance, where 10 ns pulse is used for the simulation; (f) The output voltage of the CSN for different states. Device variations are considered in the simulation. Taken (e) and (f) together, the transfer function of the proposed CSN can be obtained. Source: Zhang et al. [34].

During the reset phase, all the MTJs are set to the parallel configuration. As a result, the transfer function realized by the CSN can be characterized by the relationship of the $I_{synapse}$ and the output voltage signal V_{out}. Figure 2.12e shows the circuit-level simulations of the relationship of the resultant synaptic current $I_{synapse}$ and the resistance state of the CSN during the encode phase. As seen, the $I_{synapse}$ can be encoded into four discrete resistance states of the CSN at 69/97/123 μA. For example, if the $I_{synapse}$ is larger than 69 μA but smaller than 97 μA, the CSN is encoded in the "001" state. Figure 2.12f shows the relationship of the output voltage signal V_{out} and the resistance state of the CSN, considering the process and device variations. In the simulation, the I_{read} is set to be 1 μA to avoid the voltage bias effect on the resistance of the CSN.

Thanks to its analogue conductance feature, the spintronic memristor is also exploited to implement the non-step transfer function [41] as shown in Figure 2.13. Such "non-step" neural device, here termed as Neuron MTJ, is connected to a Reference MTJ, which is fixed in the AP state. In this way, it constitutes a resistive divider, and then it drives a transistor in saturation regime to ensure that the supplied current to its fan-out synaptic devices is independent of the magnitude of the resistances of its fan-out synaptic devices. This neuron operation also involves three phases, i.e. the write phase, the read phase, and the reset phase. During the write phase, the Neuron MTJ receives the resultant synaptic weighted current (I_{in}) that flows between the terminal T_2 and T_3, whose magnitude determines the position of the DW. The larger the magnitude of the I_{in} is, the further the DW moves to right and the lower the resistance of the Neuron MTJ is. During the read phase, the terminal T_2 is deactivated. The gate voltage of the driver transistor (i.e. PMOS) decreases with the decrease of the resistance of the Neuron MTJ (or the increase of the I_{in}). It means that the output current (I_{out}) of the PMOS transistor increases with the increase of the I_{in}. As a result, the neural transfer function can be characterized by the relationship between the I_{out} and I_{in}. And the PMOS transistor can mimic the axon functionality of the biological neuron by propagating the neural output

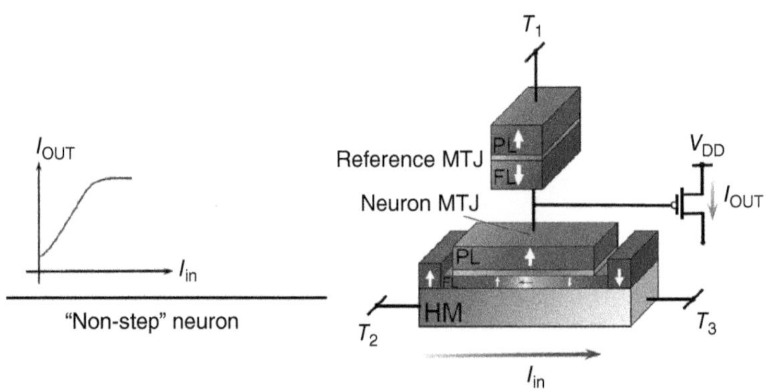

Figure 2.13 The non-step transfer function can be implemented by connecting the spintronic memristor as neuron MTJ to a reference MTJ. Source: Sengupta and Roy [41].

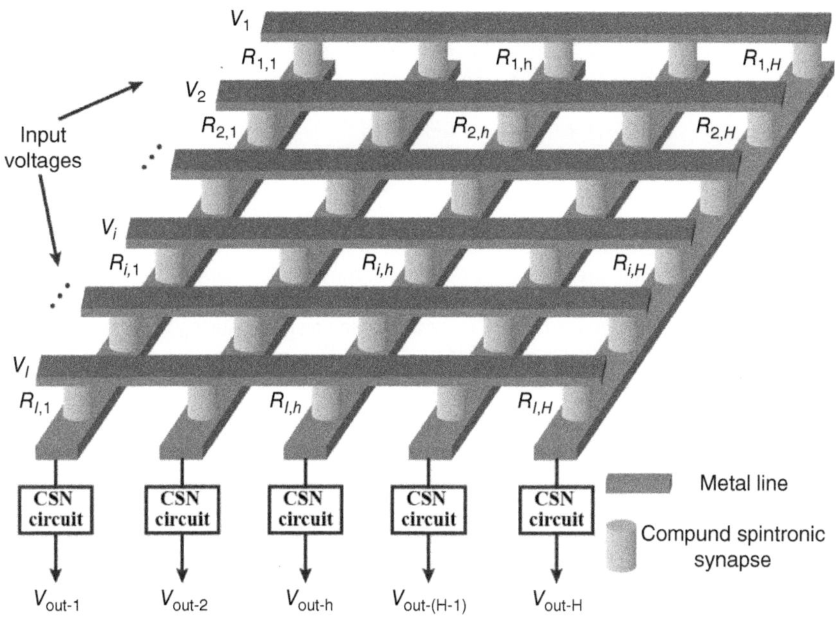

Figure 2.14 Proposed all-spin artificial neural network with proposed CSS and CSN circuit. Source: Zhang et al. [34].

signal to its fan-out synaptic devices. During the reset phase, the DW will be placed in the left edge of the "Free" layer of the Neuron MTJ for next operation cycle.

2.2.4 All-Spin Neural Network

2.2.4.1 All-Spin Artificial Neural Network with Compound Scheme

Based on the proposed CSS and CSN, an all-spin artificial neural network (ASANN) with a fully connected construction can be constructed as shown in Figure 2.14. Depending on the training data, each input neuron will generate a corresponding voltage, which will be applied to the corresponding horizontal metal lines of the CSS crossbar. Then, these input voltages will be modulated by the corresponding CSS in each vertical metal line, thus producing the weighted input signals (i.e. the synaptic currents). These weighted input signals will be summed into the CSN circuit located at the end of each vertical metal line. In the encode phase of the CSN in hidden layer, the CSN is encoded into a specific resistance state with respect to the magnitude of the summed synaptic currents. Then, in the read phase of the CSN, the CSN will output a voltage signal, the magnitude of which depends on the resistance state of the corresponding CSN. And the output voltage signal will be transferred to the next-layer network as an input signal.

Figure 2.15 shows the system-level simulation results of the proposed ASANN for hand-written digital recognition [34]. Figure 2.15a shows the examples of the handwritten digits from the database. Initially, each CSS weight was set randomly as shown in Figure 2.15b. Figure 2.15c shows the learned handwritten digital patterns.

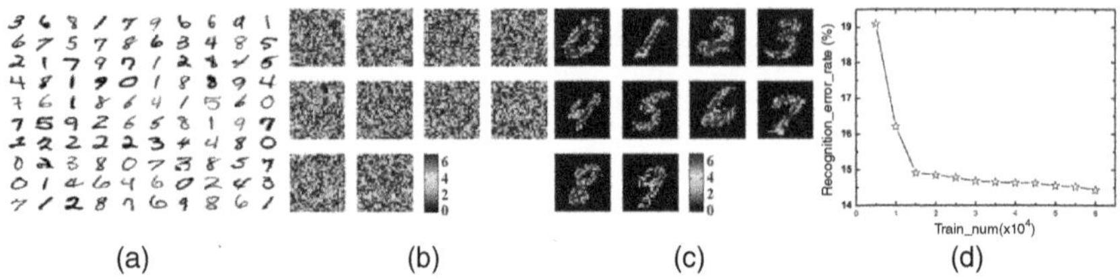

Figure 2.15 System-level simulation results of the proposed ASANN for hand-written digital recognition. (a) Examples of digital patterns in the MNIST database. (b) Initialization of the synaptic weight matrix of the network between the input and hidden layer. (c) The learned handwritten digital patterns. (d) The recognition error rate during the training process. Source: Based on Sengupta et al. [34].

Figure 2.15d shows the recognition error rate during the training process, where the recognition error rate could reach about 9.15%. Moreover, the resolution of both the CSS weight and CSN output signal can be optimized with respect to different applications. The higher the weight resolution, the greater tolerance the ASANN.

Additionally, the SOT-MTJ with the compound method has also been exploited for neuromorphic computing [57].

2.2.4.2 All-Spin Spiking Neural Network with Spintronic Memristor

The spintronic memristors in the form of the crossbar architecture can be used to construct an all-spin spiking neural network (ASSNN) [58], where the spintronic memristor works as the synaptic device and the special spintronic memristor behaves as the spiking neuron. Such crossbar topology of the spintronic memristor enables the parallel matrix-vector operation during the feedforward phase. During the training phase, such spintronic memristor can achieve a STDP learning rule [58]. By performing the system-level simulation with such learning rule, the constructed ASSNN can be used for handwritten digital recognition [58].

2.2.5 Summary and Outlook

This section mainly provides a review of various synaptic and neural functionalities implemented by the MTJ device. However, it demands for the experimental demonstration of these proposals at the network level. Moreover, device-algorithm-architecture co-design and optimization will be required in future works.

2.3 Skyrmion-Based Neuromorphic Computing

2.3.1 The Introduction of Skyrmions

Magnetic skyrmion, a topological nanoscale spin structure, has attracted the attention of the academic community ranging from physics to electronics [59]. It is expected to be the next-generation ultrahigh-density information carrier [60]. Until now, two typical skyrmions with different vorticities and helicities have been

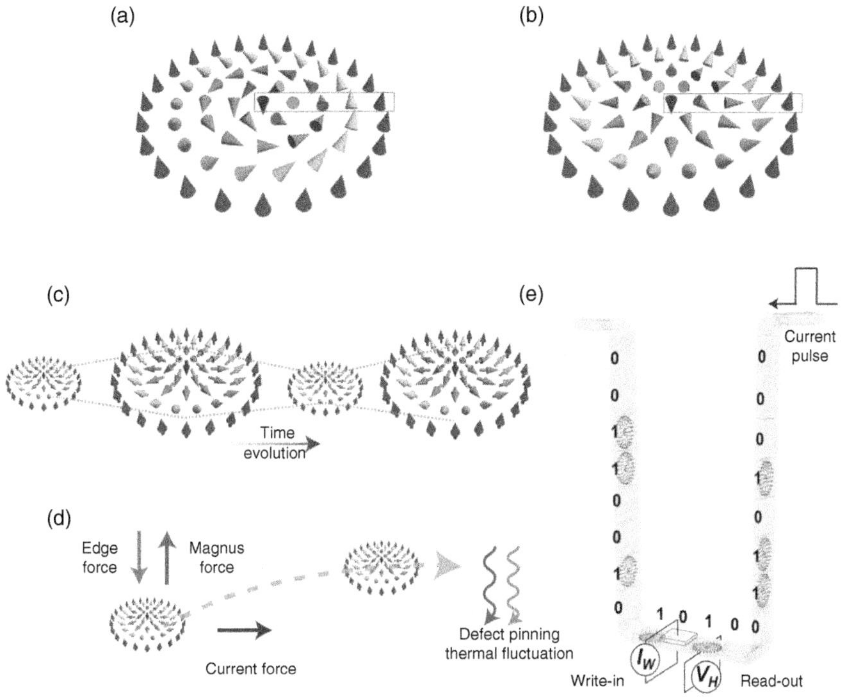

Figure 2.16 (a) Bloch skyrmion. (b) Néel skyrmion. (c) Skyrmion breathing mode with skyrmion size oscillation. (d) Skyrmion motion mode motivated by edge force, magnus force, and current force. (e) The skyrmion-based racetrack memory. Source: Li et al. [61].

mostly studied: Bloch skyrmion and Néel skyrmion (Figure 2.16a,b) [62]. In 2009, Mühlbauer et al. firstly observed skyrmions in bulk material MnSi by small angle neutron scattering (SANS) [63]. In 2013, Sampaio et al. proposed the possibility of developing the practical applications of thin-film isolated skyrmions by numerical investigations [64].

Compared to other spintronic device candidates of information technology, the skyrmion owns many advantages, such as small size, strong stability, defect-tolerance, and low threshold current density due to its topological structure [65, 66]. Recently, many experimental advances have been made in the generation, motion, and detection of skyrmions at room temperature [60, 67–71], further validating the potential of skyrmion-based electronics, called as skyrmionics. For a detailed introduction of the skyrmion and skyrmionics, readers could check Bogdanov et al. [72], Kang et al. [59], and Zhang et al. [73].

The application of skyrmion is mainly based on its controllable and abundant dynamics [74–79], where the first proposal is racetrack memory learnt from the concept designed by S. Parkin in 2008 [80], as shown in the Figure 2.16e. However, the replacement of the memory Flash or SRAM demands for the high write/read accuracy and reliability, which is very difficult to achieve by current skyrmionics.

The reasons are twofold: the skyrmion Hall effect (SkHE) induced by the topological Magnus force and skyrmion diffusion at room temperature, would enable skyrmion deviate from the specified location and annihilate possibly [81]; the ubiquitous material defects will pin skyrmions unexpectedly. Therefore, the traditional manner of skyrmionic device to compute or store the information bit is not practical for the moment. In this context, neuromorphic applications which exploit the dynamics of skyrmion has attracted the attention of researchers. In general, two behaviors of skyrmion were applied. One is the breathing-like behaviors of skyrmion, similar to the STO device, which acts as regular size variation locally (Figure 2.16c) and fluctuates under a certain frequency [82]. On the other hand, the skyrmion could be motivated by multiple forces (Figure 2.16d), where the shape and stimulations of the device could be designed to perform different functions [83]. To date, many conceptual designs of skyrmion-based neuromorphic devices have been proposed [84–87] and some of them have been implemented experimentally [88, 89]. In this section, we will introduce skyrmion-based neuromorphics according to the type of neuromorphic computing studies: synapse, neuron, reservoir computing (RC), stochastic computing (SC). Meanwhile, we will highlight the diverse computing needs for skyrmionic implementations and be more focused on possible future trends. Finally, we present the challenges and outlook.

2.3.2 Skyrmion-Based Synapse Devices

The function of synapse is connecting two neurons and transmitting nerve signals, the strength of which (so called synaptic weight) is varying with the time and excitation of neurons at both ends, as shown in the Figure 2.17. Different from the conventional resistor element, the artificial synapse device needs to have a time-dependent modification with memory to the applied excitations. If the spiking time lag $\Delta\tau$ ($\Delta\tau = t_{\text{post}} - t_{\text{pre}}$) between the two neurons is positive, the synaptic weight increases. So, it can be understood that why the STDP approach enables the devices more suitable to solve the event-driven tasks.

Magnetic skyrmions have the physical nature of pinning and stable current-driven motions, it is possible to construct the multistate resistance. Skyrmions have been demonstrated to circumvent around pinning sites or impurities in materials owing to its chiral topology [91], thus to reduce the possibility of information lost. On the other hand, skyrmion has a smaller critical current density ~10^2 A/cm^2, which is promising to significantly lower power consumption. In this context, Huang et al. proposed a skyrmion-based artificial synaptic device, as shown in Figure 2.18b, where main structure consists of two FM/HM layers [92]. The device presented in Figure 2.18a is divided into a post-synapse and a pre-synapse, the former is the detection area depending on the number of skyrmions moving in. The magnetoresistance detected by the MTJ represents the synaptic weight. During the synapse learning process, bidirectional spin currents generated by the SOT in HM will drive the skyrmion into (or out of) the postsynaptic region to increase (or decrease) the synaptic weights, thereby mimicking the potentiation/depression process of the biological synapse. The STDP scheme of the synaptic device could be

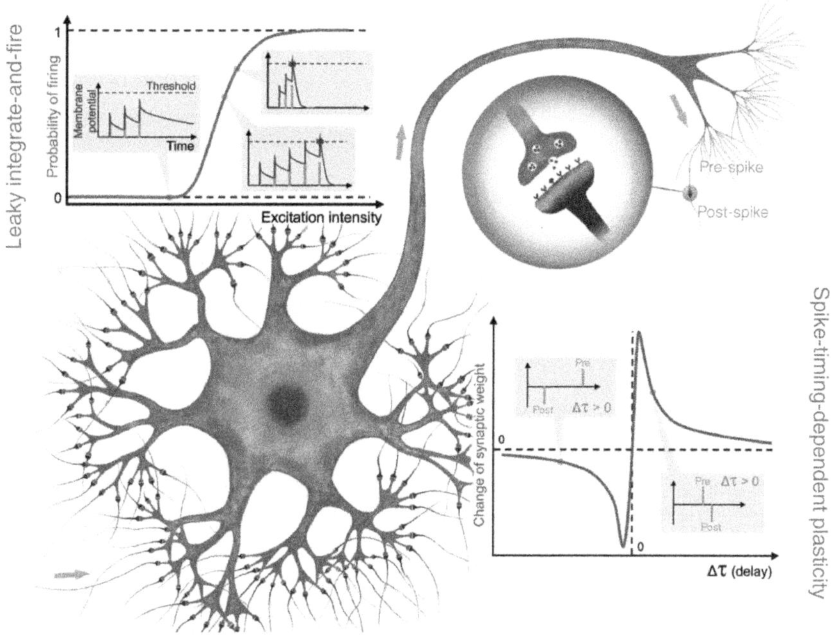

Figure 2.17 A schematic diagram of a biological neural network describing functionalities of a neuron and a synapse. Source: Kurenkov et al. [90].

achieved by adjusting the energy potential barrier in the nanotrack. Apart from the motion of skyrmion on the track, the localized skyrmion in the MTJ could also have abundant physical capabilities. In Figure 2.18c, the STT current could generate and annihilate skyrmions thereby the number variation of skyrmions in the free layer could be read out by TMR (see Figure 2.19), realizing a multi-level resistance [87]. Recently, Song et al. have experimentally demonstrated the functionality of such skyrmion-based synaptic devices by using a combination of current and magnetic field to modulate the generation, movement, and annihilation of skyrmion, where the detected Hall resistance represents the synaptic weight [88].

2.3.3 Skyrmion-Based Neuron Devices

Compared to synapse, the biological neuron is more complex to mimic for its probabilistic spiking nature required for the frequency-dependent performance [42]. The bio-fidelity of the spintronic device not only depends on the choose of proper neuron model, but also the sensitivity of the device responding to the stimulations transmitted by excitatory or inhibitory of synaptic signals. The LIF model is the most widely adopted neuron model in skyrmionics [93]. With the PMA-varying nanotrack, Li et al. proposed a skyrmion-based artificial neuron device [94] shown in Figure 2.20. To achieve a time-dependent motion on the track, the device is designed to be consisting of a thickness-gradient FM layer adjacent with a uniform HM layer to obtain a linear PMA along the nanotrack. Therefore,

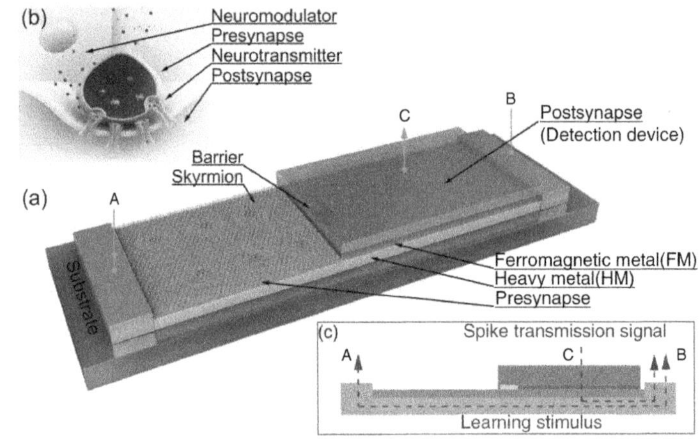

Figure 2.18 Schematic of (a) the proposed skyrmionic synaptic device and (b) a biological synapse. (b) A bidirectional current flowing through the HM from terminal A to terminal B (or vice versa) drives skyrmions into (or out of) the postsynapse region to increase (or decrease) the synaptic weight. (c) A detection device at terminal C measures the weight of the synaptic device via the magnetoresistance effect, which also modulates the post-synaptic spike current (depending on the weight of the synaptic device). Source: Huang et al. [92].

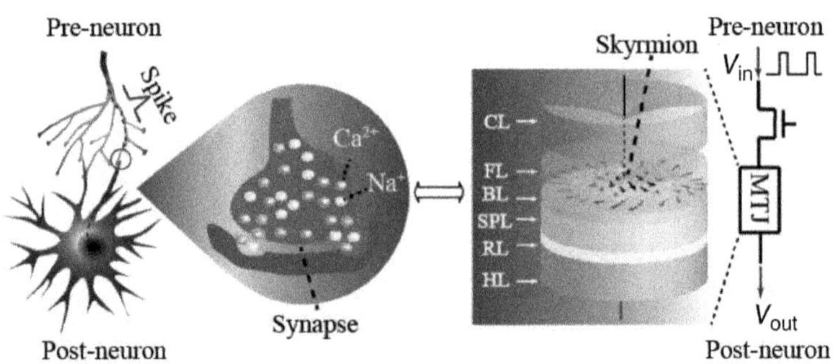

Figure 2.19 Analogy between the biological synapse and the electronic ternary synapse based on the skyrmionic MTJ device. Source: Pan et al. [87].

the pre-spike current force and the backward PMA force could compete each other and make skyrmion mobile along the nanotrack forth and back, showing a "leaky-integrate" pattern. Once the skyrmion reached the detection area, the detection MTJ will "fires" an output spike. Then, the skyrmion will be pushed back quickly to the initial position to wait for next spike, which is the reset mode of the proposed neuron device. Similarly, Chen et al. proposed a skyrmion-based neuron device using a wedge-shaped nanotrack [95], where the current force competes with the wedge repulsive force, instead of the PMA-gradient force. Except LIF, there are some other models to describe the neuron behaviors. A fixed magnetic skyrmion in an MTJ with Giga-Hertz oscillation could exactly emulate resonate

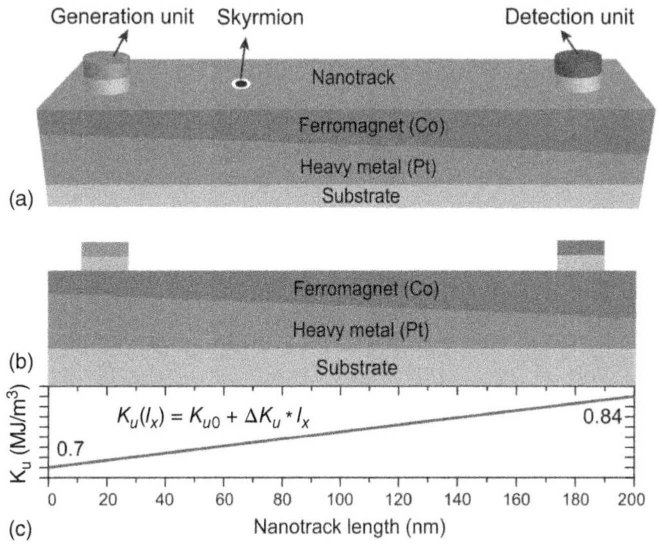

Figure 2.20 (a) Schematic of the skyrmion-based artificial neuron device; (b) The front cross-section view; (c) The linear increase of the PMA value along the nanotrack. Source: Li et al. [94].

and fire neuron [96], which could potentially work together with STO arrays in neuromorphic computing. For a brain-like neuron function, the device sensitivity to signal frequency and amplitude is desired. The spintronic devices, especially skyrmionic device could be a dense, energy efficient alternative in this regard.

2.3.4 Skyrmion-Based Reservoir Computing

RC has recently been introduced for its bio-inspired feature of fixed interconnection weights, especially applicable to solve the spatial-temporal data problems. In RC, a reservoir medium could transform the input data into spatiotemporal patterns in a high-dimensional feature space then be readout by a simple pattern analysis. Therefore, the nonlinear physical systems/devices are very appropriate to realize RC (Figure 2.21). The weights of the input-reservoir and reservoir interior are fixed, which do not need training. Only the readout weights could be trained with a simple algorithm, for instance, linear regression.

The reduced training amount makes RC outperform classical (recurrent neural networks, RNNs) in some certain tasks, such as identification, prediction, and classification of nonlinear system. Its advantages include the following: (i) improved modeling accuracy, e.g. in predicting chaotic dynamics; (ii) enlarged applied time-resolved application in continuous-time-and-value systems; (iii) biological plausibility closer to the brain computing mechanism; and (iv) lower latency [97].

Pychynenko et al. firstly proposed to use is a pinned skyrmion as a physical implementation for RC application, where the nonlinear voltage is detected by anisotropic magnetoresistance (AMR) [98]. As shown in Figure 2.22, the skyrmion is initially pinned at the center, the electrical field induced by the two contacts will lead to a

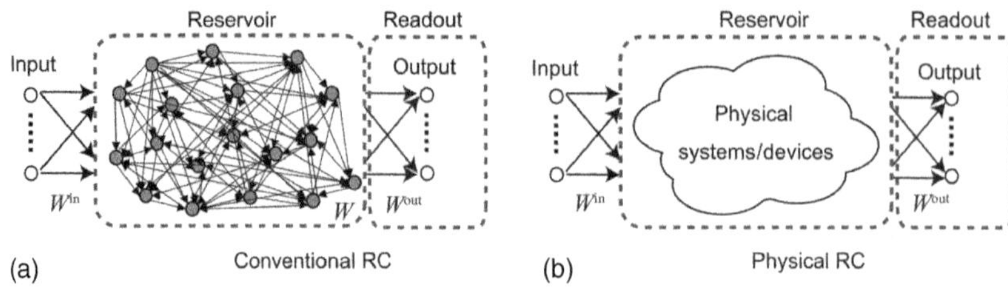

Figure 2.21 Comparative concepts of (a) Conventional RC and (b) Physical RC. Source: Tanaka et al. [97].

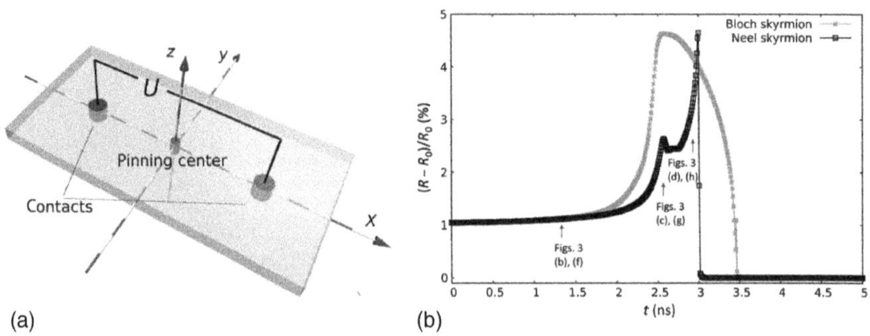

Figure 2.22 (a) Schematic of the skyrmion-based RC device. (b) Resistance of a pinned Skyrmion as a function of an applied voltage visualizing the nonlinear behavior. Source: Prychynenko et al. [98].

characteristic current flow calculated by AMR. Combining the pinning force and the STT force, the skyrmion will be deformed or stretched. The random thermal fluctuations may even enhance the nonlinearity, which could turn the temperature from a negative effect on traditional device, e.g. logic gate, to a positive effect on the performance of RC systems. Except single skyrmion, the skyrmion fabrics with more complex and confusing magnetic texture could also generate a non-linear current flow responding to variable input voltages, increasing the tunability by optimizing the material [99]. Compared to other RC candidates, skyrmions offer a rich array of topology-dependent static and dynamic properties to support an internally interconnected reservoir, such as the skyrmion–skyrmion interaction, thermal diffusion. In 2019, Jiang et al. implemented one single magnetic skyrmion memristor as a RC node, where its nonlinear responses directly perform as an output to process classification task, achieving digital image classification by skyrmionic RC system [100]. In summary, the more nonlinearity of the device, the more attractive option for RC implementation. While for a neural network, the detectable capacity of the device in each reservoir node should also be considered.

2.3.5 Skyrmion-Based Stochastic Computing

In SC, the information is processed and encoded by probabilities of random binary bit streams. So, it does not require very high reliability and low error rate as conventional computing does, providing the advantages of the hardware simplicity

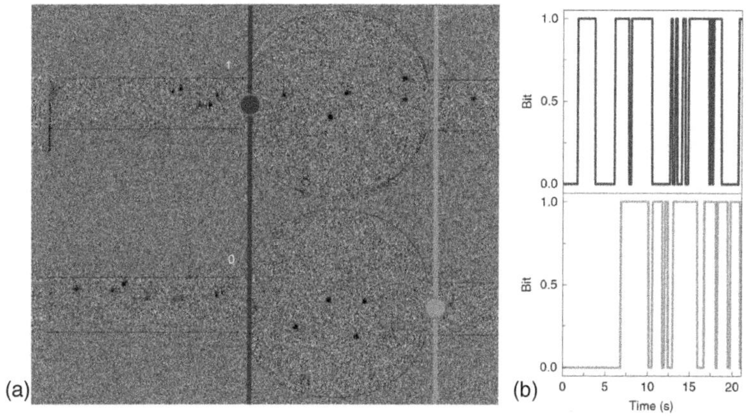

Figure 2.23 Skyrmion reshuffler device. (a) Reshuffler operation with skyrmion nucleation by a direct current. The input signal is constructed as a time frame in which the skyrmion crosses the threshold line, while the output is determined on crossing the right line. (b) The corresponding input (up) and reshuffled output signals (below). Source: Based on Wang et al. [89].

and fault tolerance [101]. It exploits the statistical nature of data metrics of some emerging applications, where devices could implement bit-wise operations to complete computations at low energy cost and high noise robustness. The first important component is the random number generator, which occupies the most part of SC energy cost. In the past, pseudorandom number generator is widely used, but its issues are low-quality of random and high energy consumption. In this context, skyrmionic devices become good options for their real stochastic physical properties.

Skyrmion has a semblable thermal-induced random phenomenon. D. Pinna et al. devised a skyrmion reshuffle [102] to wash out the correlations of the two bitstreams, improving the accuracy of the SC. As shown in Figure 2.23, the skyrmions are generated at a constant rate and driven by the two up-and-down pulses injected into the chamber. The stochastic Brownian motions of skyrmions in the chamber will keep the "1" rate of the pulse streams but change "1" location of the stream in the time frame, experimentally demonstrated by Zázvorka et al. [89] Thermal Brownian motion is also used by Yao et al. to design a TRNG device [103]. In Figure 2.24, the relative position of skyrmion inside the chamber assigns the output bits, which is detected by MTJ as a random bit sequence. The probability of "0" and "1" could be tuned by anisotropy gradient. To sum up, skyrmionic devices show the advantages of low cost, high quality of TRNG and high density for SC. The further work of peripheral circuit design and real-world task demonstration are needed.

2.3.6 Challenges and Perspectives

This subsection will highlight the two main directions of skyrmion from the viewpoint of application. As shown in Figure 2.25a, the material optimization of skyrmion and experimental demonstration of neuromorphic device are urgently

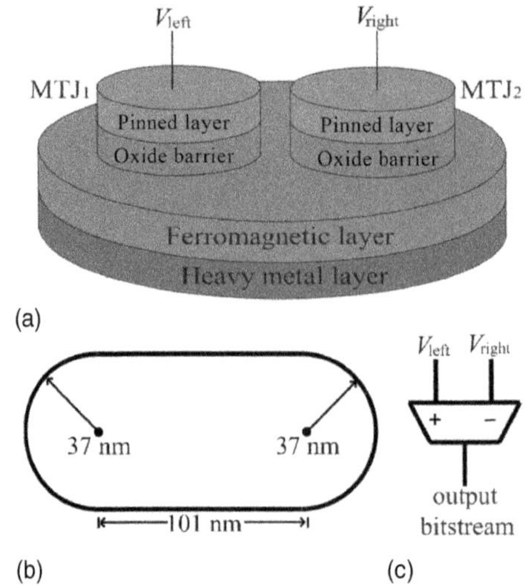

Figure 2.24 (a) The schematic structure of the skyrmion-based TRNG: the ferromagnetic layer is used for skyrmion Brownian motion and the two MTJs are used to detect the skyrmion. (b) The top view of the FM chamber (c) The schematic of the comparator. It is assigned a bit of "0" if the skyrmion topological is on the left; otherwise, a bit of "1" is indicated. Source: Yao et al. [103].

needed. To date, there are only two experimental studies to demonstrate the possibilities of unconventional skyrmion-based devices. On the other hand, the current skyrmionic neuromorphics was limited at the device level to demonstrate the functionality and analyze the various parameters. The future device-circuit-algorithm codesign can accelerate device optimization and explore design space, thereby the development of this area from concepts to practical applications. In terms of performance, energy, and reliability characteristics, hybrid simulation framework/tool could be further explored. To present the most necessary targets of future skyrmion-based unconventional computing, we show some attributes as below (Figure 2.25b):

(1) Density: The skyrmion is considered to be the smallest information carrier, and its size has been theoretically proven to be in the range of 1–300 nm, offering a great advantage in density compared to other emerging technologies.
(2) Nonvolatility: Skyrmion is spintronic object and thus physically nonvolatile, but the stability at room temperature needs more explorations.
(3) Power consumption: It mainly consists of the energy cost of skyrmion generation, detection, and manipulation. The low-power operation of skyrmionic devices is the most necessary target.
(4) Endurance: Skyrmion could be created/deleted repeatedly without damaging the device.
(5) Maturity: The potential of skyrmionics at the circuit- and architecture-level needs more investigation.
(6) Latency: The output delay of skyrmionic device and the peripheral circuitry are key determinants, which could be improved by multiple optimization methods.

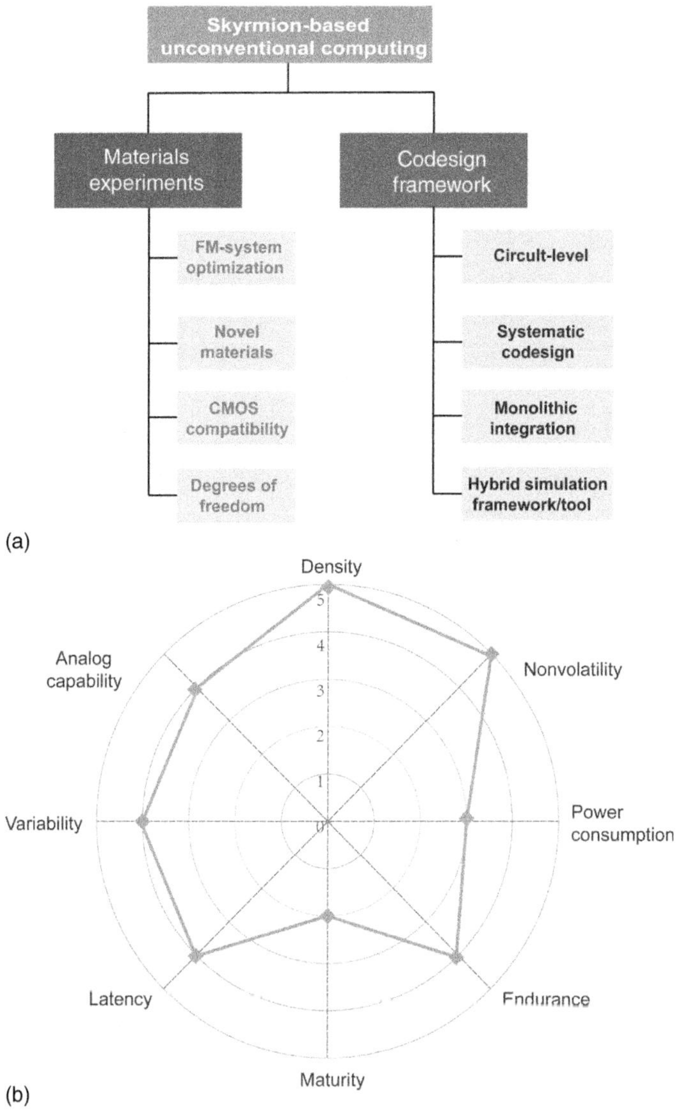

Figure 2.25 (a) An overview of the current challenges of skyrmion-based neuromorphic computing. (b) The spider chart presenting the most necessary targets for future unconventional computing and the current status of skyrmionic devices. Source: Li et al. [61].

(7) Variability: It is most relevant to the physical properties of skyrmion. In the neuromorphics computing, variability can be well tolerated or even offer the performance enhancement.
(8) Analog capability: It is the superiority of skyrmionic device, which is suitable for designing the computing devices requiring high spatial-temporal flexibility.

2.4 Spin Torque Oscillators for Neuromorphic Computing

2.4.1 Introduction to Spin Torque Oscillator

The oscillators are widely used in modern telecommunication system. The voltage controlled oscillators (VCOs) fabricated by CMOS technology can generate high quality radio frequency signal when it is coupled with a phase locked loop (PLL) circuit [104–106]. Besides, it was proposed by von Neumann that oscillators can be as logic gates, where information was represented by the phase [107]. This concept was further developed by Goto and others and fully functional Boolean computers using oscillators were realized in 1950s, which was the very early age of modern electronic computers [108, 109]. Thus, computation base on oscillators has history almost as long as electronic computers.

The slowdown of semiconductor transistors' scaling pace and the desire for low energy neuromorphic computation to perform task like recognition prompt the renaissance of the oscillator-based computation. Among novel nano-sized oscillators, STOs stand out since its small dimension (~100 nm), high tunable frequency range (~GHz) and CMOS compatible fabrication process [110–113]. A typical structure of STO is plotted in Figure 2.26a [114]. It is just an MTJ, which is already introduced in Section 2.2.1.1. Under appropriate condition, the damping term from Gilbert constant can be compensated by the spin torque term, where a stable closed loop oscillation can be sustained, as shown in Figure 2.26b.

It is revealed both theoretically and experimentally that, the STOs are highly non-linear oscillators since the change of its oscillation amplitude will cause the change of its oscillation frequency [115–117]. Such strong nonlinearity also enhances the injection locking range [115]. The injection locking phenomenon is illustrated in Figure 2.27. The STOs can be not only injection-locked to the external perturbations, but also mutually injection locked to each other. This is called synchronization [118]. In mutual synchronization scenario, the oscillation of a STO

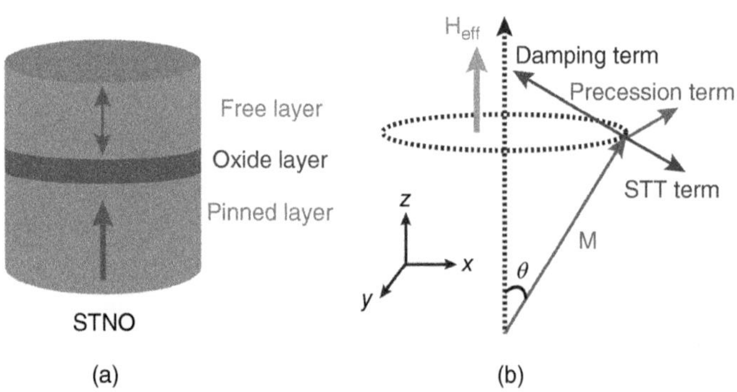

Figure 2.26 (a) The schematic of spin torque oscillator. (b) The balance between damping term and spin transfer torque term, and the magnetic dynamic caused by precession term. Source: Zeng et al. [114].

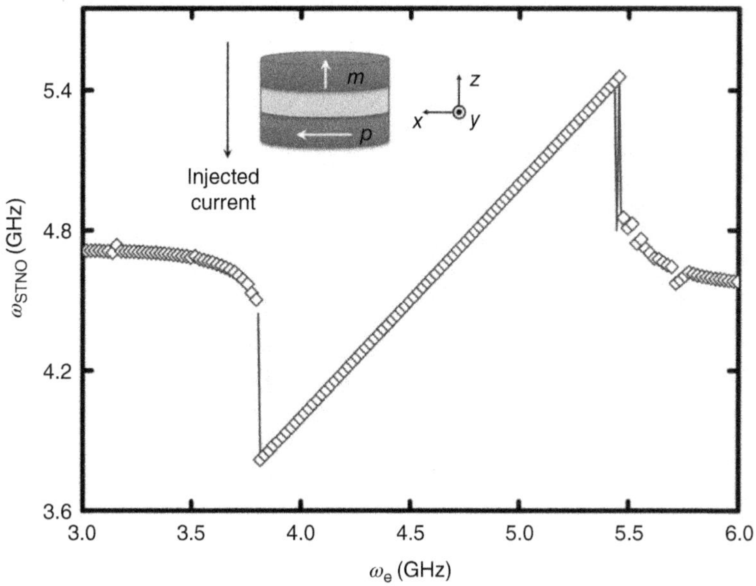

Figure 2.27 The illustration of injection locking phenomenon. ω_{STNO} is the oscillation frequency of STO and ω_e is the frequency of the injected microwave current. Source: Zeng et al. [115].

array is just like in a Chorus thus the phase of each STO is coherent. The mutual synchronization of STO is considered an efficient solution to reduce the phase noise and increase the output power of STO simultaneously. Also, there are a number of neuromorphic computation proposals utilizing STO synchronization.

When the input stimulus change, the oscillation frequency of STO will change accordingly. However, since the STO is highly non-linear oscillator, the amplitude and frequency are coupled. The oscillation amplitude (or the oscillation orbital) will change to realize frequency change. The speed of oscillation orbital change is controlled by amplitude relaxation rate which is usually tens or hundreds of MHz [119, 120]. Thus, the frequency of STO cannot change simultaneously. This implies that the STO has somewhat memory effect when the input stimulus changes continuously in time domain.

All of the above-described features of the STO including injection locking, synchronization and memory effect can be used to design specific neuromorphic computation system. In the following, we will introduce them one by one.

2.4.2 Associative Memory Based on Injection Locking of STO

The associative memory is defined as the ability to learn and remember the relationship between unrelated items in psychology. In neuromorphic computation, associative memory is often used to recall the closest pattern of the training set from the noisy input inference set [121–123]. Such associative computing is inefficient to perform on von Neumann architecture in CMOS technology. However, it can be implemented energy efficiently with injection locked STOs.

For the edge detection of an image by injection locking of STOs, each pixel corresponds to one STO [123]. In the locking period, each STO in the array is injected with the same current I_{bias}. After a short time (several nanoseconds), the STO array starts to oscillate at the same frequency since the injected current is the same. In the computing period, the additive current I_{pixel} whose strength depends on the pixel intensity is also injected into each STO. The current I_{pixel} is a maximum for the white pixels and a minimum for the black pixels. For the gray pixel, the current I_{pixel} is calculated with the intensity of pixel. Thus, the total current injected into STO in the computing period can be written as $I_{input} = I_{bias} + I_{pixel}$. It can be postulated that the STOs corresponding to the pixels with similar intensity has similar input current leading to close oscillating frequency. While the STOs corresponding to the pixels with high contrast in intensities have separated input current thus separated oscillating frequency. The STOs with similar oscillating frequency can be injection locked whereas that with separated oscillating frequency will not lock, as their frequencies differ considerably.

The locked and non-locked STOs can be detected with a proposed detector circuit [123]. The output voltage from the STO array is added and averaged first. When the STOs are locked, their output waveforms are phase coherent thus the addition is constructive. This means the output voltage for locked STO array has high peak-to-peak voltage. When the STOs are non-locked, the addition of output waveforms is destructive leading to lower averaged output voltage. The higher integrated voltage means the "non-edge" part while the lower integrated voltage means the "edge" part.

2.4.3 Reservoir Computing Based on STO

As we mentioned previously, the finite amplitude relaxation rate and strong non-linearity implied complicated magnetic dynamics of STO, which is suitable for utilizing as a reservoir.

One experimentally implementation of physical reservoir computing system using STO is shown in Figure 2.28 [124]. The structure of STO is shown in Figure 2.28a and the oscillation waveform is shown in Figure 2.28b. As the injected DC current varies, the output voltage amplitude changes accordingly as shown in Figure 2.28c. The complex non-linear response of STO to the input voltage V_{in} is observed in Figure 2.28e. The STO can be regarded as a filter to the input voltage V_{in} since the finite amplitude relaxation rate. In this experimental demonstration, oscillation amplitude is chosen to perform the calculation.

The computing is performed with single STO as single reservoir neuron and time-multiplexing should be employed in order to emulate neurons interconnected through time [124, 125]. The input audio file is pre-processed by two different methods: spectrogram and cochlear models in Figure 2.29a,b. Then the pre-processed input audio files are further filtered by STO and the recoded trace are measured and reconstructed by a computer in Figure 2.29c,d. After post-processing, the recognition rates with and without oscillator are plotted in Figure 2.29e,f. The

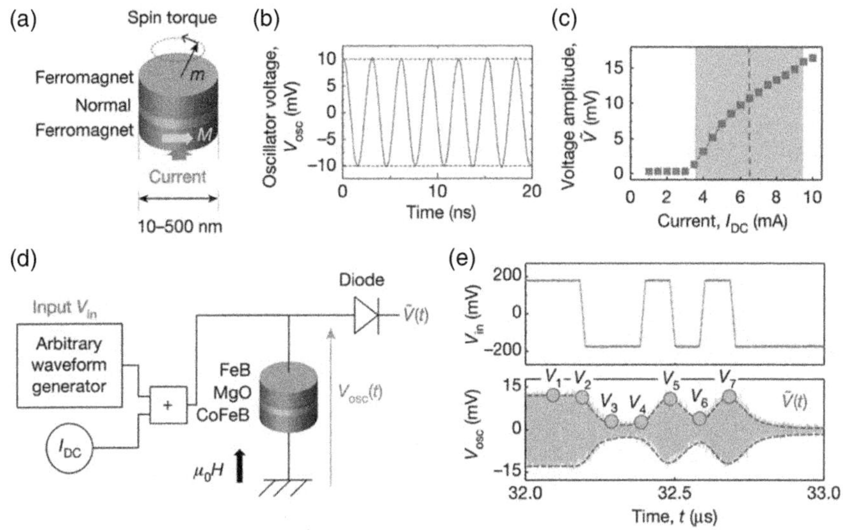

Figure 2.28 STO can be acted as a reservoir. Source: Torrejon et al. [124].

state-of-the-art 99.6% recognition rate is achieved. With oscillator by cochlear filtering. The effect of STO as reservoir on reducing the error rate is significant.

Although the significant reduction of the error rate with STO acting as reservoir is shown experimentally above, this field is still in its infancy and needs further deeper exploration. Besides oscillation amplitude, it is proved that the oscillation phase and frequency also can be used to perform reservoir computing [126]. The memory property of the STO dynamics, called short-term memory (STM), can be tuned by forced synchronization. The STM capacity can be increased about two times compared with that obtained without forced synchronization [127]. The dependence of memory function on pulse width injected into STO is investigated. A step-like dependence on the pulse width for the STM and parity-check capacities is obtained [128]. The capacities remain at 1.5 for a range of the pulse width, and drop to 1.0 for a long pulse-width. The utilization of an array of STOs as reservoir non-linear neurons is presented by theoretical analysis [129].

2.4.4 Recurrent Neural Network based on Delayed Feedback of STO

The memory function of STO can be made even more complicated by introducing time-delay feedback. With such time-delay feedback, a RNN can be implemented [130]. The connections of the RNN form a directed graph along a temporal sequence, and it is suitable to tasks such as handwriting recognition and speech recognition. An example of RNN based on delayed feedback of STO is shown in Figure 2.30.

The RNN shown in Figure 2.30 is very similar with the reservoir computing based on STO shown in Figure 2.28d. The difference is the feedback loop from the output of STO through a delay line and injecting back into the input. This time-delay feedback extends the memory function of STO and this extrinsic memory can increase the

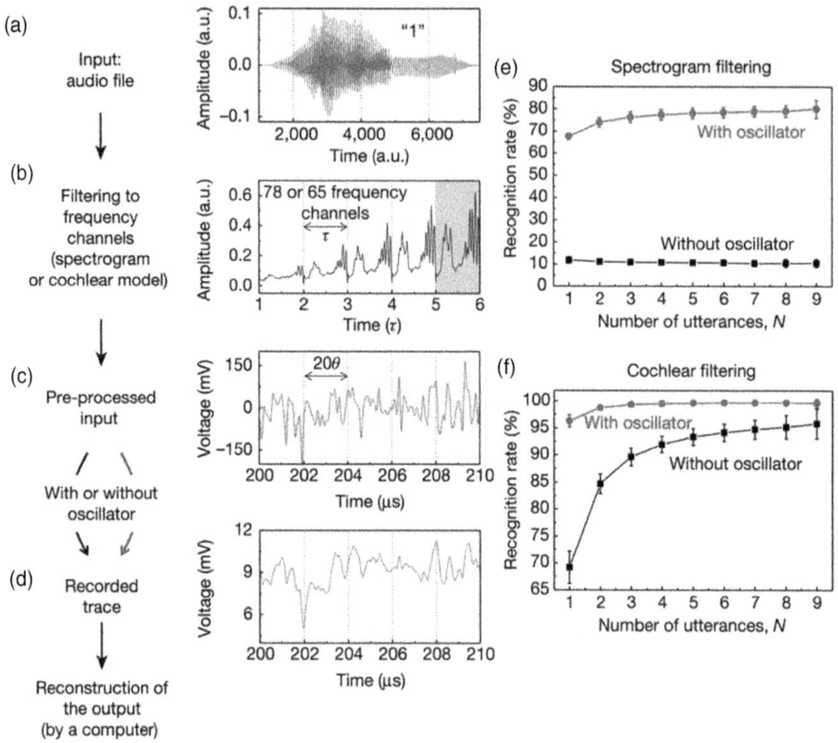

Figure 2.29 The flow chart for reservoir computing with STO. Source: Torrejon et al. [124].

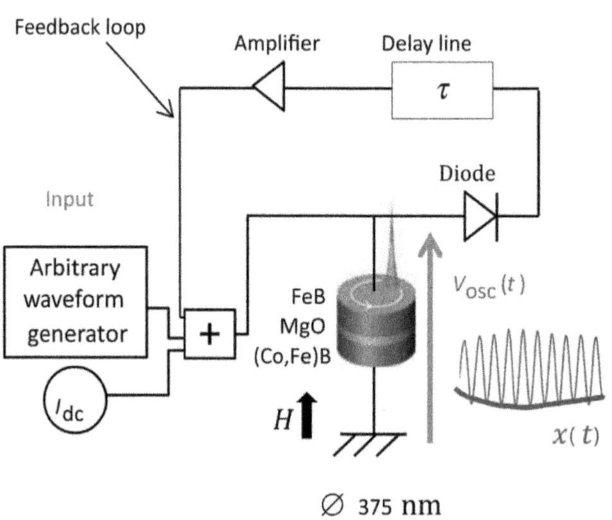

Figure 2.30 Recurrent neural network based on delayed feedback of STO. Source: Riou et al. [130].

efficiency of solving pattern recognition tasks. The successful discrimination of sine and square waveforms is demonstrated.

2.4.5 Neuromorphic Computing Based on the Synchronization of STO

To perform calculations with the synchronizations of oscillators has a long history. Ising machine is a widely used computation model which can perform generical purpose calculation as well as non-Boolean calculation. The synchronization of oscillators can be mapped into the minimizations of energy of the Ising machine [131]. There are lots of work on such topic with different kinds of oscillators, like CMOS oscillators, VO_2 oscillators as well as STOs. Besides the Ising machine, it is proposed that neuromorphic computing can be realized based on the synchronization of STO [132, 133]. The synchronization of STOs can act as neurons.

The neuromorphic computation system with the synchronization of STOs is proposed as shown in Figure 2.31. The input is f_A and f_B, and four coupled STOs with full connections are treated as neurons. The four coupled STOs are biased with different input current and oscillating at different frequency. The different combinations of input f_A and f_B can trigger different response of the four coupled STOs. Experimental synchronization map as a function of the frequencies is shown in Figure 2.31e. Each color corresponds to a different synchronization state, and each synchronization state can be regarded as recognition of a different spoken vowel. The recognition rate can be as high as 88%. If ideal noiseless oscillator is employed, the recognition rate can be boosted to 94%.

The full connections between different coupled STOs may be a problem. In another work, the connections are proposed to be realized by coupling through spin waves [134]. The illustration for the neuromorphic computing based on the spin wave synchronization is plotted in Figure 2.32a. The two-dimensional STOs array is fabrication as shown in Figure 2.32b. The synchronization characteristics for each STNO is shown in Figure. 2.32c. The corresponding experimental synchronization map is shown in Figure 2.32d.

For designing a large coupled STOs array, the frequency separation of each STO should be designed carefully, and its feasibility should be examined. A large array of STOs with n equally spaced frequency separation of 5 MHz is designed by analytical and numerical method [135]. It is proved that arrays of hundreds of STOs with ranges of several hundred MHz are possible even considering nanofabrication constraints. It is worth noting that an ANN can be implemented with a neuron of active function designed by spin wave coupled STOs [136]. Additionally, a ReLU activation function is experimentally demonstrated in STO by spin diode effect [137].

2.4.6 Problems and Perspectives

Although the neuromorphic computation based on STOs has advantages of high speed, simple structure, ample functionality and low power, this research field still is in its early stage and more mysteries need to be solved. Here we list some obstacles

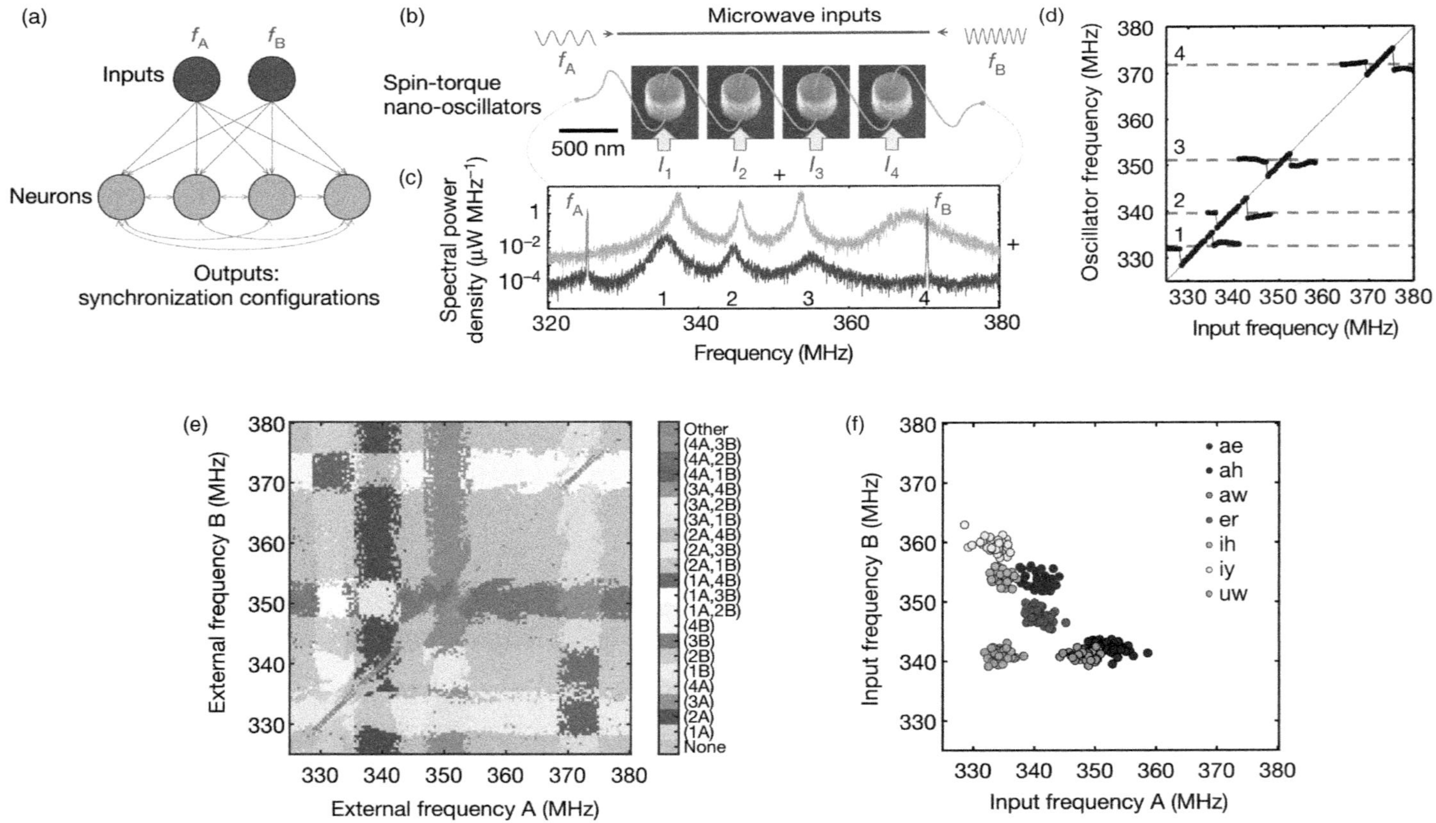

Figure 2.31 Coupled STOs act as neurons. Source: Romera et al. [132].

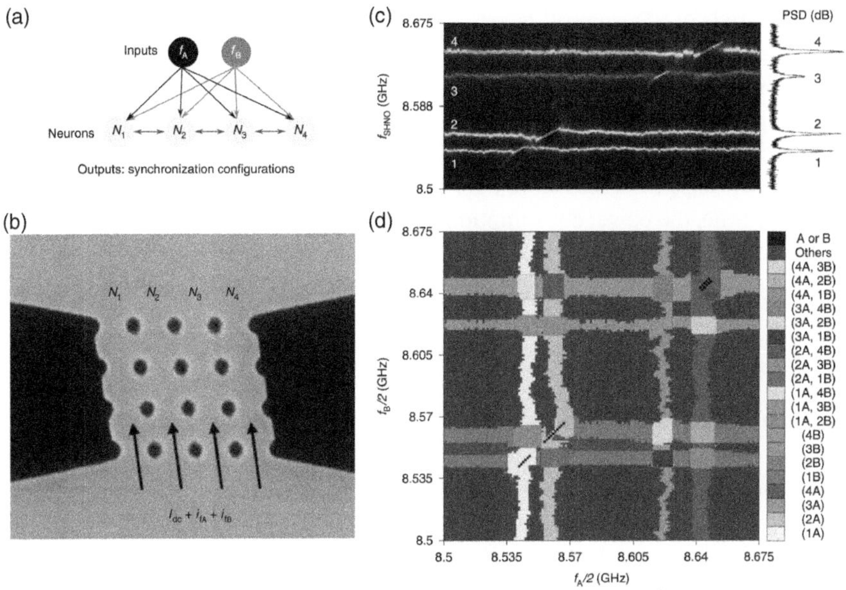

Figure 2.32 Coupling of two dimensional STOs through spin waves and its corresponding experimental synchronization map. Source: Zahedinejad et al. [134].

needed to be surpass and potential solutions and directions for further exploration in this topic.

(1) Noise: The small size of STOs indicates a strong effect of thermal noise on their magnetic dynamics. Actually, it is reported that the noise of STOs is 4–6 orders higher than that of VCOs. The noise reduces injection locking range and hinders the mutual synchronization of STOs. It can also reduce the recognition rate comparing with ideal noiseless oscillators. There are several theoretical and experimental works devoted to the reduction of noise in STOs. A recent work proposes a novel method to reduce noise and enhance mutual synchronization which introduces an additional filter in the coupling loop. [17]
(2) Interconnections: The interconnections between neurons are major challenges for almost all kinds of neuromorphic networks. The reason is the number of interconnections scales to the square of the number of neurons. Fortunately, STOs coupled through spin waves may provide a simple solution. In an array of STOs, the STOs will emit and exchange spin waves intrinsically [44]. The coupling through spin waves between STOs cannot be tuned individually in principle. However, in reservoir computing, the interconnections between neurons are random and fixed, and the computation is performed by the complicated memory function of the reservoir network. This scenario fits perfectly with STOs array coupling through spin wave and needs no additional interconnections. Again, noise will play an important and harmful role in this scenario.
(3) Scaling: The experimental work introduced above usually uses only a few STOs, sometimes even only one STO. The scaling ability of the neural network to tens

or hundreds of STOs is still questionable. Although it has been verified that a large array of STOs is possible even when nano fabrication variability is taken into account [45]. But it is just an analytical and numerical analysis and demands further experimental proof.

(4) Power: Although it is claimed that STOs can have very small power consumption down to pJ per oscillation, the result is just a theoretical prediction. In the present time, the power consumption of one single STO is in the order of several hundred μW which is not small enough [138]. The next step is to use extremely scaling STOs whose dimension should be about 10 nm to demonstrate actual low power neuromorphic calculation.

2.5 Conclusion and Outlook

In this chapter, the present status of the spintronic devices for neuromorphic computing is summarized and reviewed. Versatile spintronic devices, including the MTJ, spintronic memristor based on DWM, skyrmion device and STO device, can be employed to implement the synaptic and neural functionalities as well as the all-spin neural network for pattern recognition applications. However, most of the related work is still limited at the proof-of-concepts device simulation level. To convincingly prove the power of neuromorphic computation based on spintronics devices, experimental fabrication of the proposed synapse and neuron based on spintronics devices should be demonstrated. Furthermore, the array of such proposed devices is demanded to facilitate the device-algorithm-architecture co-design and optimization in system level. Nevertheless, it opens up a way for hardware implementation of neuromorphic computing and broadens the spectrum of material systems for the synaptic and neural device.

References

1 Fong, X., Kim, Y., Venkatesan, R. et al. (2016). Spin-transfer torque memories: devices, circuits, and systems. *Proc. IEEE* 104: 1449–1488.

2 Girard, P., Cheng, Y., Virazel, A. et al. (2020). A survey of test and reliability solutions for magnetic random access memories. *Proc. IEEE* 109: 1–21.

3 Zhang, Y., Zhao, W., Lakys, Y. et al. (2012). Compact modeling of perpendicular-anisotropy CoFeB/MgO magnetic tunnel junctions. *IEEE Trans. Electron. Devices* 59: 819–826.

4 Wang, M., Cai, W., Zhu, D. et al. (2018). Field-free switching of a perpendicular magnetic tunnel junction through the interplay of spin–orbit and spin-transfer torques. *Nat. Electron.* 1: 582–588.

5 Ikeda, S., Miura, K., Yamamoto, H. et al. (2010). A perpendicular-anisotropy CoFeB-MgO magnetic tunnel junction. *Nat. Mater.* 9: 721–724.

6 Lin, X., Yang, W., Wang, K.L., and Zhao, W. (2019). Two-dimensional spintronics for low-power electronics. *Nat. Electron.* 2: 274–283.

7 Peng, S., Zhu, D., Li, W. et al. (2020). Exchange bias switching in an antiferromagnet/ferromagnet bilayer driven by spin–orbit torque. *Nat. Electron.* 3: 1–8.

8 Wang, G., Zhang, Y., Wang, J. et al. (2019). Compact modeling of perpendicular-magnetic-anisotropy double-barrier magnetic tunnel junction with enhanced thermal stability recording structure. *IEEE Trans. Electron. Devices* 66: 2431–2436.

9 Cao, K., Li, H., Cai, W. et al. (2019). Low-temperature performance of nanoscale perpendicular magnetic tunnel junctions with double MgO-interface free layer. *IEEE Trans. Magn.* 55: 1–4.

10 Wang, Z., Zhou, H., Wang, M. et al. (2019). Proposal of toggle spin torques magnetic RAM for ultrafast computing. *IEEE Electron. Device Lett.* 40: 726–729.

11 Lee, S.-W. and Lee, K.-J. (2016). Emerging three-terminal magnetic memory devices. *Proc. IEEE* 104: 1831–1843.

12 Peng, S.Z., Lu, J.Q., Li, W.X. et al. Field-free switching of perpendicular magnetization through voltage-gated spin-orbit torque. In: *2019 IEEE International Electron Devices Meeting (IEDM)*, 28.6.1–28.6.4. IEEE.

13 Xiong, D., Peng, S., Lu, J. et al. (2020). Modulation of thermal stability and spin–orbit torque in IrMn/CoFeB/MgO structures through atom thick W insertion. *Appl. Phys. Lett.* 117: 212401.

14 Wang, Z., Zhang, L., Wang, M. et al. (2018). High-density NAND-like spin transfer torque memory with spin orbit torque erase operation. *IEEE Electron. Device Lett.* 39: 343–346.

15 Zhu, D. and Zhao, W. (2020). Threshold current density for perpendicular magnetization switching through spin-orbit torque. *Phys. Rev. Appl.* 13: 044078.

16 Peng, S., Zhu, D., Zhou, J. et al. (2019). Modulation of heavy metal/ferromagnetic metal interface for high-performance spintronic devices. *Adv. Electron. Mater.* 5: 1900134.

17 Zhang, D., Hou, Y., Zeng, L., and Zhao, W. (2019). Hardware acceleration implementation of sparse coding algorithm with spintronic devices. *IEEE Trans. Nanotechnol.* 18: 518–531.

18 Sengupta, A., Shim, Y., and Roy, K. (2016). Proposal for an all-spin artificial neural network: emulating neural and synaptic functionalities through domain wall motion in ferromagnets. *IEEE Trans. Biomed. Circuits Syst.* 10: 1152–1160.

19 Wang, C., Wang, Z., Wang, M. et al. (2020). Compact model of dzyaloshinskii domain wall motion-based MTJ for spin neural networks. *IEEE Trans. Electron. Devices* 67: 2621–2626.

20 Zhang, X., Vernier, N., Cao, Z. et al. (2018). Magnetoresistive sensors based on the elasticity of domain walls. *Nanotechnology* 29: 365502.

21 Wang, M., Cai, W., Cao, K. et al. (2018). Current-induced magnetization switching in atom-thick tungsten engineered perpendicular magnetic tunnel junctions with large tunnel magnetoresistance. *Nat. Commun.* 9: 671.

22 Pan, Y., Jia, X., Cheng, Z. et al. (2020). An STT-MRAM based reconfigurable computing-in-memory architecture for general purpose computing. *CCF Trans. High Perform. Comput.* 2: 272–281.

23 Fukami, S., Yamanouchi, M., Kim, K.-J. et al. 20-nm magnetic domain wall motion memory with ultralow-power operation. In: *2013 IEEE International Electron Devices Meeting*, 3.5.1–3.5.4. IEEE.

24 Wang, Z., Li, Z., Wang, M. et al. (2019). Field-free spin–orbit-torque switching of perpendicular magnetization aided by uniaxial shape anisotropy. *Nanotechnology* 30: 375202.

25 Xu, Y., Wu, B., Wang, Z. et al. (2020). Write-efficient STT/SOT hybrid triple-level cell for high-density MRAM. *IEEE Trans. Electron. Devices* 67: 1460–1465.

26 Devolder, T., Hayakawa, J., Ito, K. et al. (2008). Single-shot time-resolved measurements of nanosecond-scale spin-transfer induced switching: stochastic versus deterministic aspects. *Phys. Rev. Lett.* 100: 057206.

27 Qu, Y., Cockburn, B.F., Huang, Z. et al. (2018). Variation-resilient true random number generators based on multiple STT-MTJs. *IEEE Trans. Nanotechnol.* 17: 1270–1281.

28 Vincent, A.F., Larroque, J., Locatelli, N. et al. (2015). Spin-transfer torque magnetic memory as a stochastic memristive synapse for neuromorphic systems. *IEEE Trans. Biomed. Circuits Syst.* 9: 166–174.

29 Grollier, J., Querlioz, D., and Stiles, M.D. (2016). Spintronic nanodevices for bioinspired computing. *Proc. IEEE* 104: 2024–2039.

30 Grollier, J., Querlioz, D., Camsari, K.Y. et al. (2020). Neuromorphic spintronics. *Nat. Electron.* 3: 360–370.

31 Chang, L., Ma, X., Wang, Z. et al. (2019). CORN: in-buffer computing for binary neural network. In: *Proceedings of the 2019 Design, Automation and Test in Europe Conference and Exhibition, DATE 2019*. https://doi.org/10.23919/DATE.2019.8715265. Epub ahead of print 2019.

32 Wang, C., Zhang, D., Zeng, L., and Zhao, W. (2020). Design of magnetic non-volatile TCAM with priority-decision in memory technology for high speed, low power, and high reliability. *IEEE Trans. Circuits Syst. I Regul. Pap.* 67: 464–474.

33 Zhang, D., Zeng, L., Zhang, Y. et al. (2016). Stochastic spintronic device based synapses and spiking neurons for neuromorphic computation. In: *Proceedings of the 2016 IEEE/ACM International Symposium on Nanoscale Architectures, NANOARCH 2016*, 173.

34 Zhang, D., Zhao, W., Zeng, L. et al. (2016). All spin artificial neural networks based on compound spintronic synapse and neuron. *IEEE Trans. Biomed. Circuits Syst.* 10: 828–836.

35 Peng, S., Wang, M., Yang, H. et al. (2015). Origin of interfacial perpendicular magnetic anisotropy in MgO/CoFe/metallic capping layer structures. *Sci. Rep.* 5: 18173.

36 Bennett, C.H., Hassan, N., Hu, X. et al. Semi-supervised learning and inference in domain-wall magnetic tunnel junction (DW-MTJ) neural networks. In: *Spintronics XII* (eds. H.-J.M. Drouhin, J.-E. Wegrowe and M. Razeghi), 126. SPIE.

37 Jin, T., Gan, W., Tan, F. et al. (2019). Synaptic element for neuromorphic computing using a magnetic domain wall device with synthetic pinning sites. *J. Phys. D Appl. Phys.* 52: 445001.

38 Siddiqui, S.A., Dutta, S., Tang, A. et al. (2020). Magnetic domain wall based synaptic and activation function generator for neuromorphic accelerators. *Nano Lett.* 20: 1033–1040.

39 Shibata, T., Shinohara, T., Ashida, T. et al. (2020). Linear and symmetric conductance response of magnetic domain wall type spin-memristor for analog neuromorphic computing. *Appl. Phys. Express* 13: 043004.

40 Jaiswal, A., Agrawal, A., Panda, P., and Roy, K. (2020). Neural computing with magnetoelectric domain-wall based neuro-synaptic devices. *IEEE Trans. Magn.*: 1–1.

41 Sengupta, A. and Roy, K. (2017). Encoding neural and synaptic functionalities in electron spin: a pathway to efficient neuromorphic computing. *Appl. Phys. Rev.* 4: 041105.

42 Sengupta, A., Panda, P., Wijesinghe, P. et al. (2016). Magnetic tunnel junction mimics stochastic cortical spiking neurons. *Sci. Rep.* 6: 30039.

43 Wang, Z., Zhao, W., Deng, E. et al. (2015). Perpendicular-anisotropy magnetic tunnel junction switched by spin-Hall-assisted spin-transfer torque. *J. Phys. D Appl. Phys.* 48: 065001.

44 Deng, E., Zhang, Y., Klein, J.-O. et al. (2013). Low power magnetic full-adder based on spin transfer torque MRAM. *IEEE Trans. Magn.* 49: 4982–4987.

45 Kang, W., Wang, Z., Zhang, Y. et al. (2016). Spintronic logic design methodology based on spin Hall effect–driven magnetic tunnel junctions. *J. Phys. D Appl. Phys.* 49: 065008.

46 Chang, L., Wang, Z., Glova, A.O. et al. (2017). PRESCOTT: Preset-based cross-point architecture for spin-orbit-torque magnetic random access memory. In: *IEEE/ACM International Conference on Computer-Aided Design, Digest of Technical Papers*. ICCAD https://doi.org/10.1109/ICCAD.2017.8203785. Epub ahead of print 2017.

47 Ghosh-Dastidar, S. and Adeli, H. (2009). Spiking neural networks. *Int. J. Neural. Syst.* 19: 295–308.

48 Nessler, B., Pfeiffer, M., Buesing, L., and Maass, W. (2013). Bayesian computation emerges in generic cortical microcircuits through spike-timing-dependent plasticity. *PLoS Comput. Biol.* 9: e1003037.

49 Wallace, E., Benayoun, M., van Drongelen, W., and Cowan, J.D. (2011). Emergent oscillations in networks of stochastic spiking neurons. *PLoS One* 6: e14804.

50 Benayoun, M., Cowan, J.D., van Drongelen, W., and Wallace, E. (2010). Avalanches in a stochastic model of spiking neurons. *PLoS Comput. Biol.* 6: e1000846.

51 Nessler, B., Pfeiffer, M., and Maass, W. (2009). STDP enables spiking neurons to detect hidden causes of their inputs. In: *Advances in Neural Information Processing Systems 22 – Proceedings of the 2009 Conference*, 1357.

52 Liyanagedera, C.M., Sengupta, A., Jaiswal, A., and Roy, K. (2017). Stochastic spiking neural networks enabled by magnetic tunnel junctions: from nontelegraphic to telegraphic switching regimes. *Phys. Rev. Appl.* 8: 064017.

53 Sengupta, A. and Roy, K. (2018). Neuromorphic computing enabled by physics of electron spins: prospects and perspectives. *Appl. Phys. Express* 11: 030101.

54 Agrawal, A. and Roy, K. (2019). Mimicking leaky-integrate-fire spiking neuron using automotion of domain walls for energy-efficient brain-inspired computing. *IEEE Trans. Magn.* 55: 1–7.

55 Jaiswal, A., Agrawal, A., Chakraborty, I. et al. On robustness of spin-orbit-torque based stochastic sigmoid neurons for spiking neural networks. In: *2019 International Joint Conference on Neural Networks (IJCNN)*, 1–6. IEEE.

56 Sengupta, A., Choday, S.H., Kim, Y., and Roy, K. (2015). Spin orbit torque based electronic neuron. *Appl. Phys. Lett.* 106: 143701.

57 Ostwal, V., Zand, R., DeMara, R., and Appenzeller, J. (2019). A novel compound synapse using probabilistic spin–orbit-torque switching for MTJ-based deep neural networks. *IEEE J. Explor. Solid-State Comput. Devices Circuits* 5: 182–187.

58 Sengupta, A. and Roy, K. (2016). A vision for all-spin neural networks: a device to system perspective. *IEEE Trans. Circuits Syst. I Regul. Pap.* 63: 2267–2277.

59 Kang, W., Huang, Y., Zhang, X. et al. (2016). Skyrmion-electronics: an overview and outlook. *Proc. IEEE* 104: 2040–2061.

60 Soumyanarayanan, A., Raju, M., Oyarce, A.L.G. et al. (2017). Tunable room-temperature magnetic skyrmions in Ir/Fe/Co/Pt multilayers. *Nat. Mater.* 16: 898–904.

61 Li, S., Wang, K., Zhang, X. et al. (2020). Magnetic skyrmions for unconventional computing. *Mater Horizons.* https://doi.org/10.1039/d0mh01603a. Epub ahead of print.

62 Leonov, A.O., Monchesky, T.L., Romming, N. et al. (2016). The properties of isolated chiral skyrmions in thin magnetic films. *New J. Phys.*: 18. https://doi.org/10.1088/1367-2630/18/6/065003. Epub ahead of print.

63 Muhlbauer, S., Binz, B., Jonietz, F. et al. (2009). Skyrmion lattice in a chiral magnet. *Science (80-)* 323: 915–919.

64 Sampaio, J., Cros, V., Rohart, S. et al. (2013). Nucleation, stability and current-induced motion of isolated magnetic skyrmions in nanostructures. *Nat. Nanotechnol.* 8: 839–844.

65 Li, S., Xia, J., Zhang, X. et al. (2018). Dynamics of a magnetic skyrmionium driven by spin waves. *Appl. Phys. Lett.* 112: 142404.

66 Zhu, D., Kang, W., Li, S. et al. (2018). Skyrmion racetrack memory with random information update/deletion/insertion. *IEEE Trans. Electron. Devices* 65: 87–95.

67 Maccariello, D., Legrand, W., Reyren, N. et al. (2017). Electrical detection of single magnetic skyrmions in metallic multilayers at room temperature. *Nat. Nanotechnol.* https://doi.org/10.1038/s41565-017-0044-4. Epub ahead of print.

68 Legrand, W., Maccariello, D., Reyren, N. et al. (2017). Room-temperature current-induced generation and motion of sub-100 nm skyrmions. *Nano Lett.* 17: 2703–2712.

69 Raju, M., Yagil, A., Soumyanarayanan, A. et al. (2019). The evolution of skyrmions in Ir/Fe/Co/Pt multilayers and their topological Hall signature. *Nat. Commun.* 10: 696.

70 Moreau-Luchaire, C., Moutafis, C., Reyren, N. et al. (2016). Additive interfacial chiral interaction in multilayers for stabilization of small individual skyrmions at room temperature. *Nat. Nanotechnol.* 11: 444–448.

71 Wang, Y., Wang, L., Xia, J. et al. (2020). Electric-field-driven non-volatile multi-state switching of individual skyrmions in a multiferroic heterostructure. *Nat. Commun.* 11: 1–8.

72 Bogdanov, A.N. and Panagopoulos, C. (2020). Physical foundations and basic properties of magnetic skyrmions. *Nat. Rev. Phys.* 2: 492–498.

73 Zhang X, Zhou Y, Song KM, et al. (2019). A decade of skyrmionics: writing, deleting, reading and processing magnetic skyrmions toward spintronic applications, http://arxiv.org/abs/1906.04718 .

74 Kang, W., Huang, Y., Zheng, C. et al. (2016). Voltage controlled magnetic skyrmion motion for racetrack memory. *Sci. Rep.*: 6. https://doi.org/10.1038/srep23164. Epub ahead of print.

75 Zhang, X., Cai, W., Zhang, X. et al. (2018). Skyrmions in magnetic tunnel junctions. *ACS Appl. Mater. Interfaces* 10: 16887–16892.

76 Zhang, D.L., Sun, C., Lv, Y. et al. (2018). L10 Fe-Pd synthetic antiferromagnet through an fcc Ru spacer utilized for perpendicular magnetic tunnel junctions. *Phys. Rev. Appl.* 9: 44028.

77 Kang, W., Lv, W., Zheng, C. et al. (2017). Compact modeling and evaluation of magnetic skyrmion-based racetrack memory. *IEEE Trans. Electron. Devices* 64: 1060–1068.

78 Kang, W., Zheng, C., Huang, Y. et al. (2016). Complementary skyrmion racetrack memory with voltage manipulation. *IEEE Electron Device Lett.* 37: 924–927.

79 Chen, X., Kang, W., Zhu, D. et al. (2018). Complementary skyrmion racetrack memory enables voltage-controlled local data update functionality. *IEEE Trans. Electron Devices* 65: 4667–4673.

80 Parkin, S.S.P., Hayashi, M., and Thomas, L. (2008). Magnetic domain-wall racetrack memory. *Science (80-)* 320: 190–194.

81 Jiang, W., Zhang, X., Yu, G. et al. (2017). Direct observation of the skyrmion Hall effect. *Nat. Phys.* 13: 162–169.

82 Kim, J.-V., Garcia-Sanchez, F., Sampaio, J. et al. (2014). Breathing modes of confined skyrmions in ultrathin magnetic dots. *Phys. Rev. B* 90: 064410.

83 Chen, X., Kang, W., Zhu, D. et al. (2017). Skyrmion dynamics in width-varying nanotracks and implications for skyrmionic applications. *Appl. Phys. Lett.* 111: 202406.

84 Zhang, H., Zhu, D., Kang, W. et al. (2020). Stochastic computing implemented by skyrmionic logic devices. *Phys. Rev. Appl.* https://doi.org/10.1103/PhysRevApplied.13.054049. Epub ahead of print.

85 Li, S., Kang, W., Chen, X. et al. (2018). Emerging neuromorphic computing paradigms exploring magnetic skyrmions. In: *IEEE Computer Society Annual Symposium on VLSI, ISVLSI 2018*, 539–544. IEEE.

86 Bhattacharya, T., Li, S., Huang, Y. et al. (2019). Low-power (1T1N) skyrmionic synapses for spiking neuromorphic systems. *IEEE Acc.* 7: 5034–5044.

87 Pan, B., Zhang, D., Zhang, X. et al. (2019). Skyrmion-induced memristive magnetic tunnel junction for ternary neural network. *IEEE J. Electron Devices Soc.* 7: 529–533.

88 Song, K.M., Jeong, J.-S., Pan, B. et al. (2020). Skyrmion-based artificial synapses for neuromorphic computing. *Nat. Electron.* 3: 148–155.

89 Zázvorka, J., Jakobs, F., Heinze, D. et al. (2019). Thermal skyrmion diffusion used in a reshuffler device. *Nat. Nanotechnol.* 14: 658–661.

90 Kurenkov, A., Fukami, S., and Ohno, H. (2020). Neuromorphic computing with antiferromagnetic spintronics. *J. Appl. Phys.* 128: 010902.

91 Purnama, I., Gan, W.L., Wong, D.W., and Lew, W.S. (2015). Guided current-induced skyrmion motion in 1D potential well. *Sci. Rep.* 5: 1–9.

92 Huang, Y., Kang, W., Zhang, X. et al. (2017). Magnetic skyrmion-based synaptic devices. *Nanotechnology* 28: 08LT02.

93 Borisyuk, R.M. and Borisyuk, G.N. (1997). Information coding on the basis of synchronization of neuronal activity. *Biosystems* 40: 3–10.

94 Li, S., Kang, W., Huang, Y. et al. (2017). Magnetic skyrmion-based artificial neuron device. *Nanotechnology* 28: 31LT01.

95 Chen, X., Kang, W., Zhu, D. et al. (2018). A compact skyrmionic leaky-integrate-fire spiking neuron device. *Nanoscale* 10: 6139–6146.

96 Azam, M.A., Bhattacharya, D., Querlioz, D., and Atulasimha, J. (2018). Resonate and fire neuron with fixed magnetic skyrmions. *J. Appl. Phys.* 124: 152122.

97 Tanaka, G., Yamane, T., Héroux, J.B. et al. (2019). Recent advances in physical reservoir computing: a review. *Neural Netw.* 115: 100–123.

98 Prychynenko, D., Sitte, M., Litzius, K. et al. (2018). Magnetic skyrmion as a nonlinear resistive element: a potential building block for reservoir computing. *Phys. Rev. Appl.* 9: 14034.

99 Bourianoff, G., Pinna, D., Sitte, M., and Everschor-Sitte, K. (2018). Potential implementation of reservoir computing models based on magnetic skyrmions. *AIP Adv.* 8: 055602.

100 Jiang, W., Chen, L., Zhou, K. et al. (2019). Physical reservoir computing using magnetic skyrmion memristor and spin torque nano-oscillator. *Appl. Phys. Lett.* 115: 192403.

101 Jin, C., Song, C., Wang, J., and Liu, Q. (2016). Dynamics of antiferromagnetic skyrmion driven by the spin Hall effect. *Appl. Phys. Lett.*: 109. https://doi.org/10.1063/1.4967006. Epub ahead of print.

102 Pinna, D., Abreu Araujo, F., Kim, J.V. et al. (2018). Skyrmion gas manipulation for probabilistic computing. *Phys. Rev. Appl.* 9: 1–34.

103 Yao, Y., Chen, X., Kang, W. et al. (2020). Thermal brownian motion of skyrmion for true random number generation. *IEEE Trans. Electron Devices* 67: 2553–2558.

104 Krstić, D. (1998). RF microelectronics. *Microelectronics J.* 29: 1041–1042.

105 Hor, H.C. and Siek, L. Review on VCO based ADC in modern deep submicron CMOS technology. In: *2012 IEEE International Symposium on Radio-Frequency Integration Technology (RFIT)*, 86–88. IEEE.

106 Soens, C., Van der Plas, G., Wambacq, P. et al. (2005). Performance degradation of LC-tank VCOs by impact of digital switching noise in lightly doped substrates. *IEEE J. Solid-State Circuits* 40: 1472–1481.

107 Von, N.J. (1957). Non-linear capacitance or inductance switching, amplifying, and memory organs. Google Patents.

108 Goto, E. (1959). The parametron, a digital computing element which utilizes parametric oscillation. *Proc. IRE* 47: 1304–1316.

109 Muroga, S. and Takashima, K. (1959). The parametron digital computer MUSASINO-1. *IEEE Trans. Electron Comput.* EC-8: 308–316.

110 Tsoi, M., Jansen, A.G.M., Bass, J. et al. (1998). Excitation of a magnetic multilayer by an electric current. *Phys. Rev. Lett.* 80: 4281–4284.

111 Kiselev, S.I., Sankey, J.C., Krivorotov, I.N. et al. (2003). Microwave oscillations of a nanomagnet driven by a spin-polarized current. *Nature* 425: 380–383.

112 Rippard, W., Pufall, M., Kaka, S. et al. (2004). Direct-current induced dynamics in CoFe/NiFe point contacts. *Phys. Rev. Lett.* 92: 027201.

113 Locatelli, N., Cros, V., and Grollier, J. (2014). Spin-torque building blocks. *Nat. Mater.* 13: 11–20.

114 Zeng, L., Zhang, D., Zhang, Y. et al. Spin wave based synapse and neuron for ultra low power neuromorphic computation system. In: *2016 IEEE International Symposium on Circuits and Systems (ISCAS)*, 918–921. IEEE.

115 Zeng, L., Xu, X., Chen, H.-H. et al. (2020). Realization of mutual synchronization of spin torque nano-oscillators under room temperature by noise reduction technique. *Appl. Phys. Lett.* 117: 082404.

116 Zeng, L., Ma, X., Chen, H.-H. et al. (2020). Noise reduction of spin torque oscillator by phase-locked loop with combinational frequency tuning method. *Appl. Phys. Lett.* 117: 072407.

117 Zeng, L., Liu, Y., Chen, H.-H. et al. (2020). Robust phase shift keying modulation method for spin torque nano-oscillator. *Nanotechnology* 31: 375205.

118 Yan Zhou, B.S., Persson, J., and Akerman, J. (2009). Capacitance enhanced synchronization of pairs of spin-transfer oscillators. *IEEE Trans. Magn.* 45: 2421–2423.

119 Slavin, A. and Tiberkevich, V. (2009). Nonlinear auto-oscillator theory of microwave generation by spin-polarized current. *IEEE Trans. Magn.* 45: 1875–1918.

120 Zhou, Y., Tiberkevich, V., Consolo, G. et al. (2010). Oscillatory transient regime in the forced dynamics of a nonlinear auto oscillator. *Phys. Rev. B* 82: 012408.

121 Yogendra, K., Koo, M., and Roy, K. Energy efficient computation using injection locked bias-field free spin-hall nano-oscillator array with shared heavy metal. In: *2017 IEEE/ACM International Symposium on Nanoscale Architectures (NANOARCH)*, 89–94. IEEE.

122 Yogendra, K., Fan, D., and Roy, K. (2015). Coupled spin torque nano oscillators for low power neural computation. *IEEE Trans. Magn.* 51: 1–9.

123 Yogendra, K., Liyanagedera, C., Fan, D. et al. (2017). Coupled spin-torque nano-oscillator-based computation. *ACM J. Emerg. Technol. Comput. Syst.* 13: 1–24.

124 Torrejon, J., Riou, M., Araujo, F.A. et al. (2017). Neuromorphic computing with nanoscale spintronic oscillators. *Nature* 547: 428–431.

125 Riou, M., Araujo, F.A., Torrejon, J. et al. Neuromorphic computing through time-multiplexing with a spin-torque nano-oscillator. In: *2017 IEEE International Electron Devices Meeting (IEDM)*, 36.3.1–36.3.4. IEEE.

126 Marković, D., Leroux, N., Riou, M. et al. (2019). Reservoir computing with the frequency, phase, and amplitude of spin-torque nano-oscillators. *Appl. Phys. Lett.* 114: 012409.

127 Tsunegi, S., Taniguchi, T., Nakajima, K. et al. (2019). Physical reservoir computing based on spin torque oscillator with forced synchronization. *Appl. Phys. Lett.* 114: 164101.

128 Yamaguchi, T., Akashi, N., Nakajima, K. et al. (2020). Step-like dependence of memory function on pulse width in spintronics reservoir computing. *Sci. Rep.* 10: 19536.

129 Kanao, T., Suto, H., Mizushima, K. et al. (2019). Reservoir computing on spin-torque oscillator array. *Phys. Rev. Appl.* 12: 1.

130 Riou, M., Torrejon, J., Garitaine, B. et al. (2019). Temporal pattern recognition with delayed-feedback spin-torque nano-oscillators. *Phys. Rev. Appl.* 12: 024049.

131 Csaba, G. and Porod, W. (2020). Coupled oscillators for computing: a review and perspective. *Appl. Phys. Rev.* 7: 011302.

132 Romera, M., Talatchian, P., Tsunegi, S. et al. (2018). Vowel recognition with four coupled spin-torque nano-oscillators. *Nature* 563: 230–234.

133 Talatchian, P., Romera, M., Tsunegi, S. et al. Microwave neural processing and broadcasting with spintronic nano-oscillators. In: *2018 IEEE International Electron Devices Meeting (IEDM)*, 27.4.1–27.4.4. IEEE.

134 Zahedinejad, M., Awad, A.A., Muralidhar, S. et al. (2020). Two-dimensional mutually synchronized spin Hall nano-oscillator arrays for neuromorphic computing. *Nat. Nanotechnol.* 15: 47–52.

135 Talatchian, P., Romera, M., Araujo, F.A. et al. (2020). Designing large arrays of interacting spin-torque nano-oscillators for microwave information processing. *Phys. Rev. Appl.* 13: 024073.

136 Arai, H. and Imamura, H. (2018). Neural-network computation using spin-wave-coupled spin-torque oscillators. *Phys. Rev. Appl.* 10: 024040.

137 Cai, J., Zhang, L., Fang, B. et al. (2019). Sparse neuromorphic computing based on spin-torque diodes. *Appl. Phys. Lett.* 114: 192402.

138 Wei, J., Fang, B., Wu, W. et al. (2020). Amplitude and frequency modulation based on memristor-controlled spin nano-oscillators. *Nanotechnology* 31: 045202.

3

Multiterminal Neuromorphic Devices with Cognitive Behaviors

Li Qiang Zhu[1], Jia Cheng Cai[2], Zheng Yu Ren[2], Wen Xiong[3], and Qing Wan[4]

[1]*Ningbo University, School of Physical Science and Technology, 818 Fenghua Road, Ningbo, 315211, Zhejiang, P. R. China*
[2]*Chinese Academy of Sciences, Ningbo Institute of Materials Technology and Engineering, 1219 Zhongguan West Road, Ningbo, 315201, Zhejiang, P. R. China*
[3]*Fudan University, School of Microelectronics, 220 Handan Road, Shanghai, 200433, P. R. China*
[4]*Nanjing University, School of Electronic Science & Engineering, 163 Xianlin Road, P. R. China*

3.1 Introduction

Over the past half century, the developments in microelectronics and information technology follow the Moore's law [1]. It has been faced fundamental physical limitations in terms of miniaturization of conventional Si-based planar complementary metal-oxide semiconductor (CMOS) devices [2]. There are several candidate technologies to extend the Moore's law, including single-electron device, quantum device, spin-based device, etc. Alternatively, bionics also provides us potential methodologies to solve the dilemma. Our brain is a large-scale neural network with $\sim 10^{11}$ neurons and $\sim 10^{15}$ synapses [3]. With the hierarchical structure of the cerebral neocortex and parallel processing, it is an energy-efficient computation center. It can execute huge amount of parallel synaptic computation. Though von Neumann computers can solve structural mathematical problems very well, high-energy power consumption is needed for dealing with ever-growing data. Moreover, there will be limitations in dealing with real-time, nonstructural data. Compared with traditional computation based on von Neumann architectures, brain computation has the priorities including but not limited to characteristics of parallel processing, high efficiency, low power consumption, and autonomous learning. Inspired by the brain computation, neuromorphic computing attracts more and more attentions with the emergence of big data era. Consider the fact that synapses and neurons dominate the architecture of the brain, which are responsible for the massive parallelism, structural plasticity, and robustness. Thus, one of the main tasks in neuromorphic systems would be the realization of neuromorphic devices, capable of mimicking multi-fundamental synaptic computation behaviors [4]. Brain-like neuromorphic devices have great potentials in the development of energy-efficient artificial neural networks (ANNs) and artificial intelligence (AI) technology.

Neuromorphic Devices for Brain-Inspired Computing: Artificial Intelligence, Perception, and Robotics, First Edition. Edited by Qing Wan and Yi Shi.

Recently, new materials and new conceptual devices have been proposed for neuromorphic applications, including two terminal resistive switching (RS) memories and new-concept transistors. Several biological synaptic plasticity behaviors have been mimicked on these neuromorphic devices, including short-term/long-term potentiation (STP/LTP), paired-pulse facilitation (PPF), synaptic filtering, spike-timing-dependent plasticity (STDP), spiking rate-dependent plasticity (SRDP), metaplasticity, etc. Moreover, neural calculations and processing behaviors have been simulated successfully, including pattern recognition, classical conditioned reflex, reservoir computing, spatiotemporal dynamic logic, etc. Considering the dendrite synapse configurations in biological neurons, multiterminal neuromorphic devices have also been designed. Thus, more interesting neural functions have been mimicked on these multiterminal neuromorphic devices, exhibiting great potentials in neuromorphic engineering. Interestingly, our body is a natural multi-perception learning system. The sensory organs can perceive external stimuli and transmit the stimuli to the central nervous system (CNS) through afferent nerves. Then, various senses are formed including tactile sense, olfaction sense, visual sense, etc. [5, 6] Recently, there are also growing achievements in artificial multi-perception learning system by integrating neuromorphic devices with sensors, including pressure sensors, photo-sensors, biochemical sensors, etc. [7, 8] Moreover, single neuromorphic device has also been proposed for mimicking perception learning behaviors. Thus, the realization of artificial perception learning system at hardware level results in new research branch. It will provide new opportunities for extending human intelligence.

Here, we discuss about recent developments in multiterminal neuromorphic devices. Firstly, we simply overview memristor-based neuromorphic devices. Then, we discuss the newly reported multiterminal memristors for neuromorphic engineering applications. Next, we mainly discuss about the neuromorphic transistors. Thanks for the inherent priorities for transistors, it is much easier to integrate multi-gates in a single transistor. In this part, neuron transistor has been discussed. Neuron transistor for biochemical sensor applications is reviewed. Furthermore, spiking operation mode for chemical sensor based on multiterminal neuromorphic devices is discussed. Lastly, perception learning activities have been reviewed based on multiterminal neuromorphic devices. Such achievements indicate a new branch for neuromorphic engineering.

3.2 Multiterminal Neuromorphic Memristors

3.2.1 Memristor-Based Neuromorphic Devices

Memristor is the fourth type of component, except for resistor, capacitor, and inductance, theoretically proposed by Chua in 1971 [9]. It has two electrodes sandwiching an RS layer and can "remember" electrical states even when turned off. Thus, the device can demonstrate variable and nonvolatile resistances. In 2008, HP Lab observed RS effects in a TiO_x-based device [10]. The RS behavior is related to the oxygen vacancies within the TiO_x layer. The achievements indicates the coming of new era for the memristor. Basically, the internal ionic transportations induce the RS activities of the memristor. Thus, it is a kind of ionotronic devices. During the operation, the conductive path or filament within the RS layer will be formed or dissipated due to the ion dynamics under external electrical field. Depending on the RS electrolyte materials and the electrode materials, the operation mechanisms could be classified into electrochemical metallization and the valence change mechanism, corresponding to electrochemical metallization memristors (ECMMs) [11, 12] and valence change memory memristors (VCMMs) [13, 14], respectively. For ECMMs, electrochemically active electrodes and electrochemically inert electrodes always act as top electrode (TE) and bottom electrode (BE), respectively. During operation, electrochemically active TE will be ionized into metal cations under external voltage bias. Thus, they will migrate to the inert BE through the RS layer and get reduction there, resulting in the formation and growth of nanoscale metal filament. In another word, the electrically conductive state will change from a high resistance state (HRS) to a low resistance state (LRS), i.e. the SET process. While at the reverse voltage bias, the nanoscale filament will dissipate, resulting in the transition from LRS to HRS, i.e. the RESET process. Thus, in the ionotronic ECMMs, the RS layer does not take part in chemical redox processes of metal cations. For VCMMs, inert electrodes act as both TE and BE, and oxides typically act as RS electrolyte. Oxygen ions (O^{2-}) or oxygen vacancies (V_o^{2+}) usually act as active ionic species. Thus, the switching layer always participates in chemical redox processes of active species. The distribution of V_o^{2+} concentration between V_o^{2+}-poor and V_o^{2+}-rich layers results in the RS process. Morphologically, the structure of memristor is quite similar to that of a biological synapse consisting of presynapse, synaptic cleft, and postsynapse. On the other hand, there are nonlinear electrical characteristics and nonvolatile resistance modulation effects. Thus, memristors could be attractive candidates for neuromorphic platform applications [15, 16]. In 2005, Aono et al. [17] reported an atomic switch device with a Ag_2S/nanogap/Pt crossbar structure, demonstrating "ON" and "OFF" states due to the formation and annihilation of a bias-controlled Ag bridge in the nanogap. In 2011, they also demonstrate a benchmarking study where synaptic functions have been mimicked based on Ag/Ag_2S/nanogap/Pt-based atomic switch, including STP, LTP, and transition from short-term memory (STM) to long-term memory (LTM) [18].

Additionally, bilayer structure-based memristors would have better electrical performances than those of single-layer structure-based ones [19, 20]. For example, Pt/HfO_2/TiO_2/indium tin oxide (ITO)-based resistive random access memory (RRAM) devices can operate at a lower voltage than Pt/HfO_2/ITO-based RRAM device [21]. Thus, bilayer structure-based memristors have also been proposed for neuromorphic device applications [22]. As shown in Figure 3.1a, in a bilayered HfO_2/TiO_x cognitive memristor, oxygen ions (O^{2-}) generated in HfO_2 layer can drift to BE when the TiN bottom electrode biased at a positive bias. Oxygen ions can be stored within the TiO_x film, which results in the formation of oxygen vacancy-based conducting filament, bridging the Pt TE and TiN BE, as is the electroforming process. When the TiN bottom electrode biases at a negative bias, oxygen ions

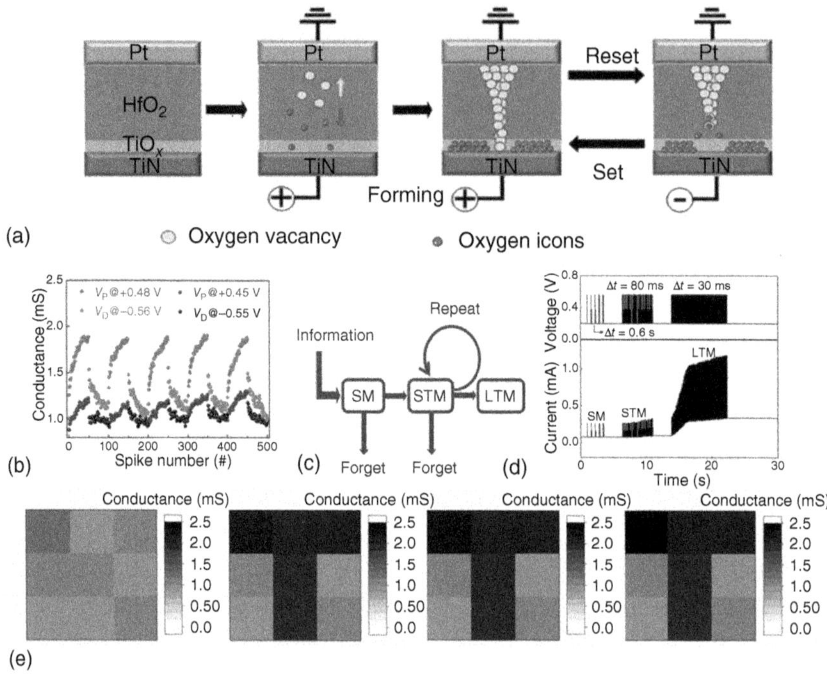

Figure 3.1 (a) Proposed resistive switching model for bilayered HfO_2/TiO_x cognitive memristor. (b) Conductance (or synaptic weight) modulation under repeated positive (potentiation) or negative (depression) spikes. (c) Schematic diagram of Atkinson–Shiffrin multi-store memory model. (d) Mimicked multi-store memory model. (e) Pattern memory of the memristor-based neuromorphic device array. Source: Xiong et al. [22].

can migrate to combine with oxygen vacancies, triggering RESET process. When a positive bias is applied again, the device will switch from HRS to LRS, i.e. SET process. Here, the formation/rupture of the oxygen vacancy-based conducting filaments get easier because of the presence of TiO_x oxygen reservoir layer. Thus, the neuromorphic device can operate at low voltage as low as 0.6 V. Two groups of potentiation/depression spikes (V_P/V_D) have been applied on the memristors, inducing the gradually increased and decreased conductances, respectively, as shown in Figure 3.1b. At higher pulse amplitudes, the changes in conductances get bigger. Furthermore, good repeatabilities have been observed, indicating good stabilities. In 1968, Atkinson and Shiffrin proposed a multi-store memory model to explain brain memory behavior [23], as schematically shown in Figure 3.1c. When we receive information, it will be stored as sensory memory (SM). Yet, it will be forgotten easily. However, SM can be converted to STM by receiving information repeatedly. Furthermore, STM can also be converted to LTM. The bilayered HfO_2/TiO_x cognitive memristor can mimic such behavior, as shown in Figure 3.1d. Due to the nonvolatile conductance, a "T"-shaped pattern is programmed on a 3×3 memristor array. Then, the state is read out in five minutes and in one hour. As shown in Figure 3.1e, the color of "T"-shaped image is barely faded, demonstrating that this pattern is stored in LTM mode.

It would also be interesting to obtain memristor with environmentally friendly materials, including starch-based flexible biomemristors [24], polydopamine-based wearable memristor device [25], etc. Moreover, flexible memristors with low power consumption is interesting for potential brain-inspired multifunctional perception learning system applications. For example, flexible cognitive memristors based on poly(vinyl alcohol)/graphene oxide (PVA-GO) hybrid nanocomposite exhibit excellent electrical performances at ultralow voltage below 0.5 V with high stabilities against mechanical stress [26]. The threshold voltage of resistance switching is only ~0.2 V. The device demonstrates ultralow power consumption of ~0.5 μW during the SET process. The value is lower than those of other reported biomaterial-based memristors during SET process. By connecting PVA-GO hybrid nanocomposite memristor together, Pavlovian conditioned reflex can be mimicked, as shown in Figure 3.2. Initially, M_1 and M_2 are in LRS (HWS, high synaptic weight state) and HRS (LWS, low synaptic weight state), respectively. The response of M_3 will reflect the salivation or not. "Food" stimuli and "Bell" stimuli can be applied on M_1 and M_2, respectively. When applying "Food" stimuli (US) on M_1, a salivation response (UR) is observed on M_3, while there is no salivation response (UR) on M_3 when

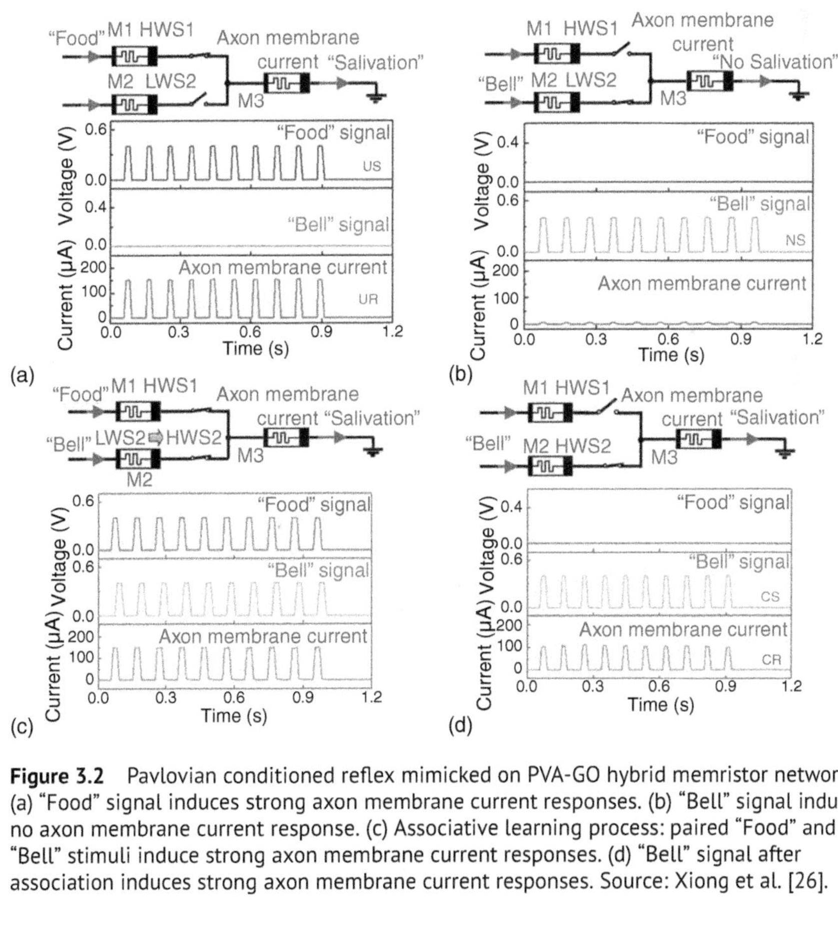

Figure 3.2 Pavlovian conditioned reflex mimicked on PVA-GO hybrid memristor network. (a) "Food" signal induces strong axon membrane current responses. (b) "Bell" signal induces no axon membrane current response. (c) Associative learning process: paired "Food" and "Bell" stimuli induce strong axon membrane current responses. (d) "Bell" signal after association induces strong axon membrane current responses. Source: Xiong et al. [26].

applying "Bell" stimuli (NS) on M_2. When paired "Food" and "Bell" stimuli are applied, the correlation is established. M_2 can be switched from LWS to HWS. Thus, a salivation response is observed on M_3. After the association learning, "Bell" stimuli alone still trigger strong salivation response.

3.2.2 Multiterminal Memristor for Neuromorphic System

In the nervous systems, a single neuron can connect with thousands of pre-neurons through thousands of dendrite synapses [27]. A neuron is an integrator. Through these dendrite synapses, the neuron can communicate with other neurons in a spatiotemporal parallel mode and execute neural computation in energy-efficient way [28, 29]. They act as computational engines in brain computation by processing inputs from thousands of dendrite synapses. Generally, spikes from multi-pre-neurons are needed to spike a post-neuron, because the strength of one spike coming from one pre-neuron is always not high enough. When the integrated signal arrives at a certain threshold, the neuron will be triggered to transfer a spike to a post-neuron. This multiterminal structure is the basis of complex neural functions, including dendrite integration and heterosynaptic plasticity. Up to date, several basic synaptic learning rules have been demonstrated by using two-terminal memristors, demonstrating homosynaptic plasticity activities [30–32]. Due to the inherent limitations for two-terminal configurations in memristors, it is difficult to realize dendrite synaptic integration on memristors. Fortunately, people have introduced additional electrodes in memristors. Compared with traditional two-terminal memristors, the multiterminal memristive devices not only exhibit memristive characteristics but also offer another degree of freedom for synaptic weight modulation [33].

3.2.2.1 Synaptic Competition and Cooperation on Multiterminal Memristor

Synaptic dynamics observed in biology is the basis of neural signal transmission [34–36]. Therefore, mimicking the synaptic dynamics is considered to be an effective way to realize the synaptic interactions on solid-state devices [37, 38]. Synaptic cooperation and competition are important components of synaptic plasticity in nervous systems [39]. They can tune synapses for the formation of associative LTP, a cellular correlate of associative LTM, and are critical for the stability and functions of the neural networks in biological systems. In general, this synaptic interaction is induced by the diffusion of plasticity-related proteins (PRPs) (for example, calmodulin-dependent kinase II [CaMKII]) that are needed for synaptic growth among adjacent synapses. Lu et al. [40] emulated the PRP exchange processes among biological synapses through ionic coupling between MoS_2-based memristive devices. Specifically, they proposed anionic coupling-based multiterminal network structural memristive device composed of multiple planar structure of $Au/Li_xMoS_2/Au$ memristor. Local 2H-1T′ (semiconductor–metal) phase transitions of MoS_2 films can be controlled by the migration of Li^+ ions under an electric field to implement the reversible modulation of conductance. Based on this memristive mechanism, the synaptic competition and synaptic cooperation have been implemented by establishing the ionic coupling effect between multiple Au/Li_xMoS_2/Au memristors. Two memristors are designed to share a common grounded electrode, resembling the postsynaptic terminals. Electrodes A and B resemble the presynaptic terminals. Initially, Li^+ ions in device 1 and device 2 were driven toward the region under electrode G to emulate the limited PRPs. During the redistribution of Li^+ ions between devices driven by an electric field, the device that received more Li^+ ions exhibited a larger conductivity. Thus, increasing the degree of potentiation (depression) in one Li_xMoS_2-based synaptic device reduces (increases) the activity of the other, emulating the synaptic competition behavior in biological networks. In addition, four memristors sharing a common electrode (G electrode) were integrated into a multiterminal network structural memristive device to simulate synaptic cooperation. Initially, Li^+ ions under the common electrode G were driven into near electrodes A, B, C, and D. When the conductance of device 1 was modulated by electric field stimulus, the Li^+ ions under the common electrode G cooperatively increase/decrease. Therefore, similar synaptic potentiation/depression was also observed in other devices, indicating the realization of synaptic cooperation behaviors.

3.2.2.2 Heterosynaptic Plasticity on Multiterminal Memristor

Heterosynaptic plasticity is a fundamental form of synaptic response behavior in brain cognitive activities. Under the heterosynaptic plasticity mechanism, synaptic connection between pre- and postsynaptic terminals can be effectively modulated by an additional terminal [41, 42]. As schematically shown in Figure 3.3a,b, heterosynaptic plasticity requires another modulatory terminal to regulate synaptic facilitation/depression behavior between the pre- and postsynaptic neurons. Yang et al. [43] proposed a conductive bridge-based multiterminal memristive device to emulate heterosynaptic plasticity. As shown in Figure 3.3c, Ag nanoclusters are dispersed randomly within the dielectric between presynaptic and postsynaptic terminals (Pt electrodes). Beside the device, there is an additional modulatory terminal (Pt electrodes). The gap between the modulatory terminal (Pt electrodes) and the nanoclusters is slightly larger. Thus, filaments can be prevented from being connected to the modulatory terminal. As shown in Figure 3.3d, the modulation of synaptic weight is based on the self-organization of Ag nanoclusters under an electric field. Interestingly, the modulatory terminal can affect the local electric field between the presynaptic and postsynaptic terminals. Thus, the formation and destruction of conductive filaments can be effectively modulated. When presynaptic input is paired with the modulation input, the number of pulses required to reach the same facilitation level (e.g. weight change by a ratio of 10^2) was significantly reduced, as shown in Figure 3.3e. Such effect becomes more significant when the paired modulatory input becomes stronger, demonstrating the heterosynaptic facilitation (Figure 3.3f). Furthermore, the modulatory behavior of the modulatory terminal was also verified during the destruction of conductive filaments, demonstrating the heterosynaptic depression (Figure 3.3g).

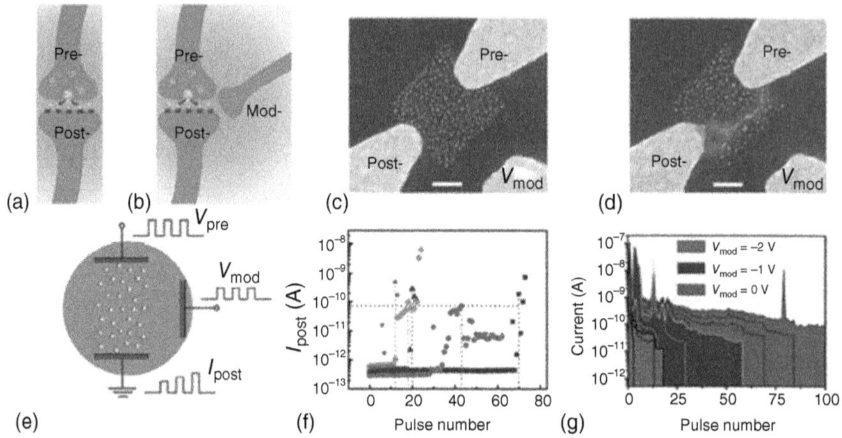

Figure 3.3 Schematic diagram of (a) homosynaptic plasticity and (b) heterosynaptic plasticity. (c) SEM image of as-fabricated heterosynaptic device. (d) SEM image of the device after heterosynaptic facilitation. (e) Schematic illustration of the input and output signals for the measurement. (f) Heterosynaptic facilitation. (g) Heterosynaptic depression. Source: Yang et al. [43].

3.2.2.3 Multiterminal Memtransistor

For the conductive bridge-based multiterminal memristive device, filaments are needed to be formed across the channel under certain biases. Thus, it is difficult to accurately and repeatedly adjust synaptic weights when more modulatory terminals need to be integrated. To solve such dilemma, transistor configurations have also been integrated within the multiterminal memristive devices. The device can operate typical memristive behaviors. Additionally, gate coupling effects also play a key role. Thus, such devices have been named as memtransistor [33, 44–46]. Thus, a multiterminal memtransistor can combine the nonvolatility characteristics of a memristor with the gate tunability of a transistor to mimic multiple synaptic connections in neurons. Sangwan et al. [44] fabricated a multiterminal memtransistor by using monolayer molybdenum disulfide (MoS_2) in a scalable fabrication process. The two-dimensional (2D) planar geometry of the device allows multiterminals to be integrated. The conductances could be tuned through the modulation of the local Schottky barrier at each terminal, resulting in the simulation of multiple synaptic connections in neurons. The conductance between any two of the four inner electrodes can be modulated by high bias pulses applied to the two outer electrodes, demonstrating the heterosynaptic plasticity activities. It should be noted that the switching ratio for any pair of side electrodes can be increased by about 2–10 times by varying V_G from 50 to 20 V. In another word, the fabricated multiterminal memtransistors allow further modulation of heterosynaptic plasticity through a gate electrode. Furthermore, LTP/long-term depression (LTD) and STDP were also mimicked on the multiterminal memtransistors. Overall, the multiterminal memtransistors implementing memristive switching in the transistor could enable complex neuromorphic learning rules.

3.3 Multiterminal Neuromorphic Transistors

There are three terminals in a field effect transistor, including gate, drain, and source. Due to the unique characteristics of transistor, it has certain advantages over two-terminal devices in terms of neuromorphic device applications [8]. Firstly, gate spiking and drain detecting can execute synchronously. Thus, transistors have advantages in simultaneously realizing learning and processing functions. Secondly, multi-gates can be integrated in a single transistor. Thus, synaptic performances can be regulated by applying potentials on additional gates to modulate signals. Moreover, energy consumptions could also be adjusted accordingly due to the capacitive coupling. Thirdly, with multi-gates, complex synaptic functions and synaptic arithmetic can be executed, including dendrite integrate and dendrite algorithm. Recently, there are great achievements in neuromorphic transistors for neuromorphic platform applications. Several synaptic functions have been mimicked, including STM, LTP, STDP, synaptic learning, dendrite algorithm, etc. Here, multiterminal neuromorphic transistors are discussed.

3.3.1 Neuron Transistors

As early as 1992, Shibata et al. proposed a highly functional neuron MOS (νMOS) transistor. The device possesses multiple input control gates, which can capacitively couple with a continuous floating gate [47]. Thus, the device can calculate weighted sum of all input signals at the gate level with an ultralow power dissipation, controlling the "ON" and "OFF" states of the transistor. The operation mode is quite similar to those in biological neurons, as is the reason why the device can be called as neuron MOS transistor. Due to the low energy consumption, the device is suitable for energy-efficient computation. It has been designed for different applications, including D/A and A/D converter [48], sensors [49, 50], analog multiplier [51], and so on. In conventional field-effect transistors, dielectrics such as SiO_2, Al_2O_3, and high dielectric constant transition metal oxide (high-k oxide) have been applied. Because of the limitations of electrostatic modulation capabilities for these conventional dielectrics, the operation voltages are relatively large, generally above 5 V. In the 1980s, electrolyte-gated transistors (EGTs) have been reported [52, 53]. In these devices, ion coupling has provided an additional method to modulate electric properties for solid-state materials. Due to the ion transfer under external electric field, phase transformations from one form to another are expected, which greatly enrich materials functionality [54]. Basically, there are two types of operation mechanism for ionic/electronic hybrid EGTs, i.e. electrical-double-layer (EDL) modulation and electrochemical doping/de-doping, corresponding to electrical-double-layer transistors (EDL-Ts) and electrochemical transistors (ECTs), respectively [55]. Figure 3.4a shows schematic diagram of operation mechanism of EDLTs. For EDL-Ts, the movements of anions/cations in the electrolyte toward the positively/negatively charged electrode form an EDL. As a result, a high EDL capacitance of up to ~500 $\mu F/cm^2$ is expected, as is highly desirable for low energy consumption applications. When the gate voltages end, the accumulated cations will return back to their equilibrium

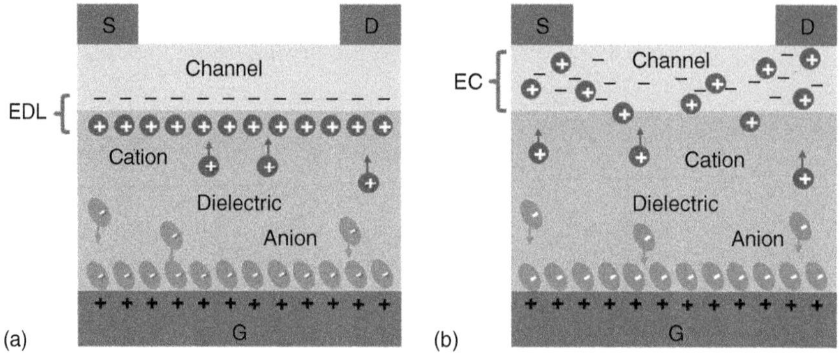

Figure 3.4 Schematic diagram of (a) electrical double-layer (EDL) mechanism and (b) electrochemical mechanism.

position. Figure 3.4b shows schematic diagram of operation mechanism in ECTs. Under gate biases, cations will migrate across the dielectric/channel interface and dope the channel. Thus, the channel conductances will change in a nonvolatile manner. When the gate voltage ends, the cations will not return back to their previous equilibrium position. Because of the strong interfacial ionic/electronic coupling, EGT can operate at low voltage, typically below 1.5 V.

Recently, neuron MOS (νMOS) transistors have been fabricated based on EGT configurations. Zhu et al. fabricated a dual in-plane gate oxide neuron transistor with a self-assembling process [56]. Logical arithmetic operation was implemented on the solid-state electrolyte-gated neuron transistors at quite low operation voltage. Firstly, P-doped nanogranular SiO_2 films were deposited by plasma-enhanced chemical vapor deposition (PECVD) using SiH_4 (95% SiH_4 + 5% PH_3) and O_2 as reactive gases. A high EDL capacitance of ~2.5 μF/cm^2 is observed, as shown in Figure 3.5a. The film also demonstrated a high room-temperature proton conductivity (σ) in the order of 1×10^{-4} S/cm, which is much higher than that of the porous SiO_2 without exposure to the atmosphere of ~10^{-8} S/cm, strongly indicating the protonic transport behavior [57, 58]. Then, the neuron transistor was fabricated with a self-assembled process through a shadow mask, as schematically shown in Figure 3.5b. Pattered indium–zinc oxide (IZO) films were deposited on the nanogranular SiO_2 film-coated fluorinated tin oxide (FTO) glass substrate acting as dual in-plane gates and source/drain electrodes. A thin IZO channel could be self-assembled between the IZO source/drain electrodes due to the reflection effect at the edge of the metal mask. The FTO layer on the glass substrate can work as a floating gate. Figure 3.5c schematically shows the operation mechanism of the neuron transistor. The current is controlled by depletion region created by the two gates. In OFF state, the depletion regions fill the channel completely, leading to the low OFF current. In ON state, the depletion region disappears, and a conducting path is created for the current to flow in the channel. As a result, the device can execute "AND" logic, as shown in Figure 3.5d. Furthermore, the logic operation can also be modulated by the operation mode of the neuron transistor. Dou et al.

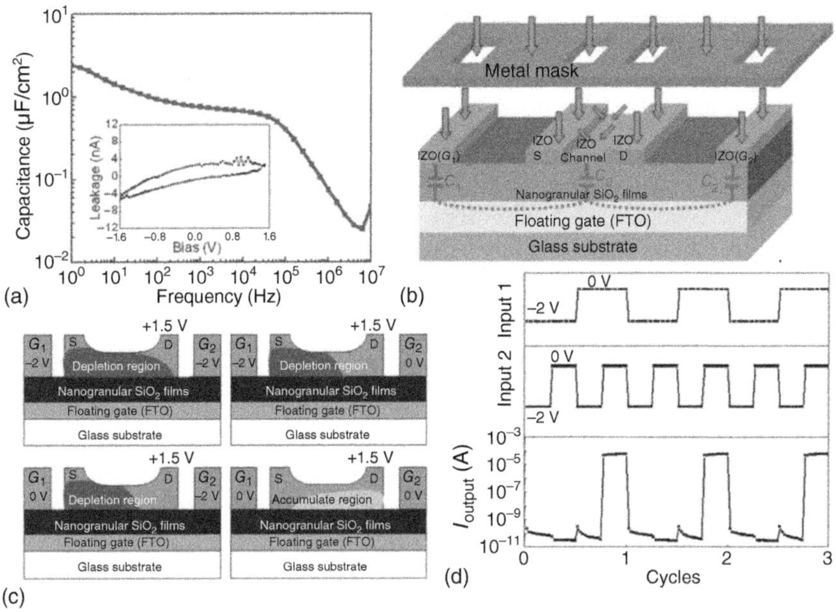

Figure 3.5 (a) Specific capacitance–frequency curves for the nanogranular SiO_2 films. (b) Schematic diagram for obtaining the neuron transistor. (c) Operation mechanism of the neuron transistor. (d) "AND" logic operation. Source: Zhu et al. [56].

proposed a flexible organic/inorganic hybrid proton/electron-coupled neuron transistors on paper substrates [59]. The device can operate at both depletion mode and enhancement mode with channel thickness of ~30 and ~10 nm, respectively. When logic states have been applied on the dual in-plane gates, "OR" and "AND" logic operations are demonstrated on devices operated in enhancement mode and depletion mode, respectively.

Due to the inherent priorities for multi-gate neuron transistors, multifunctions could be realized in terms of logic circuits. For example, multifunctional logic applications have been exhibited on PVA/GO hybrid electrolyte-gated oxide neuron transistors [60]. Firstly, PVA/GO hybrid electrolyte films have been obtained with a solution process. The hybrid film demonstrates a high EDL capacitance of ~1.4 μF/cm^2. Thus, the PVA/GO-gated neuron transistor can operate at low voltage. Furthermore, resistor-loaded inverter have been obtained by connecting a neuron transistor with a resistor in series. By doing so, "NOT," "AND," and "NAND" logic circuits were demonstrated by using one-transistor/one-resistor structures. Unified logical definitions of low potential of 2 and −2 V are used for logic state of "1" and "0," respectively.

3.3.2 Neuromorphic Devices for Chemical Biosensor Applications

3.3.2.1 Chemical Biosensors

Chemical biosensors have been developed for a long period of time in the field of bioanalytical applications, especially for Si-based ones because of their properties including sensitivity, speed, miniaturization, and cost. The operation mechanisms are protein–protein interactions, antigen–antibody binding, enzyme–substrate reactions, etc. [61, 62] For example, dopamine is an important neurotransmitter involved in human activities. It regulates multiple physiological functions of CNS and plays an important role in regulating synaptic transmission efficiency from neuron to neuron [63]. Especially, LTP and LTD are always regulated with dopamine activities [64]. On the other hand, dopamine concentration should be maintained at certain value. Otherwise, dopamine regulation disorders will happen, inducing diseases including Parkinson's disease, schizophrenia, Tourette's syndrome, attention deficit hyperactivity syndrome, etc. Liu et al. proposed a low-voltage IZO homojunction EDL transistor to detect dopamine [65]. Nanogranular SiO_2-based electrolyte acts as dielectrics. ITO bottom gate is treated by surface functionalization and biomolecule immobilization. When dipping dopamine solution on the ITO gate, the surface potential of the ITO gate will be modulated by the preferential binding of dopamine molecule to the phenylboronic acid functionalized on the gate. To compensate the decrease of the surface potential, the absolute gate voltage needs to be increased, which shifts the transistor performances. Thus, a low dopamine detection limit down to 0.1 nM level is obtained. With similar device configuration free of gate surface functionalization, N. Liu et al. reported a low-cost pH sensor with pH sensitivity of 58.1 mV/pH [66].

3.3.2.2 Neuron Transistors for Chemical Biosensor

Especially, ion-sensitive field-effect transistors, proposed by Bergveld in 1970, have been intensively investigated to probe chemical and biological species [67, 68]. The surface potential variation induced by charged species will result in a detectable electrical change in channel through field gating effect. However, dual-gate structure will amplify the sensing signal by a capacitive coupling between the sensing gate and the primary gate [69]. Conventionally, an ultrathin channel layer is usually sandwiched between the top sensing gate and the bottom primary gate, which complicates the device processing. Normally, costly silicon-on-insulator (SOI) wafers are needed in Si-based chemical biosensors [70]. Neuron transistors have been designed for chemical biosensor applications owing to its flexible device design scheme. In a typical conventional neuron transistor, the floating gate voltage depends not only on the voltages applied on the multi-gates but also on the charges present within the floating gate. M. A. R. Barranca et al. proposed a gas sensing system using a neuron transistor [50]. At proper temperature, reducing or oxidizing gases will react with a floating gate, free electrons will be either released to or taken from the gate. Thus, the floating gate voltage will be modulated accordingly, resulting in the drift of the electrical performances. B. C. Jacquot et al. proposed a time-resolved ion and molecule transport sensing with microfluidic integration by chemoreceptive neuron MOS transistors [49]. Remote input sensing gate has been designed by using an extended floating gate in a neuron transistor architecture. The device demonstrates characteristics of high sensitivity, large bandwidth, and ease of CMOS/microfluidic integration. Ion and molecular transport in various fluidic compositions can be monitored.

3.3.2.3 Multi-gate Neuromorphic Transistor for pH Sensor

N. Liu et al. designed an IZO-based EDL transistor with two laterally coupled gates for pH sensing application [71]. Figure 3.6a schematically shows the measurement layout of the device. An Al gate electrode is used as the primary gate (G1). An Ag/AgCl reference electrode immersed into a pH buffer solution droplet on nanogranular SiO_2 electrolyte film is used as sensing gate (G2). IZO channel conductivities can be modulated by Al gate electrode and Ag/AgCl reference electrode. The EDL capacitances at 1 Hz are ~4.8 and ~2.0 μF/cm^2 for the solution electrode and Al electrode, respectively, as shown in Figure 3.6b. Figure 3.6c shows transfer curves of the pH sensors measured at dual-gate synergic modulation mode with different V_{REF} on solution electrode. With the decreased V_{REF} values, the sensitivities increase. The sensitivity in a single-gate mode is only ~ 48 mV/pH. Due to the dual-gate synergic coupling effects, the sensitivity increases. It arrives at ~168 mV/pH when the sensing gate is biased at −0.6 V. The value is much bigger than the Nernst limit of ~59.2 mV/pH in ISFETs at room temperature. Thus, such dual-gate laterally synergic modulation provides a promising direction for high-performance biochemical sensors. They also design a flexible oxide-based neuromorphic transistor with multiple input gates on plastic substrates for pH sensor applications, as schematically shown in Figure 3.6d [72]. IZO channel conductivity can be modulated by the weighted sum of all inputs of sensing gate and control gates. When the device is operating at static state, dual-gate laterally synergic modulation improves the sensitivities for reference electrode biased at negative potentials. Furthermore, the device also operated at transient state. The reference electrode (G1) is biased at 0 V while the primary gate is spiked. With the decreased pH values, the drain current increased significantly, as shown in Figure 3.6e. The sensitivity is relatively low for the device operated at static state. When the device operates at spike state, the sensitivities increased significantly, as shown in Figure 3.6f. Moreover, the bending stress has no significant negative effects on the sensing performances. The energy consumption is also quite low. The power consumption reduces from 103 to 15.6 pJ/spike when the pH value increases from 4 to 10. Interestingly, when the reference electrode is biased at negative voltages, the sensitivity can be improved further, and the energy consumption can be decreased further. These results provide a novel strategy for fabricating biochemical sensors with high sensitivity and rapid response. The depression voltage applied on the sensing gate could improve the sensitivity and decrease power consumption while the transient mode improved the performances further.

3.3.3 Dendritic Algorithm on Multiterminal Neuromorphic Transistors

3.3.3.1 EGT-Based Neuromorphic Transistors

Ions play a vital role in neural system. Cognitive behaviors of our brain are realized through ion migrations, which involves physical alterations in neural substrates that modulate neuron activities and communications [73, 74]. Such ionic flux-related information storages and retrievals occur by rewiring the neuronal networks. A chemical synapse, connecting presynaptic neuron and postsynaptic neuron, can

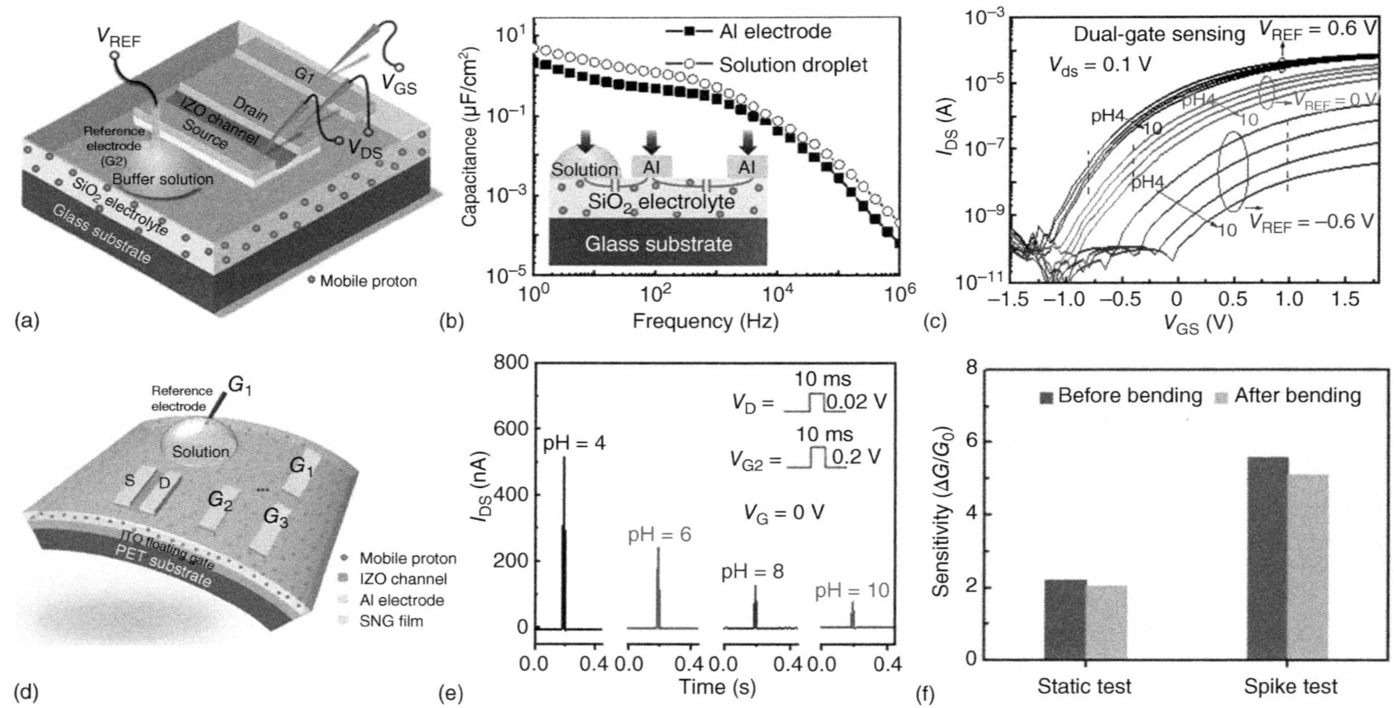

Figure 3.6 (a) Schematic measurement layout of pH sensor based on dual-gate IZO-based EDL transistor. (b) Frequency-dependent specific capacitance of SiO_2 electrolyte measured with an Al electrode and a solution droplet of pH = 6.0. (c) Transfer curves of the pH sensors measured at dual-gate synergic modulation mode with different V_{REF} on solution electrode. Source: (a–c) Liu et al. [71]. (d) Schematic of flexible pH sensor based on an IZO neuromorphic transistor. (e) Single-spike response for pH value increases from 4 to 10. (f) The influence of 1000 times bending on the sensitivity of the neuromorphic transistor for both quasi-static and single-spike sensing modes. Source: (d–f) Liu et al. [72].

transport information via an action potential. For example, voltage-gated calcium (Ca^{2+}) channels will be activated when action potential arrives at presynaptic neuron, inducing the release of neurotransmitters in presynaptic neuron. Subsequently, these released neurotransmitters will bind with receptors on postsynaptic membrane, which changes postsynaptic membrane potential and generate postsynaptic current (PSC) [75]. Interestingly, ion gating behaviors in EGTs are quite similar to ionic flux-related cognitive behaviors in biological neurons and synapses. Thus, unique ion modulation behavior in EGTs could establish bridges between ionotronic devices and neuromorphic platforms. For EGTs operated in EDL-T mode, ions within the electrolyte will migrate to electrolyte/channel interface under external electrical field. Then, a compact EDL layer will form at the interface, inducing changes in channel conductance. However, channel conductance will return back to its initial value when the external electrical field is removed due to the fact that ions will migrate back to their initial equivalent positions. In another word, channel conductance is modulated in a nonvolatile mode. The behaviors are quite similar to the STP behaviors observed in biological synapses. For EGTs operated in ECT mode, ions will react with channel layer under a relatively high external electrical field. Thus, channel conductance will be modulated in a volatile mode. These are quite similar to the LTP behaviors observed in biological synapses. Thus, EGTs have great potentials in neuromorphic engineering. Up to now, several important synaptic functions and neural algorithms have been simulated on EGTs, including STP/LTP, paired-pulse facilitation (PPF)/paired-pulse depression (PPD), STDP, synaptic filtering, etc.

3.3.3.2 Multi-gate Neuromorphic Transistor

When designing multi-gates, complicated synaptic functions have also been mimicked, including synaptic modulation, dendrite integration, brain-like neural functions, etc. Zhu et al. proposed an artificial synapse network for neuromorphic system applications using nanogranular SiO_2 proton conductor [76]. Figure 3.7a schematically shows the synapse network. Multi-gates are deemed as presynapses to receive presynaptic spikes, while multichannels with drain and source electrodes are deemed as postsynapses. Due to the lateral proton gating effects of nanogranular SiO_2 proton conductor, all the presynapses and postsynapses are intercoupled without the needs of complicated intentional hard-wire connection. Furthermore, with multi-gate synergic modulation, the transistor performances could be improved significantly [77, 78]. The distances between presynapses and postsynapses can affect the synaptic responses too, as shown in Figure 3.7b. Such distance (d) value-dependent synaptic weights are suitable for synaptic network construction. Distance-dependent PPF behaviors have also been observed. With the increased d value, PPF index decreases correspondingly, as shown in Figure 3.7c.

In nerve system, heterosynaptic plasticity is an essential synaptic response behavior [41, 42]. It plays a significant role in stabilizing synaptic weights and promoting synaptic competition. There are two types of synaptic inputs correlated with neural transmission in a neuron with branched dendrites, i.e. the driving inputs and the modulatory input. The driving inputs can make the relevant neuron fire

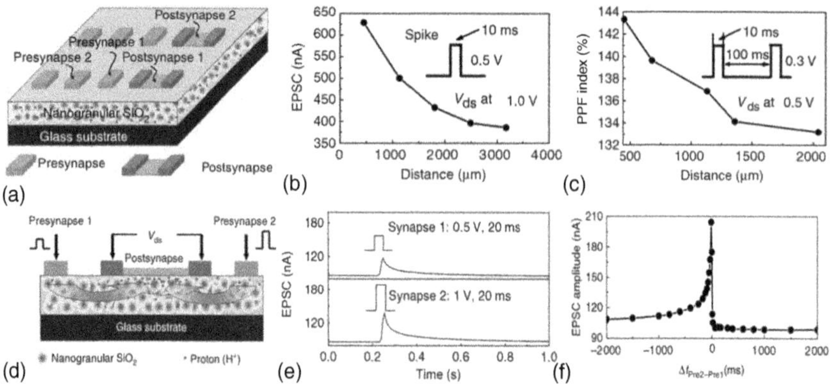

Figure 3.7 (a) Schematic diagram of synaptic network with multiple in-plane gates (presynapses) and multiple IZO channels (postsynapses). (b) Presynapse to postsynapse distance-dependent EPSC. (c) Presynapse to postsynapse distance-dependent PPF index. (d) Schematic diagram of the method to measure spatiotemporal integration. (e) EPSCs triggered by presynaptic spike 1 and presynaptic spike 2. (f) Spatiotemporal integration behavior. Source: Zhu et al. [76].

strongly, while the modulatory input can tune the effectiveness of the driving input [79, 80]. Thus, in heterosynaptic plasticity, synaptic connectivity can be reinforced or weaken by another modulatory terminal. In neuromorphic transistor, one of the lateral gates can be deemed as a modulatory terminal, while others could be deemed as presynapses. Thus, the modulatory biases would affect the synaptic responses, including excitatory postsynaptic current (EPSC) and PPF behavior. Therefore, the neuromorphic transistor with multi-gates could be deemed as a heterosynapse [81]. Figure 3.8a shows a schematic diagram of a heterosynapse in biological systems. There are dendritic inputs and modulatory input. Figure 3.8b shows schematic diagram of the method to mimic the heterosynaptic plasticity. When there is no modulatory pulse, the synapse shows excitory activities. Interestingly, when the modulatory pulse and the dendrite input terminal are triggered simultaneously, the transition between inhibitory synaptic behavior and excitory synaptic behavior can be observed, as shown in Figure 3.8c. Thus, typical heterosynaptic responses have been mimicked. Due to the modulation of EPSC current with a modulatory bias, the energy consumption could also be modulated effectively for a single synaptic response.

3.3.3.3 Dendrite Neuron and Dendritic Algorithm

In the nerve system, there are three functional parts of a neuron, including dendrites, soma, and axon. Dendrites are branched protoplasmic extensions of a nerve cell. Signals can be transmitted to soma through the dendrite synapses. Thus, a fundamental computation of neurons is the transformation of incoming synaptic information into specific synaptic output patterns [83]. Basically, soma is a central processing unit to deal with the input signals in a spatial–temporal manner, including the addition of unitary events occurring simultaneously in

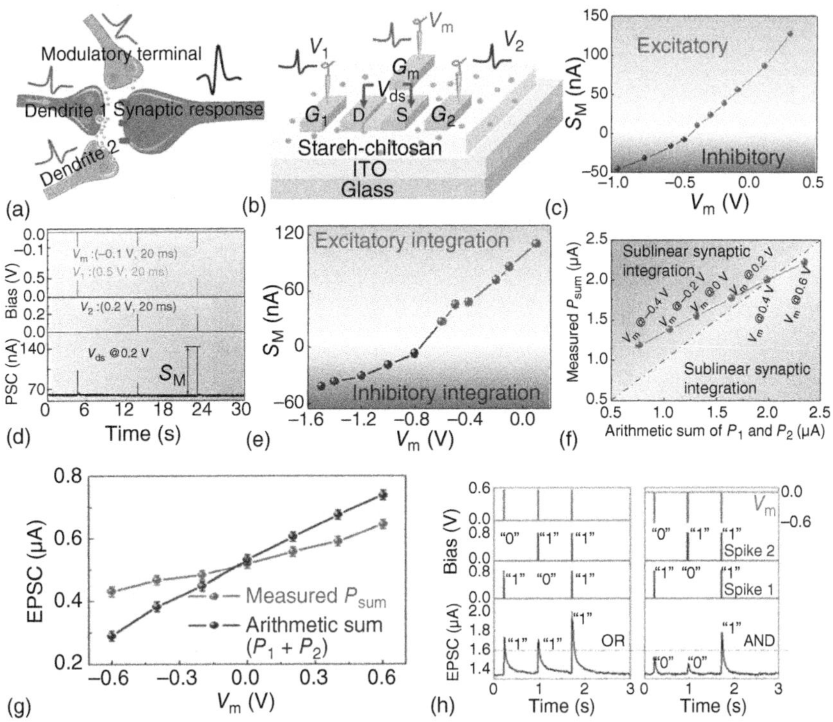

Figure 3.8 (a) Schematic diagram of a biological heterosynapse. (b) Schematic diagram of the method to mimic the heterosynaptic plasticity. (c) Modulation of synaptic activities between inhibitory response and excitatory response. (d) PSC responses triggered by synchronous and asynchronous dendrite spikes. (e) Modulation between inhibitory synaptic integration and excitatory synaptic integration. Source: (a–e) Long et al. [81]. (f) Modulation between superlinear synaptic integration and sublinear synaptic integration. Source: Zhu et al. [78]. (g) Measured EPSC sum (P_{sum}) and arithmetic EPSC sum ($P_1 + P_2$) as a function of modulatory spike (V_m). (h)"OR" logic operation and "AND" logic operation regulated by a modulatory spike (V_m). Source: (g, h) Zhu et al. [82].

separate dendrite synapse (spatial summation) and addition of non-simultaneous unitary events (temporal summation). Such spatial–temporal summation behavior is an important synaptic response to establish a dynamic logic [83]. Multi-gate neuromorphic transistor can mimic such dendrite neurons and execute dendritic algorithm. Figure 3.7d schematically shows the method to measure spatiotemporal integration. When applying spatiotemporal spikes on two presynapses, postsynaptic currents are detected on the channel with a constant V_{ds}. Figure 3.7e shows the EPSCs triggered by presynaptic spike 1 and presynaptic spike 2, and Figure 3.7f shows the measured spatiotemporal integration behavior. Heterosynaptic plasticity will also affect dendritic integration activities. When only excitory synaptic spikes are applied, the dendritic integration will exhibit excitory behavior, as shown in Figure 3.8d. When introducing modulatory terminal, dendritic integration can be modulated between excitory behavior and inhibitory behavior, as shown in Figure 3.8e. The behavior indicates great potentials of heterosynaptic plasticity in synaptic computation. In neural systems, specific synaptic algorithm results in specific output patterns. The heterosynaptic plasticity also results in the modulation of synaptic integration. When the modulatory terminal is biased with modulatory spike, both superlinear integration and sublinear integration are demonstrated, as shown in Figure 3.8f. When the modulatory terminal is biased at certain spikes, device operation would be modulated. Thus, dynamic logic "AND" and logic "OR" would be observed, as shown in Figure 3.8g,h [82]. Y. Fu et al. [84] also fabricated multiterminal organic transistors. Two presynaptic spikes applied on two gates are deemed as condition stimulus (−3 V in amplitude simulating "bell ringing") and unconditioned stimulus (−5 V in amplitude simulating "sight of food"), respectively. Thus, Pavlovian learning behaviors can be simulated. In addition, John et al. [85] also demonstrated the Pavlovian classical conditioning experiment on a dual-gated EDL transistor. The bottom gate spikes and top gate spikes are deemed as conditioned stimulus (CS) and unconditioned stimulus (US). Persistent simultaneous training results in effective association between CS and US. After training, the salivation will be activated when applying bottom spikes alone. Moreover, classical conditioning extinction, retraining, and recovery were also demonstrated.

More neural functions could also be developed. In nerve system, neuronal arithmetic includes multiplicative and additive modulation on input–output (I–O) relationship [86, 87]. C. J. Wan et al. [88] proposed proton-conducting graphene oxide-coupled neuron transistors with multi-in-plane gates to mimic such neuronal arithmetic. Voltage pulses (0.5 V, 10 ms) were applied on G1 and G2 as driving inputs, and a low bias voltage (e.g. −0.1 and 0.1 V) was applied on Gm as modulatory input. The triggered EPSC, combining effects of these inputs, was defined as neuronal output. Figure 3.9a shows schematic diagram of the multiplicative and additive operations. In multiplicative operation, neuronal response can be written as $R(d, m) = f(d) \times g(m)$, where d and m are the driving input parameters and modulatory input parameters, respectively. Similarly, in additive operation, neuronal response can be written as $R(d, m) = f(d) + g(m)$. During simulation, the driving input parameter is defined by rate of the driving input, and the modulatory input parameter is defined by the modulatory voltage, i.e. $d = f$ and $m = V_m$. Figure 3.9b shows typical neural output in response to a Poisson distribution spike

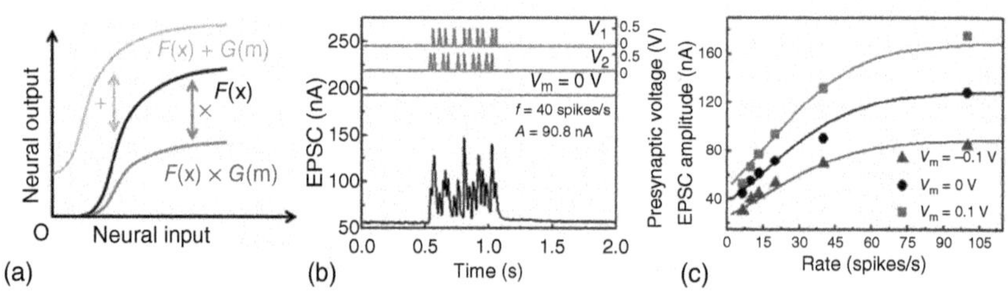

Figure 3.9 (a) Schematic diagram of multiplicative and additive operations. (b) Typical neural output in response to a Poisson distribution spike train as driving inputs. (c) Neural I–O relationship tuned with modulatory input voltages. Source: Wan et al. [88].

train as driving inputs (0.5 V, 10 ms) with rate of 40 spikes/s and a modulatory input of 0 V. Figure 3.9c shows the neural *I*–*O* relationship tuned with modulatory input voltages. With the increased f, EPSC amplitude increases gradually and gets saturated at high f values. The slopes of the *I*–*O* curves also increase with the increased V_m values, indicating the multiplicative operation. Similarly, the neural arithmetic based on temporal-correlated coding scheme was also investigated, demonstrating multiplicative operation. Thus, the artificial neuron could be potentially applied in pattern recognition and sensory processing applications. G. G. Malliaras et al. [89] also designed PEDOT:PSS ECT with 3 × 3 gates. The orientation is defined at certain spiking condition. Voltage pulses have been applied concurrently at three different gates to reproduce spatial pulse, producing orientation-resolved outputs. Thus, an obvious orientation selectivity is observed, analogous to that in visual system. Similarly, C. Qian et al. [90] also defined variable orientation on multi-gate P3HT ECTs. The polar diagram demonstrates the orientation-dependent EPSC. These functions would have potential applications, including intelligent electronic eyes, multifunctional robotics, and auxiliary equipment for visually disabled people, etc.

3.3.3.4 Multi-gate Neuromorphic Transistor for Pattern Recognition

There are also reports on pattern recognition applications with multi-gate synaptic transistors. Kim et al. [91] conducted a system-level study on pattern recognition by using neuron circuits with double-gate synaptic transistors. A single-layer spiking neural network (SNN) is proposed for pattern recognition. Furthermore, they demonstrated pattern recognition of modified national institute of standards and technology (MNIST) images based on three different kinds of SNN. The classification rate of SNN for excitatory synaptic transistors trained according to STDP rules is ~60% of the total accuracy. As for SNN with transferred synaptic weights from an ANN using a rectified linear unit (ReLU), the classification rate increases to ~87.6% of the total accuracy.

3.4 Neuromorphic Transistors for Perception Learning Activities

3.4.1 Artificial Tactile Device

The skin is an important organ for sensing external pressure stimuli. Tactile sensation, one of five important human sensations, can transform pressure information received from skin into inner feelings [6, 92, 93]. Figure 3.10 shows the schematic diagram of human tactile afferent nerves, consisting of tactile receptors (i.e. mechanoreceptors), nerve fibers, and processing components (i.e. neurons and synapses). External touch/press stimulus on the skin can be converted by tactile receptors into electrical signals. Electrical signals can be transmitted to CNS and be processed. Thus, terminal somatosensory cortex can be activated to form tactile sensation. Inspired by such process, it has attracted widespread attention recently to integrate artificial synapses with sensor devices to imitate the tactile perception

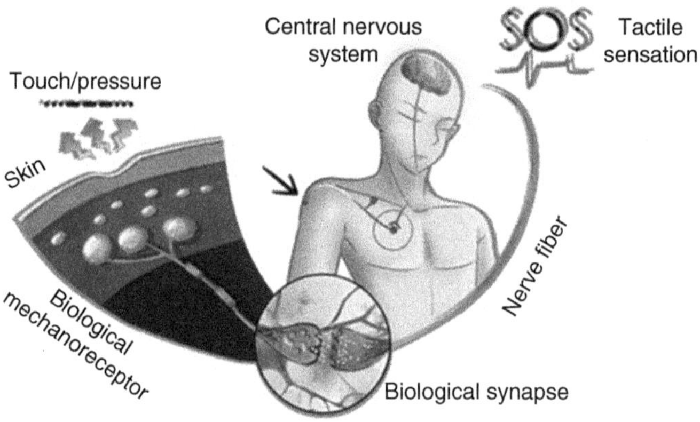

Figure 3.10 Schematic diagram of human tactile afferent nerve. Source: Yu et al. [94].

behavior of organisms. Initially, tactile sensors have been demonstrated in the field of E-skins [95]. Due to the unique ionic relaxation behaviors and ionic–electronic effects, great achievements have been made for ionotronic devices in neuromorphic engineering applications. Recently, brain-inspired tactile afferent nerve has been realized by integrating ionotronic neuromorphic devices with E-skin devices [96–99].

Piezoelectric nanogenerators (PENGs) have also been proposed for artificial sensory synapse applications [98]. This is integrated with ion gel-gated transistors, as schematically shown in Figure 3.11a. In the piezotronic artificial synapse, piezopotential triggered by external strain, ions in the ion gel dielectric, and channel current are deemed as the presynaptic input, neurotransmitters, and postsynaptic output in the biological sensory synapse, respectively. The piezopotential from PENG efficiently power the ion gel-gated synaptic transistor due to the formation of electrical double layer. Various synaptic behaviors have been simulated in this device, such as excitation/inhibition, synaptic plasticity, PPF, responses to multiple strain spikes, and dynamic modulation behaviors. Furthermore, two parallel PENGs are coupled with one synaptic transistor. Thus, two artificial presynaptic sensor neurons convert mechanical strains to piezopotential pulse inputs, linking the spatiotemporal strain information with the PSC, as shown in Figure 3.11b. The piezotronic artificial sensory synapse can be considered as an artificial afferent nerve system, enabling parallel strain signal recognition and processing.

By combining pressure sensors with artificial synapses, it is also possible to mimic the sensory neuron and is capable of perceptual learning. Wan et al. demonstrated a neuromorphic tactile processing system (NeuTap) composed of piezoresistive sensor, soft ionic cable, and synaptic transistor, as shown in Figure 3.11c [100]. The system can mimic the functions of receptor, axon, and synapse. The pressure signal of the pressure sensor is converted into an electrical signal and can be transmitted to the synaptic transistor through the ion cable for information processing. Combining two pressure sensors with dual-gate synaptic transistors can integrate and distinguish the temporal and spatial characteristics of tactile patterns, enabling parallel

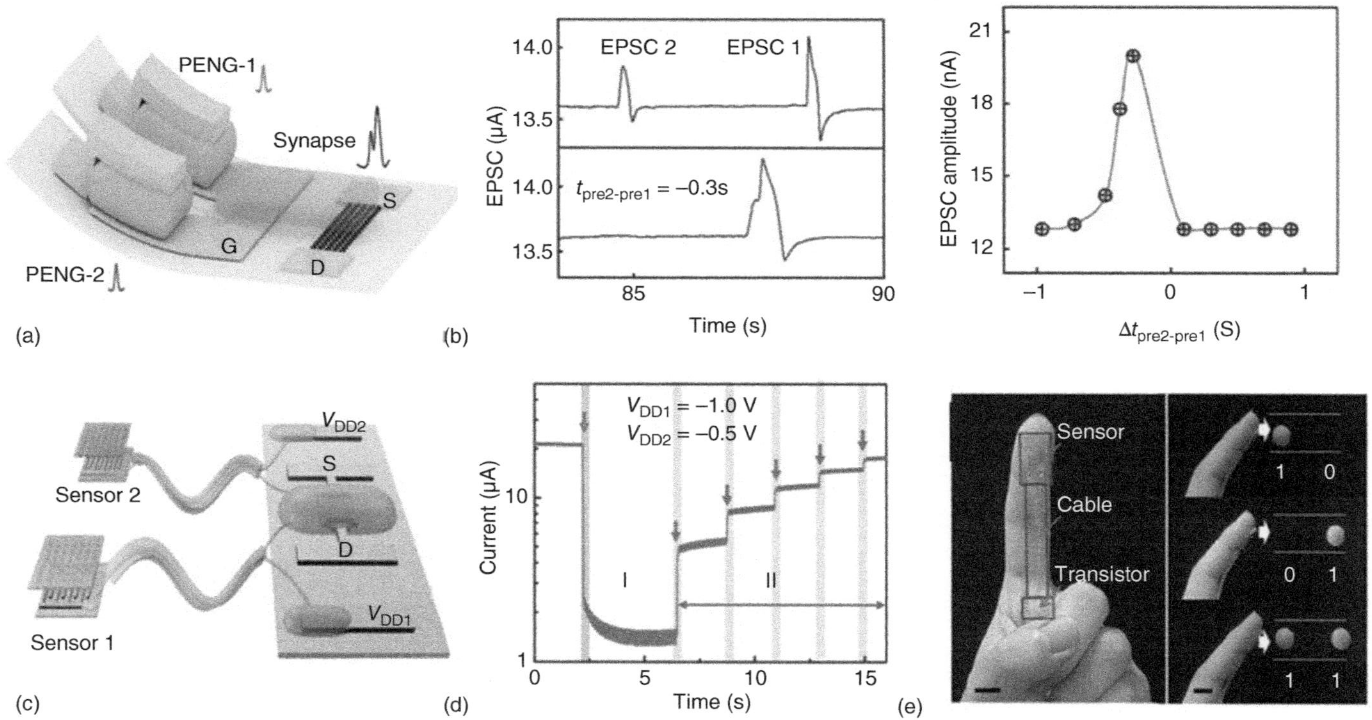

Figure 3.11 (a) Schematic diagram of artificial sensory synapse based on piezoelectric nanogenerators (PENG). (b) Spatiotemporal strain information integration. Source: (a, b) Chen et al. [98]. (c) Schematic diagram of a neuromorphic tactile processing system (NeuTap) composed of resistance pressure sensor, soft ion cable, and synaptic transistor. (d) Integration of the spatiotemporal correlated tactile stimuli. (e) Tactile pattern recognition. Source: (c–e) Wan et al. [100].

sensory signal processing. V_{DD1} is set at −1.0 V and V_{DD2} is set at 0.5 V. Sensor 1 was pressed firstly, followed by sensor 2. Upon pressing sensor 1, current decreases dramatically. By repeatedly pressing of sensor 2, the current could be tuned stepwise back to the initial level. These results indicate that the NeuTap could also integrate the spatiotemporal correlated tactile stimuli and could also implement tactile pattern recognition, as shown in Figure 3.11e.

3.4.2 Artificial Vision Device

External stimuli of 80% are perceived through vision [101, 102]. So, the retina can be deemed as one of the most important parts of the human perception system. In fact, this is the extension of the brain. It is a thin layer of tissue lining the inner surface of the eye. It contains millions of light-sensitive photoreceptor cells and other nerve cells. Photoreceptor cells include rods and cones and are differentiated structurally by their distinctive shapes. Thus, they will sense different kinds of light. When light passes through the retina, cone cells can detect specific wavelengths of light and preprocess its intensity. Then, signals will be forwarded through optic nerve to the brain, forming the vision.

In field-effect transistor-based neuromorphic devices, synaptic performances can be regulated by adding additional gates to amplify signals and reduce energy consumption. When illuminating the devices, synaptic performances could also be modulated correspondingly. Thus, photo illumination could be deemed as an additional gating, named as photogating. Thus, photogating transistors would also have great potentials in artificial visual neuromorphic system. HL Park et al. [103] proposed an organic photonic synapse, using carbon nitride (C_3N_4) as an UV-responsive floating gate layer with transistor configuration. C_3N_4 nanodots dominantly absorb UV light. Thus, the device can selectively detect ultraviolet light and process various optical stimuli. The device has extremely low power consumption, consuming only 18.06 fJ of energy per synaptic event and demonstrating good synaptic plasticity. Y. B. Guo et al. [104] proposed a starch-based bio-polysaccharide electrolyte-gated indium–gallium–zinc oxide (IGZO) phototransistor, as shown in Figure 3.12a. Photoelectric synergic effects have been measured on such neuromorphic transistor. The EPSC is obtained with a constant V_{ds} of 0.5 V. A single light spike induces an EPSC with absolute EPSC value of ~92 nA, while a single electric spike triggers an EPSC with absolute EPSC value of ~44 nA. As shown in Figure 3.12b, when there is a time interval between the light spike and the electric spike, a spatial–temporal coupling behavior has been triggered. When the two spikes are triggered synchronously, superlinear integration behaviors have been observed. Pavlovian conditioned reflex is a typical form of associative learning in the nervous system [105]. Due to the photoelectronic synergic coupling behavior for the neuromorphic transistor, Pavlovian conditioned reflex behaviors have also been simulated. Light spikes are deemed as feeding food (unconditioned stimulus, US), while gate voltage spikes are deemed as bell ringing (neutral stimulus, NS). Figure 3.12c shows the imitation of Pavlovian conditioning. Similarly, this has also been mimicked on synergistic gating of electro-iono-photoactive 2D

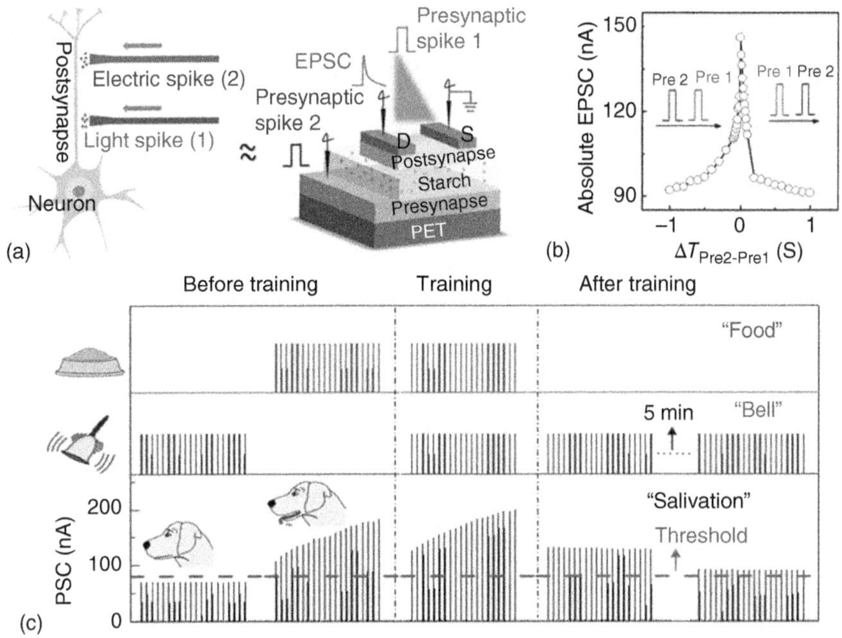

Figure 3.12 (a) Schematic diagram of synergic effect in neurons activated by electric and light stimuli and schematic diagram of the photoelectric synergic coupled IGZO neuromorphic transistor. (b) Photoelectric synergic spatial–temporal coupling. (c) Imitation of Pavlovian associative learning by associating gate voltage spikes with light spikes. Source: Guo et al. [104].

chalcogenide neuristors [106] and synaptic devices under all-light stimulation [107, 108]. Furthermore, multi-input light-stimulated synaptic transistor has also been fabricated. The device integrates perovskite quantum dot (QD). Owing to the persistent photoconductivity (PPC) effect of the semiconductor layer and synaptic behaviors, the device has been proposed to mimic the Pavlovian classical conditioning, using all-light spikes. G1 spikes ($150\,mW/cm^2$, 100 ms) is deemed as food (unconditioned stimulus, US) to activate salivation (unconditioned response, UR). G3 ($30\,mW/cm^2$, 100 ms) spikes are considered as bell (conditioned stimulus, CS). Association between G1 and G3 could be established. Thus, Pavlovian classical conditioning has been mimicked. Such multi-input optical signal synaptic devices do not need additional electrodes and thus have great potentials in neuromorphic engineering.

3.4.3 Artificial Auditory Device

Biological auditory system is a sensory system for the sense of hearing [109, 110]. The system consists both the sensory organs and the auditory parts of the sensory system. Most creatures have the abilities to localize sounds and perceive the location of the origin of the sound at certain frequencies. Mimicking of auditory on hardware-based platforms would be meaningful for artificial multifunctional perception learning system. Wan group proposed a capacitively coupled multiterminal oxide-based neuro-transistor [111]. The device demonstrates dynamic spatiotemporal information processing. Figure 3.13a schematically shows the multiterminal EDL neuro-transistor using chitosan-based electrolyte as gate dielectric. Due to the enormous EDL gate capacitance, the device can work at a low voltage of 2.0 V, leading to the low power consumption. Right panel of Figure 3.13a shows the corresponding ANN with multiple presynaptic neurons (PRENs) and postsynaptic neurons (POSTNs). Spatiotemporal electrical pulse sequences have been applied on the multiple gate electrodes in one direction with different gate-to-channel distances. It is observed that the EPSC amplitude decreases with the increased gate-to-channel distance, though the same presynaptic spike has been applied. Furthermore, sequential activation in the IN direction can produce higher response than that of the OUT direction. The behavior is similar to that in neural system [112, 113]. In biological system, sound location is achieved by detecting the interval time difference (ΔT), as schematically shown in Figure 3.13b. Sound azimuth detecting has also been realized by designing a neural system consisting of two gate electrodes and two pairs of source/drain terminals, as shown in Figure 3.13c. PREN1 and PREN2 are deemed as the left ear and right ear sound-sensing neurons, respectively. When spikes are applied on PREN1 and PREN2 with certain interval time, responses on POSTN1 and POSTN2 have carried the sound azimuth information. By doing so, sound azimuth detection function of the human brain has been emulated, as shown in Figure 3.13d,e. The results indicate that the neuro-transistor can perform spatiotemporal information decoding. Capacitively coupled neuro-transistors with multi-gates can greatly enhance the computation power in the future neuromorphic systems [112].

3.5 Conclusion and Outlook

With the hierarchical structure of the cerebral neocortex and parallel processing, brain computation is more robust, plastic, and fault tolerant than any current digital computer. Inspired by the brain computation, neuromorphic system has been proposed. It would help to process in-time nonstructural big data in an energy-efficient way, much superior to conventional digital computation based on von Neumann configuration. Considering the fact that synapses and neurons are the basic unit in brain computation, one of the main tasks in neuromorphic systems would be the realization of neuromorphic devices, capable of mimicking multiple fundamental synaptic computation behaviors. Recently, the study on solid-state neuromorphic devices leads to the emergence of a new branch of neuromorphic engineering because of its great potentials. With the developments of new materials technology and new conceptual devices, several kinds of neuromorphic devices have been proposed, including two terminal resistance switch devices and three terminal transistors.

Additionally, mimicking dendrite integration behaviors on a single neuromorphic device would be highly interesting for neuromorphic engineering. When

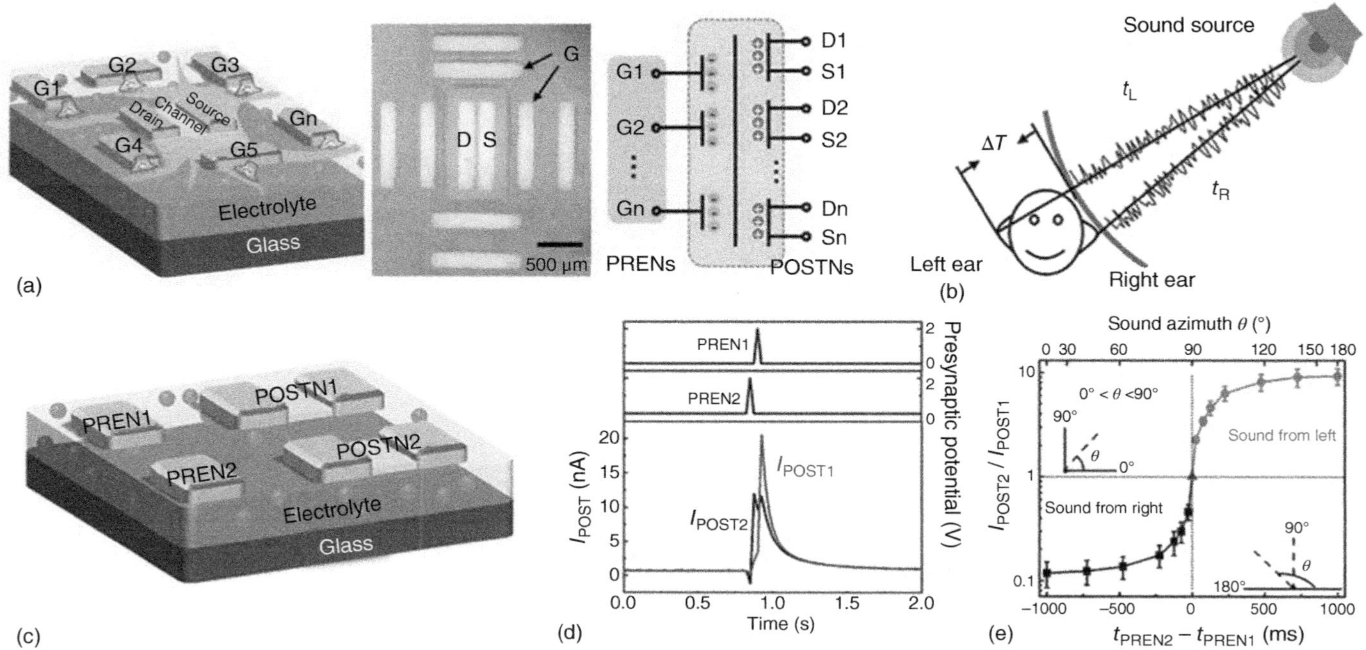

Figure 3.13 (a) Schematic diagram of multiterminal synaptic transistor structure, a photograph of the neuro-transistor, and artificial neural system based on multiterminal neuro-transistor with multiple PRENs and multiple POSTNs. (b) Schematic diagram of sound localization. (c) Schematic diagram of neural system for detecting sound azimuth. (d) Postsynaptic current of POSTN1 and POSTN2 after the sound comes. (e) Ratio of the amplitude of I_{POST2} and I_{POST1} at the time of the last PREN signal as a function of time interval and the sound azimuth. Source: He et al. [111].

multi-electrodes are integrated into a single neuromorphic device, more complex neuromorphic learning activities could be mimicked by copying biological neurons with multiple synapses, demonstrating great potentials in high-density neuromorphic platform. For memristors, multiterminal memristors have been proposed, demonstrating the advanced neural function activities. Additionally, memtransistors have been reported with interesting neuromorphic functions. For transistor-based neuromorphic devices, multi-gates could be assembled in a single device, exhibiting interesting dendritic integration functions and dendritic algorithm. Interestingly, photogating could also be deemed as additional pseudo-gating terminal and would have potentials in modulating electrical performances of memristors and transistors. With unique device configurations, visual discrimination activities could be mimicked. Thus, pseudo-multiterminal neuromorphic device could be fabricated, demonstrating great potentials in visual neuromorphic platforms. All these results indicate their great potentials in neuromorphic engineering and cognitive human–machine interfaces.

Acknowledgments

Authors would like to acknowledge the financial support from National Natural Science Foundation of China (51972316) and Zhejiang Provincial Natural Science Foundation of China (LR18F040002).

References

1 Moore, G.E. (1998). Cramming more components onto integrated circuits. *Proc. IEEE.* 86 (1): 82–85.
2 He, G., Zhu, L.Q., Sun, Z.Q. et al. (2011). Integrations and challenges of novel high-k gate stacks in advanced CMOS technology. *Prog. Mater. Sci.* 56 (5): 475–472.
3 Drachman, D.A. (2005). Do we have brain to spare? *Neurology* 64 (12): 2004–2005.
4 Zidan, M.A., Strachan, J.P., and Lu, W.D. (2018). The future of electronics based on memristive systems. *Nat. Electron.* 1 (1): 22–29.
5 Quiroga, R.Q., Reddy, L., Kreiman, G. et al. (2005). Invariant visual representation by single neurons in the human brain. *Nature* 435 (7045): 1102–1107.
6 Lumpkin, E.A. and Caterina, M.J. (2007). Mechanisms of sensory transduction in the skin. *Nature* 445 (7130): 858–865.
7 Yu, F. and Zhu, L.Q. (2019). Ionotronic neuromorphic devices for bionic neural network applications. *Phys. Status Solidi RRL* 6 (13): 1800674.
8 Guo, Y.B. and Zhu, L.Q. (2020). Recent progress in optoelectronic neuromorphic devices. *Chin. Phys. B* 29 (7): 078502.
9 Chua, L. (1971). Memristor-the missing circuit element. *IEEE Trans. Circuit Theory.* 18 (5): 507–519.

10 Strukov, D.B., Snider, G.S., Stewart, D.R., and Williams, R.S. (2008). The missing memristor found. *Nature* 453 (7191): 80–83.

11 Nandakumar, S.R., Minvielle, M., Nagar, S. et al. (2016). A 250 mV Cu/SiO_2/W memristor with half-integer quantum conductance states. *Nano Lett.* 16 (3): 1602–1608.

12 Wu, C.X., Kim, T.W., Guo, T.L. et al. (2017). Mimicking classical conditioning based on a single flexible memristor. *Adv. Mater.* 29 (10): 1602890.

13 Ting, Y.H., Chen, J.Y., Huang, C.W. et al. (2018). Observation of resistive switching behavior in crossbar core–shell Ni/NiO nanowires memristor. *Small* 14 (6): 1703153.

14 Wu, Q.T., Wang, H., Luo, Q. et al. (2018). Full imitation of synaptic metaplasticity based on memristor devices. *Nanoscale* 10 (13): 5875–5881.

15 Yang, J.J., Strukov, D.B., and Stewart, D.R. (2012). Memristive devices for computing. *Nat. Nanotech.* 8 (1): 13–14.

16 Yu, S.M. and Chen, P.Y. (2016). Emerging memory technologies: recent trends and prospects. *IEEE Solid-State Circuits Magazine* 8 (2): 43–56.

17 Terabe, K., Hasegawa, T., Nakayama, T., and Aono, M. (2005). Quantized conductance atomic switch. *Nature* 433 (7021): 47–40.

18 Ohno, T., Hasegawa, T., Tsuruoka, T. et al. (2011). Short-term plasticity and long-term potentiation mimicked in single inorganic synapses. *Nat. Mater.* 10 (8): 591–595.

19 Tsai, T.L., Chang, H.Y., Lou, J.J.C., and Tseng, T.Y. (2016). A high performance transparent resistive switching memory made from ZrO_2/AlON bilayer structure. *Appl. Phys. Lett.* 108 (15): 153505.

20 Yoon, J.H., Kwon, D.E., Kim, Y. et al. (2017). The current limit and self-rectification functionalities in the TiO_2/HfO_2 resistive switching material system. *Nanoscale* 9 (33): 11920–11928.

21 Ye, C., Deng, T.F., Zhang, J.C. et al. (2016). Enhanced resistive switching performance for bilayer HfO_2/TiO_2 resistive random access memory. *Semicond. Sci. Technol.* 31 (10): 105005.

22 Xiong, W., Zhu, L.Q., Ye, C. et al. (2019). Bilayered oxide-based cognitive memristor with brain-inspired learning activities. *Adv. Electron. Mater.* 5 (8): 1900439.

23 Atkinson, R.C. and Shiffrin, R.M. (1968). Human memory: a proposed system and its control processes. *Psychol. Learn. Motiv.* 2: 89–195.

24 Raeis-Hosseini, N. and Lee, J.S. (2016). Controlling the resistive switching behavior in starch-based flexible biomemristors. *ACS Appl. Mater. Interfaces* 8 (11): 7326–7332.

25 Bae, H., Kim, D., Seo, M. et al. (2019). Bioinspired polydopamine-based resistive-switching memory on cotton fabric for wearable neuromorphic device applications. *Adv. Mater. Technol.* 4 (8): 1900151.

26 Xiong, W., Zhu, L.Q., Ye, C. et al. (2020). Flexible poly(vinyl alcohol)-graphene oxide hybrid nanocomposite based cognitive memristor with pavlovian-conditioned reflex activities. *Adv. Electron. Mater.* 6 (5): 1901402.

27 Calabrese, B., Wilson, M.S., and Halpain, S. (2006). Development and regulation of dendritic spine synapses. *Physiology* 21 (1): 38–37.

28 Clapham, D.E. (2007). Calcium signaling. *Cell* 131 (6): 1047–1058.

29 Krueppel, R., Remy, S., and Beck, H. (2011). Dendritic integration in hippocampal dentate granule cells. *Neuron* 71 (3): 512–518.

30 Wan, Q., Zeng, F., Yin, J. et al. (2019). Phase-change nanoclusters embedded in a memristor for simulating synaptic learning. *Nanoscale* 11 (12): 5684–5692.

31 Yan, X.B., Zhao, J.H., Liu, S. et al. (2018). Memristor with Ag-cluster-doped TiO_2 films as artificial synapse for neuroinspired computing. *Adv. Funct. Mater.* 28 (1): 1705320.

32 Guo, R., Zhou, Y.X., Wu, L.J. et al. (2018). Control of synaptic plasticity learning of ferroelectric tunnel memristor by nanoscale interface engineering. *ACS Appl. Mater. Interfaces.* 10 (15): 12862–12869.

33 Wang, L., Liao, W.G., Wong, S.H. et al. (2019). Artificial synapses based on multiterminal memtransistors for neuromorphic application. *Adv. Funct. Mater.* 29 (25): 1901106.

34 Zucker, R.S. (1999). Calcium-and activity-dependent synaptic plasticity. *Curr. Opin. Neurobiol.* 9 (3): 305–313.

35 Markram, H., Pikus, D., Gupta, A., and Tsodyks, M. (1998). Potential for multiple mechanisms, phenomena and algorithms for synaptic plasticity at single synapses. *Neuropharmacology* 37 (4–5): 489–480.

36 Tsodyks, M.V. and Markram, H. (1997). The neural code between neocortical pyramidal neurons depends on neurotransmitter release probability. Proc.Natl.Acad.Sci. *USA.* 94 (10): 5495–5495.

37 Wang, Y.Y., Zhang, Z.Y., Xu, M.K. et al. (2019). Self-doping memristors with equivalently synaptic ion dynamics for neuromorphic computing. *ACS Appl. Mater. Interfaces.* 11 (27): 24230–24240.

38 Zhu, J.D., Yang, Y.C., Jia, R.D. et al. (2018). Ion gated synaptic transistors based on 2D van der Waals crystals with tunable diffusive dynamics. *Adv. Mater.* 30 (21): 1800195.

39 Shetty, M.S., Gopinadhan, S., and Sajikumar, S. (2016). Dopamine D1/D5 receptor signaling regulates synaptic cooperation and competition in hippocampal CA1 pyramidal neurons via sustained ERK1/2 activation. *Hippocampus* 26 (2): 137–130.

40 Zhu, X.J., Li, D., Liang, X.G., and Lu, W.D. (2018). Ionic modulation and ionic coupling effects in MoS_2 devices for neuromorphic computing. *Nat. Mater.* 18 (2): 141–148.

41 Bailey, C.H., Giustetto, M., Huang, Y.Y. et al. (2000). Is heterosynaptic modulation essential for stabilizing Hebbian plasticity and memory? *Nat. Rev. Neurosci.* 1 (1): 11–10.

42 Chistiakova, M., Bannon, N.M., Bazhenov, M., and Volgushev, M. (2014). Heterosynaptic plasticity: multiple mechanisms and multiple roles. *Neuroscientist* 20 (5): 483–488.

43 Yang, Y.C., Chen, B., and Lu, W.D. (2015). Memristive physically evolving networks enabling the emulation of heterosynaptic plasticity. *Adv. Mater.* 27 (47): 7720–7727.

44 Sangwan, V.K., Lee, H.S., Bergeron, H. et al. (2018). Multi-terminal memtransistors from polycrystalline monolayer molybdenum disulfide. *Nature* 554 (7693): 500–504.

45 Yang, Y., Du, H.Y., Xue, Q. et al. (2019). Three-terminal memtransistors based on two-dimensional layered gallium selenide nanosheets for potential low-power electronics applications. *Nano Energy.* 57: 566–573.

46 Zhang, Z.Y., Li, T.R., Wu, Y.J. et al. (2019). Truly concomitant and independently expressed short-and long-term plasticity in a Bi_2O_2Se-based three-terminal memristor. *Adv. Mater.* 31 (3): 1805769.

47 Shibata, T. and Ohmi, T. (1992). A functional MOS transistor featuring gate-level weighted sum and threshold operations. *IEEE Trans. Electron Devices.* 39 (6): 1444–1455.

48 Wang, P.J., Lu, J.G., and Xu, J. (2007). Application of neuron MOS in multiple-valued logic. *Neural. Comput. Appl.* 17 (2): 139–133.

49 Jacquot, B.C., Lee, C., Shen, Y.N., and Kan, E.C. (2007). Time-resolved charge transport sensing by chemoreceptive neuron MOS transistors (CvMOS) with microfluidic channels. *IEEE Sens. J.* 7 (9–10): 1429–1424.

50 Barranca, M.A.R., Acevedo, S.M., Nava, L.M.F. et al. (2010). Using a floating-gate MOS transistor as a transducer in a MEMS gas sensing system. *Sensors* 10 (11): 10413–10434.

51 Mehrvarz, H.R. and Kwok, C.Y. (1996). A novel multi-input floating-gate MOS four-quadrant analog multiplier. *IEEE J. Solid-State Circuits.* 31 (8): 1123–1131.

52 White, H.S., Kittlesen, G.P., and Wrighton, M.S. (1984). Chemical derivatization of an array of three gold microelectrodes with polypyrrole: fabrication of a molecule-based transistor. *J. Am. Chem. Soc.* 106 (18): 5375–5377.

53 Tardella, A. and Chazalviel, J.N. (1985). Highly accumulated electron layer at a semiconductor/electrolyte interface. *Phys. Rev. B.* 32 (4): 2439–2448.

54 Lu, N.P., Zhang, P.F., Zhang, Q.H. et al. (2017). Electric-field control of tri-state phase transformation with a selective dual-ion switch. *Nature* 546 (7656): 124–128.

55 Kim, S.H., Hong, K., Xie, W. et al. (2013). Electrolyte-gated transistors for organic and printed electronics. *Adv. Mater.* 25 (13): 1822–1846.

56 Zhu, L.Q., Sun, J., Wu, G.D. et al. (2013). Self-assembled dual in-plane gate thin-film transistors gated by nanogranular SiO_2 proton conductors for logic applications. *Nanoscale* 5 (5): 1980–1985.

57 Nogami, M., Nagao, R., Wong, C. et al. (1999). High proton conductivity in porous P_2O_5-SiO_2 glasses. *J. Phys. Chem. B* 103 (44): 9468–9462.

58 Yue, W.B., Xu, X.X., Su, Z.X. et al. (2011). Syntheses and proton conductivity of mesoporous Nd_2O_3-SiO_2 and NdOCl-SiO_2 composites. *J. Mater. Sci.* 47 (5): 2146–2154.

59 Dou, W., Zhu, L.Q., Jiang, J., and Wan, Q. (2013). Flexible protonic/electronic coupled neuron transistors self-assembled on paper substrates for logic applications. *Appl. Phys. Lett.* 102 (09): 093509.

60 Han, H., Zhu, L.Q., Ren, Z.Y. et al. (2020). Poly(vinyl alcohol)/graphene oxide hybrid electrolyte gated oxide neuron transistors for multifunctional logic applications. *J. Phys. D: Appl. Phys.* 53 (11): 115106.

61 Schöning, M.J. and Poghossian, A. (2002). Recent advances in biologically sensitive field-effect transistors (BioFETs). *Analyst* 127 (9): 1137–1151.

62 Lee, C.S., Kim, S.K., and Kim, M. (2009). Ion-sensitive field-effect transistor for biological sensing. *Sensors* 9 (9): 7111–7111.

63 Burns, M.E. and Augustine, G.J. (1995). Synaptic structure and function: dynamic organization yields architectural precision. *Cell* 83 (2): 187–194.

64 Calabresi, P., Picconi, B., Tozzi, A., and Filippo, M.D. (2007). Dopamine-mediated regulation of corticostriatal synaptic plasticity. *Trends Neurosci.* 30 (5): 211–219.

65 Liu, Y.H., Li, B.J., Zhu, L.Q. et al. (2016). Dopamine detection based on low-voltage oxide homojunction electric-double-layer thin-film transistors. *IEEE Electron Device Lett.* 37 (6): 778–771.

66 Liu, N., Liu, Y.H., Zhu, L.Q. et al. (2014). Low-cost pH sensors based on low-voltage oxide-based electric-double-layer thin film transistors. *IEEE Electron Device Lett.* 35 (4): 482–484.

67 Bergveld, P. (2003). Thirty years of ISFETOLOGY – what happened in the past 30 years and what may happen in the next 30 years. *Sens. Actuators B Chem.* 88 (1): 1–20.

68 Stern, E., Klemic, J.F., Routenberg, D.A. et al. (2007). Label-free immunodetection with CMOS-compatible semiconducting nanowires. *Nature* 445 (7127): 519–522.

69 Spijkman, M., Myny, K., Smits, E.C.P. et al. (2011). Dual-gate thin-film transistors, integrated circuits and sensors. *Adv. Mater.* 23 (29): 3231–3232.

70 Jang, H.J., Gu, J.G., and Cho, W.J. (2013). Sensitivity enhancement of amorphous InGaZnO thin film transistor based extended gate field-effect transistors with dual-gate operation. *Sens. Actuators B Chem.* 181: 880–884.

71 Liu, N., Liu, Y.H., Feng, P. et al. (2015). Enhancing the pH sensitivity by laterally synergic modulation in dual-gate electric-double-layer transistors. *Appl. Phys. Lett.* 106 (07): 073507.

72 Liu, N., Zhu, L.Q., Feng, P. et al. (2015). Flexible sensory platform based on oxide-based neuromorphic transistors. *Sci. Rep.* 5: 18082.

73 Voglis, G. and Tavernarakis, N. (2006). The role of synaptic ion channels in synaptic plasticity. *EMBO Rep.* 7 (11): 1104–1110.

74 Catterall, W.A. and Few, A.P. (2008). Calcium channel regulation and presynaptic plasticity. *Neuron* 59 (6): 882–901.

75 Arnold, A.J., Razavieh, A., Nasr, J.R. et al. (2017). Mimicking neurotransmitter release in chemical synapses via hysteresis engineering in MoS_2 transistors. *ACS Nano* 11 (3): 3110–3118.

76 Zhu, L.Q., Wan, C.J., Guo, L.Q. et al. (2014). Artificial synapse network on inorganic proton conductor for neuromorphic systems. *Nat. Commun.* 5 (1): 3158.

77 Zhu, L.Q., Chao, J.Y., and Xiao, H. (2014). Lateral protonic/electronic hybrid oxide thin-film transistor gated by SiO_2 nanogranular films. *Appl. Phys. Lett.* 105 (24): 243508.

78 Zhu, L.Q., Xiao, H., Liu, Y.H. et al. (2015). Multi-gate synergic modulation in laterally coupled synaptic transistors. *Appl. Phys. Lett.* 107 (14): 143502.

79 Sherman, S.M. and Guillery, R.W. (1998). On the actions that one nerve cell can have on another: distinguishing "drivers" from "modulators". *Proc. Natl. Acad. Sci. USA.* 95 (12): 7121–7126.

80 Crick, F. and Koch, C. (1998). Constraints on cortical and thalamic projections: the no-strong-loops hypothesis. *Nature* 391 (6664): 245–250.

81 Long, T.Y., Zhu, L.Q., Ren, Z.Y., and Guo, Y.B. (2020). Global modulatory heterosynaptic mechanisms in bio-polymer electrolyte gated oxide neuron transistors. *J. Phys. D: Appl. Phys.* 53 (43): 435105.

82 Zhu, L.Q., Wan, C.J., Gao, P.Q. et al. (2016). Flexible proton-gated oxide synaptic transistors on Si membrane. *ACS Appl. Mater. Interfaces* 8 (33): 21770–21775.

83 Muller, C., Beck, H., Coulter, D., and Remy, S. (2012). Inhibitory control of linear and supralinear dendritic excitation in CA1 pyramidal neurons. *Neuron* 75 (5): 851–854.

84 Fu, Y., Kong, L.A., Chen, Y. et al. (2018). Flexible neuromorphic architectures based on self-supported multiterminal organic transistors. *ACS Appl. Mater. Interfaces* 10 (31): 26443–26450.

85 John, R.A., Tiwari, N., Chen, Y.Y. et al. (2018). Ultralow power dual-gated subthresholdoxide neuristors: an enabler for higher order neuronal temporal correlations. *ACS Nano* 12 (11): 11263–11273.

86 Salinas, E. and Thier, P. (2000). Gain modulation: a major computational principle of the central nervous system. *Neuron* 27 (1): 15–21.

87 Silver, R.A. (2010). Neuronal arithmetic. *Nat. Rev. Neurosci.* 11 (7): 474–489.

88 Wan, C.J., Zhu, L.Q., Liu, Y.H. et al. (2016). Proton conducting graphene oxide coupled neuron transistors for brain-inspired cognitive systems. *Adv. Mater.* 28 (18): 3557–3563.

89 Gkoupidenis, P., Koutsouras, D.A., Lonjaret, T. et al. (2016). Orientation selectivity in a multi-gated organic electrochemical transistor. *Sci. Rep.* 6: 27007.

90 Qian, C., Kong, L.A., Yang, J.L. et al. (2017). Multi-gate organic neuron transistors for spatiotemporal information processing. *Appl. Phys. Lett.* 110 (8): 083302.

91 Kim, H. and Park, B.G. (2020). Solving overlapping pattern issues in on-chip learning of bio-inspired neuromorphic system with synaptic transistors. *Electronics* 9 (1): 13.

92 Gallace, A. and Spence, C. (2010). The science of interpersonal touch: an overview. *Neurosci. Biobehav. Rev.* 34 (2): 246–259.

93 Abraira, V.E. and Ginty, D.D. (2013). The sensory neurons of touch. *Neuron* 79 (4): 618–639.

94 Yu, F., Cai, J.C., Zhu, L.Q. et al. (2020). Artificial tactile perceptual neuron with nociceptive and pressure decoding abilities. *ACS Appl. Mater. Interfaces* 12 (23): 26258–26266.

95 Hammock, M.L., Chortos, A., Tee, B.C.K. et al. (2013). 25th Anniversary article: The evolution of electronic skin (E-Skin): a brief history, design considerations, and recent progress. *Adv. Mater.* 25 (42): 5997–5998.

96 Kim, Y., Chortos, A., Xu, W.T. et al. (2018). A bioinspired flexible organic artificial afferent nerve. *Science* 360 (6392): 998–903.

97 Zhang, C., Ye, W.B., Zhou, K. et al. (2019). Bioinspired artificial sensory nerve based on nafion memristor. *Adv. Funct. Mater.* 29 (20): 1808783.

98 Chen, Y.H., Gao, G.Y., Zhao, J. et al. (2019). Piezotronic graphene artificial sensory synapse. *Adv. Funct. Mater.* 29 (41): 1900959.

99 Zang, Y.P., Shen, H.G., Huang, D.Z. et al. (2017). A dual-organic-transistor-based tactile-perception system with signal-processing functionality. *Adv. Mater* 29 (18): 1606088.

100 Wan, C.J., Chen, G., Fu, Y.M. et al. (2018). An artificial sensory neuron with tactile perceptual learning. *Adv. Mater.* 30 (30): 1801291.

101 Wang, G.Z., Wang, R.B., Kong, W.Z., and Zhang, J.H. (2018). Simulation of retinal ganglion cell response using fast independent component analysis. *Cogn. Neurodyn.* 12 (6): 615–614.

102 Lee, G.J., Choi, C., Kim, D.H., and Song, Y.M. (2018). Bioinspired artificial eyes: optic components, digital cameras, and visual Prostheses. *Adv. Funct. Mater.* 28 (24): 1705202.

103 Park, H.L., Kim, H., Lim, D. et al. (2020). Retina-inspired carbon nitride-based photonic synapses for selective detection of UV light. *Adv. Mater.* 32 (11): 1906899.

104 Guo, Y.B., Zhu, L.Q., Long, T.Y. et al. (2020). Bio-polysaccharide electrolyte gated photoelectric synergic coupled oxide neuromorphic transistor with pavlovian activities. *J. Mater. Chem. C.* 8 (8): 2780–2789.

105 Fanselow, M.S. and Poulos, A.M. (2005). The neuroscience of mammalian associative learning. *Annu. Rev. Psychol.* 56: 207–234.

106 John, R.A., Liu, F.C., Chien, N.A. et al. (2018). Synergistic gating of electro-iono-photoactive 2D chalcogenide neuristors: coexistence of hebbian and homeostatic synaptic metaplasticity. *Adv. Mater.* 30 (25): 1800220.

107 Ahmed, T., Kuriakose, S., Mayes, E.L.H. et al. (2019). Optically stimulated artificial synapse based on layered black phosphorus. *Small* 15 (22): 1900966.

108 He, W.X., Fang, Y., Yang, H.H. et al. (2019). A multi-input light-stimulated synaptic transistor for complex neuromorphic computing. *J. Mater. Chem. C.* 7 (40): 12523–12531.

109 Moerel, M., De Martino, F., and Formisano, E. (2014). An anatomical and functional topography of human auditory cortical areas. *Front. Neurosci.* 8: 225.

110 Klinke, R., Kral, A., Heid, S. et al. (1999). Recruitment of the auditory cortex in congenitally deaf cats by long-term cochlear electrostimulation. *Science* 285 (5434): 1729–1723.

111 He, Y.L., Nie, S., Liu, R. et al. (2019). Spatiotemporal information processing emulated by multiterminal neuro-transistor networks. *Adv. Mater.* 31 (21): 1900903.

112 Branco, T., Clark, B.A., and Häusser, M. (2010). Dendritic discrimination of temporal input sequences in cortical neurons. *Science* 329 (5999): 1671–1675.

113 Branco, T. and Häusser, M. (2010). The single dendritic branch as a fundamental functional unit in the nervous system. *Curr. Opin. Neurobiol.* 20 (4): 494–502.

4

Neuromorphic Devices Based on Chalcogenide Materials

Jia Chen, Yifan Lu, Zhe Yang, Yi Li, and Xiangshui Miao

Huazhong University of Science and Technology, Wuhan National Laboratory for Optoelectronics, School of Optical and Electronic Information, Luoyu Road 1037, Wuhan 430074, P. R. China

4.1 Introduction

With the ability to execute cognitive information processing operations with high energy efficiency, brain-inspired neuromorphic computing has been emerging as an alternative paradigm for non-von Neumann computing in the incoming artificial intelligence of things (AIoT) era [1].

Electronic artificial neurons and synapses are two essential blocks for constructing a hardware neuromorphic system, regardless of which technical routes are adopted, such as various artificial neural networks (ANNs), including multilayer perceptron (MLP), convolutional neural network (CNN), and spiking neural network (SNN). Over the past few years, various nanodevices, including resistive switching devices or memristors, ferroelectric devices, and magnetic devices, have been proposed to emulate the functionalities of their biological counterparts: neurons and synapses [2–6]. Among these candidates, chalcogenide devices are preferred owing to their ability to fulfill the requirements of fully hardware neuromorphic systems [7–9].

In this chapter, we will first introduce the nonvolatile memory switching (MS) and volatile threshold switching (TS) phenomena in chalcogenide materials and their corresponding physical mechanisms. The former is related to the reversible structural transition between crystalline phase and amorphous phase induced by thermal mechanisms, whereas the latter is thought to be probably a pure electronic effect. Then, we will discuss an overview on the implementations of chalcogenide-based electronic synapses and neurons, utilizing the MS and TS behaviors, respectively. The progress in hardware neural network accelerators using these neuromorphic devices will also be reviewed and discussed from the perspective of system and application levels. The chapter concludes with a discussion and presentation of future research directions.

Neuromorphic Devices for Brain-Inspired Computing: Artificial Intelligence, Perception, and Robotics, First Edition. Edited by Qing Wan and Yi Shi.

4.2 Ovonic Memory Switching (OMS) and Threshold Switching (OTS) in Chalcogenide Materials

The study of electrical-field-induced resistive switching phenomena in chalcogenide materials can be tracked back to the pioneering work of Stanford R. Ovshinsky from Energy Conversion Devices in the 1960s [10]. He firstly observed a rapid and reversible transition between a highly resistive and a conductive state in disordered semiconducting materials. Specifically, a volatile TS behavior was measured in $Ge_{10}Si_{12}As_{30}Te_{48}$, and a nonvolatile MS effect was observed when the content of arsenic was reduced to 5%. Besides, in 1971, Feinleib et al. reported fast laser-induced amorphous to crystalline transformations in $Te_{81}Ge_{15}Sb_2S_2$ [11]. These interesting behaviors rapidly gained attention from academia and industry.

In 1970, R. G. Neal and D. L. Nelson from Energy Conversion Devices and Gordon E. Moore from Intel demonstrated the first 256-bit "read-mostly memory" using Ovonic devices [12]. Typical configurations for SET, RESET, and READ driving are developed. In the SET operation, a critical voltage of the memory must be exceeded, and a current of 5 mA must flow for several milliseconds to ensure stable conversion to the low resistance state (LRS) (crystalline). In the RESET operation, the procedure is similar except that the current is increased to 200 mA and the pulse width is reduced to 5 μs to switch the device back to amorphous high resistance state (HRS). Later in 1978, Burroughs Corp. demonstrated a 1024-bit nonvolatile 15 ns bipolar read–write programmable memory using the ovonic memory switching (OMS) phase change behavior as the storage mechanism [13]. However, in the 1970s, with a semiconductor feature size of about 10 μm, the area and power consumption of the two prototype memories both could not compete with the electrically programmable read-only memory (EPROM), so the development of phase change memory (PCM) gradually fell into obscurity. Only the phase change disk technology was further developed.

Until 1987, N. Yamada et al. of Matsushita Electric Industrial Corporation (later Panasonic Corporation) discovered that the GeTe-Sb_2Te_3 pseudo-binary alloy system has a fast phase transition speed, large optical property contrast, and advantages in the endurance cycles [14]. This finding have greatly facilitated the application of phase change storage disk technology, such as CDs, DVDs and, later, Blu-ray disks. This has also led to a renewed focus on chalcogenide materials and the straightforward application of their electrical properties: phase change random access memory (PCRAM or PCM).

At the end of the twentieth century, with the rapid development of semiconductor technology, the feature size of lithography entered the nanometer level, the size of PCM was drastically reduced, and the power consumption of device operation was also reduced to an acceptable range. In 2015, Micro and Intel announced the co-development of a new class of high density standalone memory called 3D XPoint™ as a high-performance alternative to flash. So far, they have already commercialized the tech and brought product to market. Based on publicly available information, we can conclude that in the stacked cross-point array, the OMS device acts as the memory unit, and the ovonic threshold switching (OTS) device is utilized

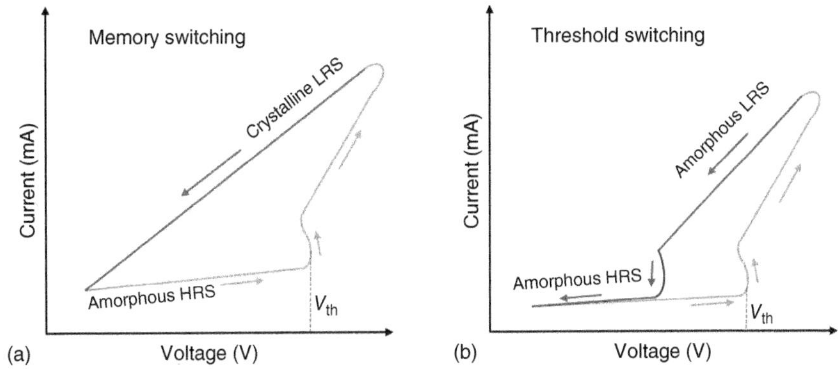

Figure 4.1 (a) Current–voltage curve of a typical memory switching device. (b) Current–voltage curve of an ovonic threshold switching device.

as a compact two-terminal selector, which eliminates the need for three-terminal transistors and increases the integration capacity.

From the point of view of the stability of LRS, the resistive switching behavior of chalcogenide materials can be divided into two types: nonvolatile MS and volatility TS. Figure 4.1 shows the comparison between the current–voltage characteristics of the MS and TS effects. The *I–V* curve in Figure 4.1a shows a typical MS behavior of a PCM. HRS state is the initial state; when the applied voltage exceeds a threshold voltage (V_{th}), the device suddenly switches to an LRS. When the applied voltage is removed, the device remains in the LRS. The two distinct nonvolatile states can be maintained for a long time and used to store two logical variables 0 and 1. This memory effect is believed to be associated with a transition from an amorphous phase to a crystalline phase [15]. Figure 4.1b shows the electrical performance of an OTS device. At a particular V_{th}, the device switches abruptly from an HRS to an LRS. However, the device LRS recovers to the HRS as the applied voltage decreases to less than the hold voltage (V_{hold}), indicating a volatile feature.

Chalcogenide materials are at the core of various applications, such as PCM and neuromorphic computing, the latter of which is the focus of this chapter. At current state, there are three main classes of chalcogenide materials that can stably exhibit nonvolatile MS behaviors [16, 17]. The first class is the nucleation-dominant GeTe-Sb_2Te_3 pseudo-binary alloy system, such as $Ge_1Sb_4Te_7$, $Ge_1Sb_2Te_4$, $Ge_2Sb_2Te_5$, $Ge_3Sb_2Te_6$, and $Ge_8Sb_2Te_{11}$. This class of materials has a sub-stable rock salt-type crystalline structure with Te atoms occupying a sublattice position (anionic position) and randomly arranged Ge and Sb atoms and about 25% vacancies occupying another sublattice position (cation position). Among them, $Ge_2Sb_2Te_5$ shows advantages of fast crystallization speed, good amorphous stability, and strong two-phase contrast. The second type is growth-dominant Sb_2Te-based materials doped with Ag, In, Ge, Ga, W, or other elements. Among them, the $Ag_5In_5Sb_{60}Te_{30}$ is the most representative, and its faster phase transition rate and large optical differences lead to the successful commercialization in phase change storage disks. The third category is Sb-based materials, such as $Ge_{15}Sb_{85}$, GaSb, AlSb, etc. This class of materials does not contain the toxic element Te and has also attracted more research attention in the recent years. In addition, since Sn is in the same primary family as Ge and As and Bi are in the same primary family as Sb, hence, Ge and Sb can be substituted to form phase change compounds with similar properties, such as the SnTe-Sb_2Te_3 subset Sn-Sb-Te, which has similar properties to GeSbTe.

In 1968, Ovshinsky reported the OTS behavior of $Te_{48}As_{30}Si_{12}Ge_{10}$ chalcogenide, which attracted widespread attention. After that, significant research interest in threshold switch behaviors of amorphous chalcogenide developed, and various OTS materials were proposed. However, the physical mechanism of the OTS device remains unclear. Up to date, many models have been proposed to explain this phenomenon.

Kroll et al. proposed the thermal runaway model in the 1970s [18, 19]. In this model, increasing applied voltage leads to the increasing current flowing through the OTS device. According to the Joule effect, the induced Joule heat increases the temperature of the device and triggers more carriers, which further increases the current. In this way, the current and temperature both increase and promote each other, resulting in a positive feedback. Once the voltage reaches V_{th}, the conductivity exponentially increases, giving rise to the threshold switch behaviors. However, convincing experimental evidence shows that this model does not match the experimental conductivity–temperature relationship [20].

In 2007, the field-induced nucleation model was introduced by Karpov et al. [21, 22]. This model attributes the threshold switching behavior to a structural phase transition. The applied electrical field reduces the energy barrier for nucleation and the critical size of the nucleus. As a result, subcritical nuclei with low resistance grow at the chalcogenide/electrode interface, increasing the conductivity of the device. Once the electrical field is removed, these subcritical nuclei become unstable and quickly disappear, resulting in the relaxation to HRS. The field-induced nucleation model explains the statistical distribution of switching times and the V_{th} temperature relationship. However, this model does not totally comply with the experimental facts. According to this model, the ON-state conductivity of the device should be the same as the crystalline phase of the alloy, which is not consistent with the experimental data.

In 1980, Adler et al. [23] proposed a comprehensive electronic model. This model describes the threshold switch event as a changing process of carrier generation and recombination under different electric fields. In this model, the conductivity of the device is limited by lone-pair-induced traps. Under a low electrical field, the field-induced carriers are captured by these traps, leading to the low conductivity. Under a high field, more carriers are generated and more traps are filled. Once all the traps are filled, the carriers can move freely in the film, forming a conductive filament and eventually leading to threshold switching behaviors. If the voltage drops below the V_{hold}, the recombination process dominates again, and the conductive filament disappears, causing the device to switch to HRS. This model successfully explained some experimental data reported by other research groups. However, there exist some issues in this model; for example, the temperature dependence of the threshold field has not been explained.

Ielmini et al. proposed another electronic model in 2007 [24]. It is known that the band gap in the amorphous semiconductors is filled by localized trap states. In the Ielmini model, the carriers are easily trapped in these traps, and extra energy is needed to get the carriers out of the traps. The applied voltage bias can reduce the energy barrier and decrease the activation energy for conduction. Therefore, under a strong field, the carriers can cross the energy barrier and jump from deep trap states into shallow trap states at higher energies. When sufficient carriers occupy shallow trap states, the threshold switching behaviors appears.

This model could well explain both the temperature dependence of the threshold voltage and thickness dependence of sample I–V characteristics. However, according to the Ielmini model, a large number of defects exist at the Fermi level of the OTS device, while other experiment data shows that the density of trap states at Fermi level is very small in some chalcogenide materials. In addition, Guo et al. proposed that there are some homopolar filaments formed during the threshold switching, which is beyond the Ielmini model [25].

Though both PCM and OTS materials belong to chalcogenides, they can be well discriminated from the perspective of their underlying bonding that is generally quantitatively characterized by bond orbital coordinates, namely, hybridization and ionicity calculated from orbital radii. As shown in Figure 4.2, these two coordinates originally helped to build a phase map for structure classification. It was further used by Lencer et al. [17] for PCM materials tracing as they were clustered in a certain region with moderate hybridization and a slightly low ionicity, supporting the existence of the metavalent bonding, a novel type of chemical bond in crystalline PCM materials. However, OTS materials differ [26, 27] that were mostly confined in another region with lower ionicity and higher hybridization than PCM materials, indicating that they were more covalently bonded and good glass former within silica-like phase. Such distinct intrinsic material properties contribute a lot to a deep understanding of the volatile and nonvolatile switching behaviors in OTS and PCM devices, respectively.

4.3 Artificial Synapses Based on MS Behaviors

At the early state, the MS behaviors enabled by the transition between the amorphous and crystalline states were intensively exploited to develop PCM, which has demonstrated perceived advantages, such as simple cell structure, fast read–write latency, 3D integration capability, multi-bit storage, and fab-friendly processes. PCM might lead among emerging memory technologies. During the research history of this technology, several troublesome issues have been the focus of research in this area [1]. The switching speed is limited by the crystallization process [2]. The thermal process, especially the amorphization, requires a large operation current and thus leads to the large power consumption [3]. Repeatable cycling always results in the reliability issue, such as failure in LRS or HRS.

In a classical phase change materials, $Ge_2Sb_2Te_5$ (GST) cell, Huang et al. realized ultrafast amorphization of only 0.2 ns, which was conjectured to be a nonthermal

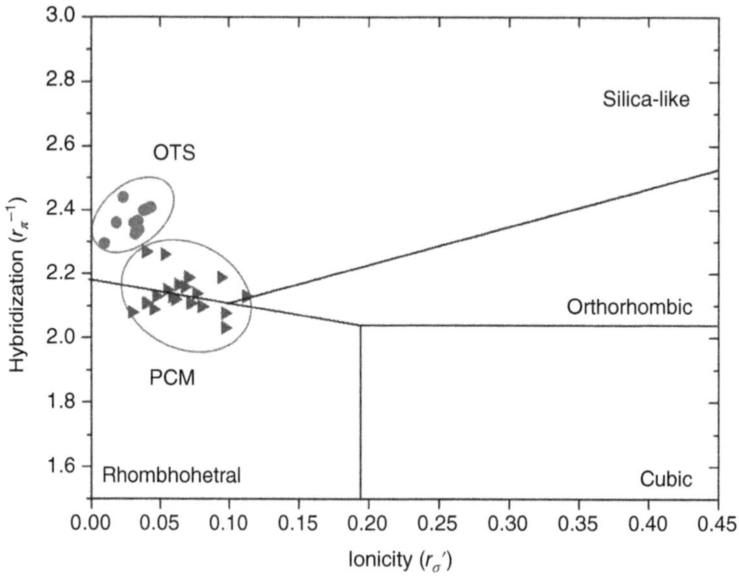

Figure 4.2 A phase map expressed by ionicity and hybridization parameters where OTS and PCM were clustered in different regions and characterized by different structures. Source: Li et al. [26].

physical process [28]. Loke et al. proposed a prestructural ordering (incubation) method to control the crystallization kinetics and achieved a crystallization speed of 500 ps [29]. More recently, a novel $Sc_{0.2}Sb_2Te_3$ material was predicted to show faster crystallization speed than GST and experimentally demonstrated with a real 0.7 ns write time in a fabricated PCM [30]. Interfacial PCM with $GeTe/Sb_2Te_3$ superlatice-like layers was developed to reduce the switching energy, since the interfaces control the local atomic switching of Ge atoms and substantially reduce entropic losses [31].

Besides, to extend the memory cycling toward unlimited endurance, various techniques including symmetric cell geometry, densification of as-deposited materials, and self-void healing have been employed to address the endurance failure issues, such as atomic migration, element segregation, and void formation [32]. A confined-type phase change memory with a metallic liner and atomic layer-deposited GST yielded a high endurance record of up to 2×10^{12} cycles [33]. It seems a great opportunity for PCM to compete with conventional floating gate-embedded nonvolatile memory at 28 nm and beyond CMOS nodes both in fully depleted silicon on insulator (SOI) and FinFET advanced technologies [34].

In the course of PCM investigation, researchers have gained sufficient understanding of the fundamental properties and physical mechanisms of the chalcogenide material and have achieved breakthroughs in device and integration techniques, paving the way for the transfer of this technology to neuromorphic computing. Over the past few decades, PCM and other emerging nonvolatile memories have been widely researched for the realization of electronic synapses due to the potential analog memory and high-density integration.

For a PCM cell, the device resistance could be tuned as multilevel intermediate states between HRS and LRS by designing the applied pulse conditions, which has been extensively applied for multilevel memory chip. Additionally, the analog resistance tuning behavior, that is, a continuous transition between amorphous and crystalline, also allows the PCM cells to be used as electronic synapses. Synapses are the connections between neurons, while the connection strength, namely, synaptic plasticity, could be continuously modulated in terms of various neural activities that serve as the basic rules for brain learning and memory. Hence, the potentiation and depression of synaptic weights could be implemented by the decrease and increase of resistance in PCM cells, respectively, which are also needed for the demonstration of biological resemble plasticity, including activity-dependent long-term potentiation (LTP) and long-term depression (LTD), spike-timing-dependent plasticity (STDP), and so on. Unlike traditional multilevel memory, which requires fewer bits and tighter state distinguishability, gradual programming and fine precise resistance modulation of the PCM cells are necessary to emulate the analog nature of synaptic weight modulation in biological synapses.

To mimic the analog nature of biological synapses, in 2011, GST phase change electronic synapse with continuous resistance transition was investigated by Kuzum et al. with the implementation of the STDP learning rule (Figure 4.3) [35]. To probe the intermediate resistance levels of GST cells, the gradually increasing voltage pulses with a custom-engineered timing level were applied. Gradual reset of the cell resistance was implemented by using pulses with increasing amplitude, and gradual set of the cell resistance was achieved by using staircase pulses with an increasing step of 0.1 V. An order of magnitude change in the phase change cell resistance was achieved through 100 steps for both set and reset transitions. Finite element simulations performed using COMSOL have helped to understand and validate the gradual set/reset in GST cells while the amorphous region grows gradually at the top of the nanoscale bottom electrode, leading to the numerous intermediate resistance states with the gradually increasing voltage pulses. Furthermore, a single GST cell with the capability of both the modulation of the time constant and the realization of the different STDP types was first demonstrated by designing the inducing pre/post spike pairs.

Suri et al. firstly discussed the consequence of LTP and LTD on a GST PCM synaptic devices by applying identical potentiating and depressing pulses (Figure 4.4a–c) [36]. To emulate the increase in synaptic efficacy (potentiation), the device was initialized to an HRS by applying a strong reset pulse. After that, identical potentiating pulses were applied. The device resistance showed the gradual decrease with the potentiating pulses until reaching a saturating plateau. On the contrary, for the synaptic depression, the device was first initialized to an LRS by applying a potentiation pulse. Then the identical depression pulses were also applied to measure the LTD. According to the experimental results, it was hard to obtain the gradual depression behavior using identical depressing pulses in a GST cell and showed almost abrupt increase in the resistance. This is because of the fact that crystallization in PCM is an energy-cumulative process, and the impact of applying several short crystallizing pulses is almost the same as applying a single long crystallizing pulse of the same amplitude. For amorphization, a region first needs to be melted and then

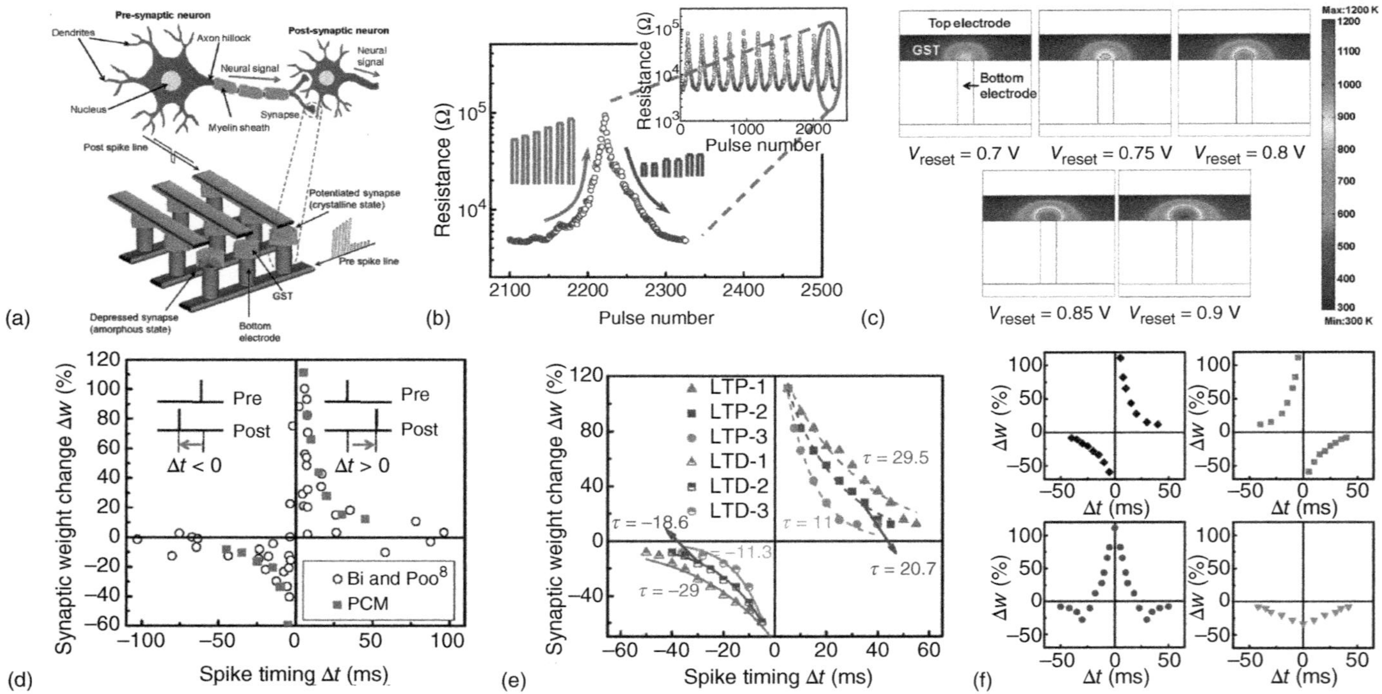

Figure 4.3 (a) Interconnection scheme of PCM synapses to reach ultrahigh density and compactness of brain. (b) Gradual reset and set of the cell resistance with the gradual increasing voltage pulses. (c) Finite element simulations performed with COMSOL. Different colors represent the temperatures across the GST region. (d) Implementation of spike-timing-dependent plasticity (STDP) with PCM cells. (e) Time constants are modulated for the STDP curve with the spike-timing delay in the range of 10–30 ms for potentiation and −10 to −30 ms for depression. (f) Different forms STDP learning rules. Source: Kuzum et al. [35].

quickly cooled. The volume of the molten region created inside the chalcogenide layer increases with the amplitude of the applied reset pulse. Thus a strong depressing pulse would melt a larger region than a weak depressing pulse, creating a larger amorphous cap and a higher value of final device resistance. When identical depressing pulses are applied, each pulse provides almost the same amount of energy to the cell, creating nearly the same amount of amorphous region every time and same value of final device resistance. To address this problem for abrupt depression, one of the solutions would be gradually increasing voltage pulse scheme. However, implementing such pulses with varying amplitudes can lead to practical problems, such as capacitive line charging and high power dissipation in large-scale neuromorphic systems involving crossbar arrays. Suri et al. further proposed a new energy-efficient synapse circuit consisting of two identical PCM devices (the "*2-PCM synapse*") with a realistic and simplified programming pulse scheme (Figure 4.4d–f) [37]. One PCM device (LTP) has a positive current contribution, while the other (LTD) contributes negatively toward the output CMOS neuron current. In the "*2-PCM synapse*," crystallizing the LTP PCM device produces synaptic LTP-like effect, while crystallizing the LTD PCM device produces synaptic LTD like effect. Since gradual crystallization can be obtained by applying simple identical pulses, the approach simplifies the pulse schemes for implementing LTP and LTD synaptic behaviors. Moreover, the "*2-PCM synapse*" is highly energy efficient, as majority of synaptic conductance changes (or weight update) are based on crystallization and not on amorphization of the PCM devices. A modified form of the biological STDP learning rule was also implemented on the "*2-PCM synapse*" to be applied in SNN capable of complex visual pattern extraction.

Different from the methods used above, Li et al. demonstrated an approach which focused on the use of the crystalline state of GST (*c*-GST) to emulate the synaptic plasticity (Figure 4.5a–c) [38]. To study the memristance of *c*-GST, the device was first set to the crystalline state at several kilo-ohms using a DC voltage sweep. Next, the hysteresis and asymmetry memristive characteristics were observed under the clockwise voltage sweep with a current compliance of 100 mA. Although bipolar resistive switching phenomena have been reported in GST, they are all based on the conductive filament theory. The memristive characteristics of *c*-GST are intrinsic characteristics. Lattice defects, mainly vacancies, and defects at grain boundaries appear as a natural cause of the memristance. To emulate this synaptic weight modification, the pulses with gradually increasing amplitudes were applied. The pulse widths and rise and fall times were fixed at 30, 10, and 10 ns, respectively. The resistance exhibited a gradual repeatable change ranging from 2.2 to 3.0 kΩ, and the mean resistance change per pulse was approximately 40 Ω, which indicates a relative programming precision of 1.8%. Based on the fine synaptic weight modulation that occurred in response to pulses of different polarities, four different STDP forms with flexible time windows were successfully implemented in the *c*-GST synapse by following a typical spike pairing protocol used in biological synaptic studies and in other electronic synapses. The feasibility of STDP time window tuning in this material was demonstrated from 50 ms to 500 ns, the latter of which is 10^5 times faster than the speed of STDP in human brain. In principle, the scaling

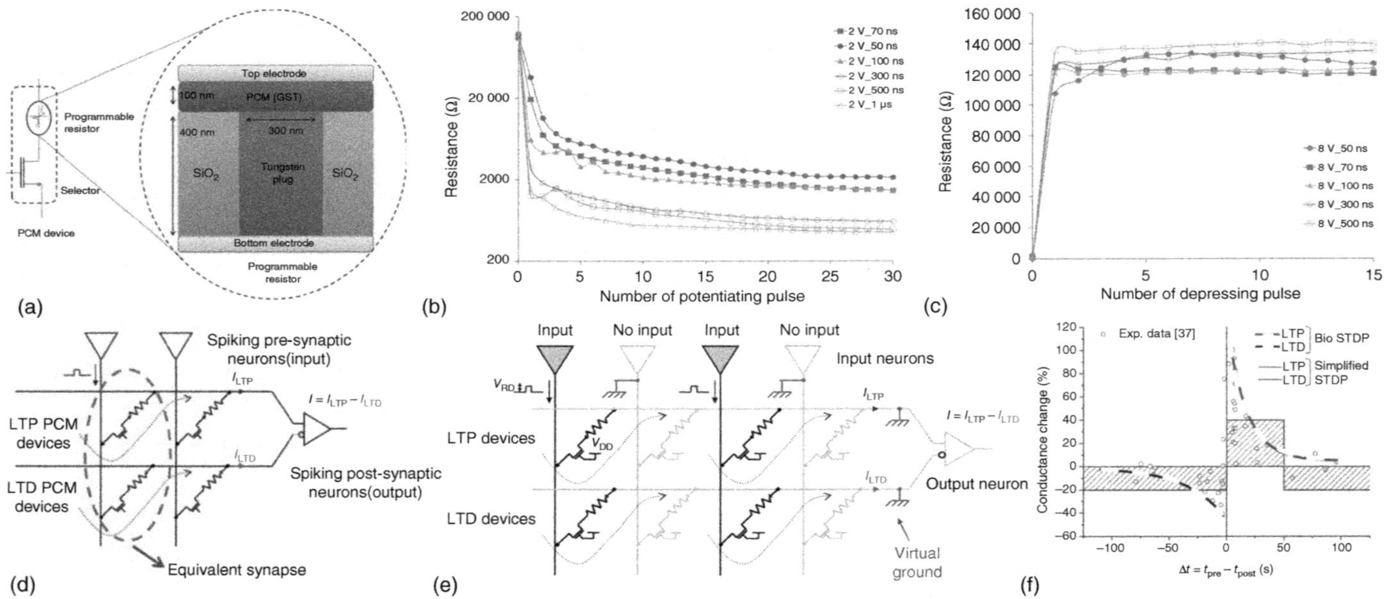

Figure 4.4 (a) Illustration showing the PCM device made of a selector and a resistor and zoomed resistor element illustrating the PCM device. (b) Demonstration of potentiating synaptic behavior on PCM device. Resistance gradually decreases with the application of each crystallizing pulse. (c) Demonstration of depressing synaptic behavior on PCM device. Resistance increases abruptly with the application of amorphizing pulses. Source: (a–c) Suri et al. [36]. (d) Circuit schematic of the "2-PCM synapse." (e) Read circuit scheme. Current from both LTP and LTD PCM devices is integrated in the output neuron, with a positive and negative contribution, respectively. (f) Biological STDP and simplified STDP. Source: (d–f) Suri et al. [37].

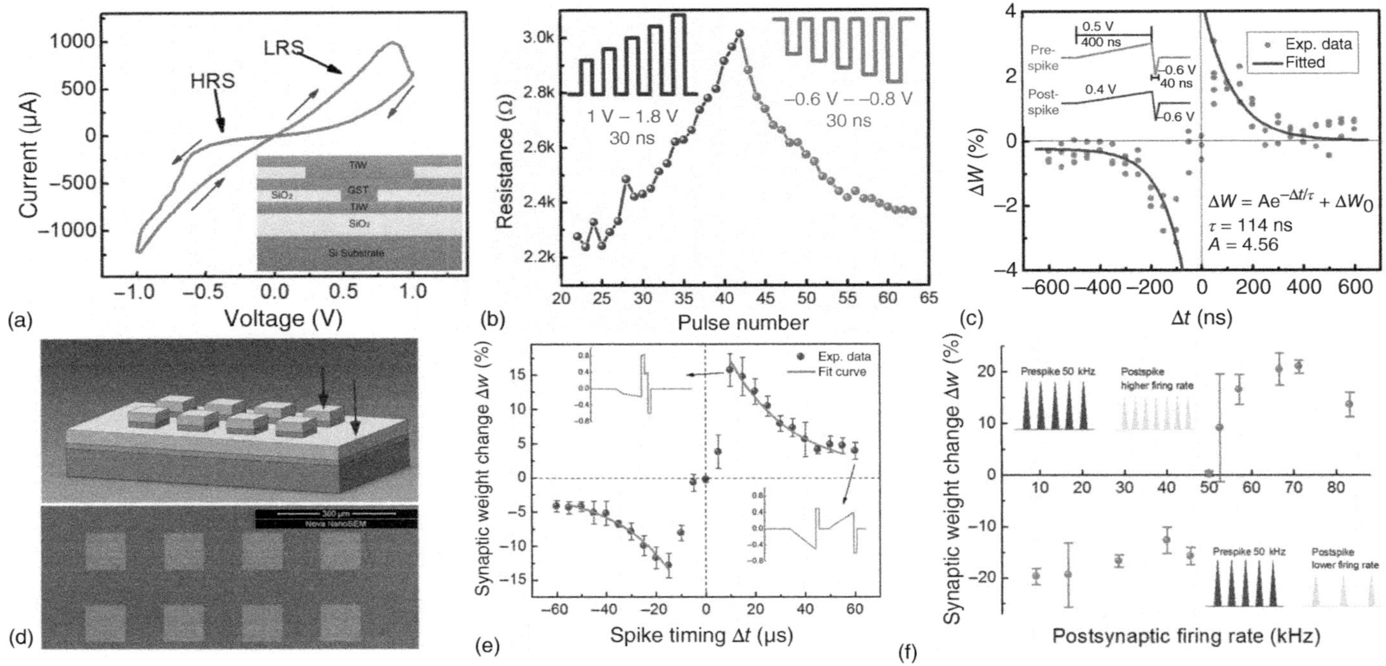

Figure 4.5 (a) $I-V$ characteristics of the c-GST cell measured by DC double sweeping. (b) The conductance increases or decreases in response to negative or positive pulses, respectively, representing synaptic weight modulation due to potentiating or depressing pulses. (c) Implementation of STDP with nanosecond-scale time windows in the c-GST chalcogenide synapse. Source: (a–c) Li et al. [38]. (d) Schematic of the Ag/AgInSbTe/Ag stack structures deposited on the Si/SiO_2 substrate. (e) The asymmetric Hebbian learning STDP rule of the AIST cell. (f) Implementation of SRDP in the AIST cell. Source: (d–f) Li et al. [39].

up of the number of analog/plastic synapses with low power consumption and ultrafast response times is in urgent demand for further evolution of neuromorphic chips. Not only that, the same group conducted another AgInSbTe (AIST)-based chalcogenide material as an electronic synapse (Figure 4.5d–f) [39]. By designing the applied pulse schemes, the activity-dependent synaptic plasticity of biological synapses was reproduced in the AIST-based electronic synapse, including four STDP forms, spike-rate-dependent plasticity (SRDP) and even synaptic saturation. The demonstrated synaptic operation together with the booming development of sub-ns synaptic devices with high density and low power consumption will contribute to the construction of next-generation neuromorphic computing architecture that requires plastic electronic synapses.

It is worth noting that the stability of the conductance has significant impact on the network training and inference performances. However, the well-known state drift of phase change materials upon aging are commonly observed. It could be another main obstacle that up to now hinders the development of artificial synapses and neural network accelerators. Ding et al. proposed a phase change heterostructure (PCH) that has low noise, small drift, long endurance, fast switching, and high energy efficiency [40]. They replaced the active phase change layer with a multilayer PCH composed of alternately deposited confinement materials (~3 nm $TiTe_2$) and phase change materials (~5 nm Sb_2Te_3). By using an iterative RESET operation (fixed pulse width of 20 ns and varied pulse amplitude from 1.95 to 2.15 V), the device achieved a varied crystalline/amorphous ratio, and nine-level states were obtained with consistently low drift coefficients ($\nu < 0.005$). Unlike the iterative RESET process, where the melt-quench amorphization was abrupt, a cumulative SET operation led to a nonlinear reduction in resistance, since it is accomplished by incubation of crystal nuclei and subsequent growth. Forty identical pulses (0.8 V and 100 ns) resulted in high-accuracy conductance states with small fluctuation (~10%).

4.4 Artificial Neurons Based on TS Effects

Artificial neurons are another basic building blocks of neuromorphic circuits. Conventional artificial neurons based on CMOS circuits have very low energy efficiency and integration density [41–43]. Therefore, it is necessary to develop new electronic devices suitable for electronic neurons. Among various candidates, memristors show unique performance, which can be used to simulate biological neurons. To date, various artificial neurons based on TS devices have been constructed, such as metal-insulator transition (MIT) devices [44], OTS devices [45], and conductive filament-type threshold conversion devices [46]. In the TS device-based neuron, the threshold switch behavior of TS device removes the need of auxiliary circuits for setting the threshold value and reset operation, which is necessary in the nonvolatile device-based neuron. Therefore, the TS device-based neuron becomes the mainstream of current researches.

To describe how a neuron works, many models have been proposed. Currently, many artificial neurons constructed with OTS devices are implemented based on

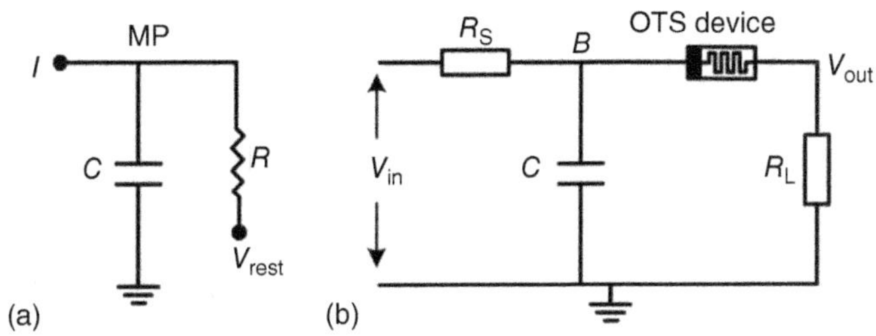

Figure 4.6 (a) Integrate-and-fire (IF) neuron model. (b) Schematic illustration of the IF neuron based on OTS device.

the integrate and fire (IF) neuron model. In the IF model, the state of the neuron is determined by its membrane potential (MP). The neuron receives input from presynaptic neuron, and then its MP will increase. When MP reaches the threshold, an action potential is triggered. The equivalent circuit of the IF model is shown in Figure 4.6a. In this circuit, the voltage across the capacitor represents the MP of the neuron. Supra-threshold MP will trigger an action potential. After that, the MP returns to its initial value.

The TS behavior of the OTS device can be used to simulate the threshold-driven spiking behavior of the IF neuron. The OTS device-based artificial neuron circuit is shown in Figure 4.6b. In response to synaptic input, the capacitor accumulates charge, and the potential at node B (V_B) increases, which mimics the integration process of the MP. Initially, the OTS device is in HRS and the leakage current is negligible. When the V_B reaches the threshold value, the OTS device switches to LRS, and the capacitor discharges through R_O, resulting in an output spike, which mimics the threshold-driven spiking behavior of a biological neuron. When the voltage across the OTS device decreases to a certain value, the device returns to HRS to prepare the next fire action.

Many artificial neurons have been proposed based on similar circuit structure. Hyungkwang Lim et al. successfully developed an IF neuron utilizing Pt/GeSe/Pt device, as shown in Figure 4.7a [47]. The input end of the circuit is a nonideal operational amplifier. The input voltage is amplified by the amplifier and applied to the circuit responsible for generating the action potential. This circuit can fire oscillation output under the square voltage pulses, as shown in Figure 4.7b. At the same time, due to the existence of amplifier at the input end, the circuit can also realize the amplification of the input signal and unidirectional signal transmission. In addition, the randomness of the threshold switch process of the OTS devices, including the randomness of threshold voltage and OFF-state resistance, will also introduce random Poisson noise to the circuit, which is very important for related applications in stochastic analog electronics.

Milim Lee et al. demonstrated an IF neuron based on Pt/GeSe/Pt device using the same circuit structure in Figure 4.8b, which shows the basic integrate-and-fire function and the rate coding ability, as shown in Figure 4.8a [45]. Besides, in the electrical switching process of the OTS device, the filled trap states will change, resulting in an increasing threshold voltage of the device. Therefore, the output spike frequency

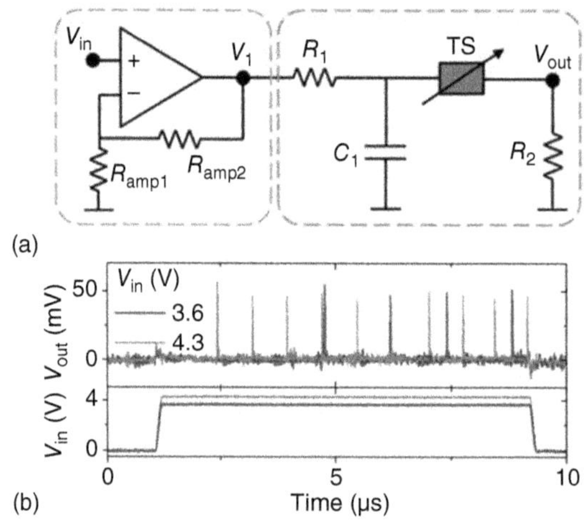

Figure 4.7 (a) Equivalent circuit of the LIF neuron. (b) Oscillating outputs of the neuron under the square voltage pulses. Source: Lim et al. [47].

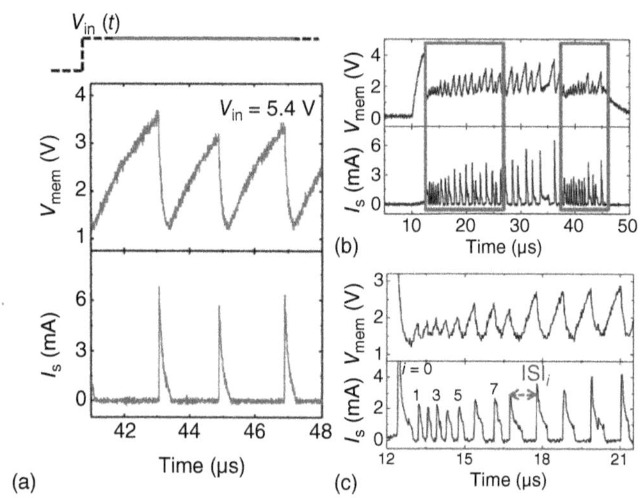

Figure 4.8 (a) The integrate-and-fire function of the OTS device-based artificial neuron. (b) Response of an artificial neuron circuit under a constant excitation input pulse. (c) The spike-frequency adaptation property of the neuron, where i and ISI_i indicate the spike count and the interspike interval, respectively. Source: Lee et al. [45].

of the neuron will gradually decrease under a constant stimulus, which shows the spike-frequency adaptation property of biological neuron, as shown in Figure 4.8b,c.

Donguk Lee et al. compared three types of threshold switch devices, including the active metal conductive filament mechanism-based TS devices, MIT devices, and OTS devices, and summarized the influence of various parameters of TS device on the neuron circuit performance [48]. The switching speed of the device will affect the functional relationship between input pulse frequency and output pulse frequency, and the OFF-state resistance and threshold voltage of the device will affect the power consumption of the neuron. OTS devices have faster switching speeds, so the activation function of the OTS device-based neuron was fitted with rectified linear unit

(ReLU) function. In addition, the OTS device has a high OFF-state resistance, which can make the leakage current of the neuron in the integration process extremely low. At the same time, the OTS device can obtain a smaller threshold voltage by reducing the thickness of the functional layer and adjusting the element ratio and reduce the amplitude of the neuron's pulse. Therefore, a low-power neuron circuit can be realized based on the OTS device.

4.5 Hardware Neural Networks

In addition to building single artificial synapses with MS behaviors and artificial neurons with TS behaviors, array-level PCM holds a great promise to contribute to the construction of energy-efficient, highly parallel in-memory neuromorphic systems, breaking the traditional von Neumann bottleneck. The PCM-based crossbar array shows huge potential to accelerate the vector–matrix multiplication operations, which is indeed computing-intensive and energy-consuming operations for traditional ANNs, including Hopfield neural network (HNN), multilayer perceptron (MLP), convolution neural network (CNN), long short-term memory (LSTM), and so on. As shown in Figure 4.9, a typical single fully connected synaptic layer is illustrated to be mapped on the PCM array. The m input nodes are fully connected with n output nodes, while an $m \times n$ array is programmed to store the same-size matrix of synaptic weights. In both forward inference and backward update phases, the vector–matrix multiplication (VMM) operations are executed in one-clock parallelized step.

Burr et al. firstly experimentally demonstrated a three-layer neural network based on 164 885 PCM synapses on a subset (5000 examples) of the MNIST database of handwritten digits using a backpropagation variant suitable for PCM crossbar arrays, while a 2-PCM unit was used as one synaptic connection that showed symmetric, linear conductance response of high dynamic range (Figure 4.10a,b) [49]. Using a neural network simulator matched to the experimental demonstrator in this work, it performed extensively tolerant with respect to some inevitable nonideal cases, including device variability, yield, stochasticity, linearity, and asymmetry of the PCM conductance tuning behavior. The bidirectional PCM synaptic unit with a symmetric, linear conductance response of high dynamic range was capable of delivering acceptable classification accuracies (82.2% for training, 82.9% for inference). However, the classification accuracies of such in situ training using nonvolatile memory hardware have generally been less than those of software-based training, owing to the insufficient dynamic range and excessive weight-update asymmetry. Hence, the same group demonstrates a "2PCM + 3T1C" unit per synapse with both more dynamic range and better update symmetry [52]. The dynamic range was increased by using two pairs of conductance of varying significance. By applying different scale factors on the read current from the conductance pairs, the total read current per synapse is proportional to $F \times (G^+ - G^-) + (g^+ - g^-)$, where F is a small gain factor, G^+ and G^- (2 PCM) are the conductance of the higher-significance conductance pair, and

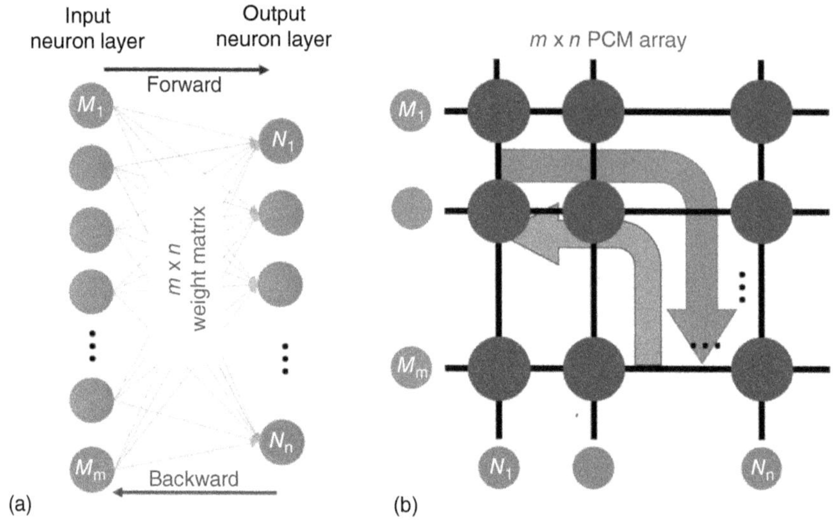

Figure 4.9 (a) A fully connected layer with *m* input nodes and *n* output nodes. (b) The mapping method of the fully connected layer on a PCM array.

g^+ and g^- (3T1C) are the conductance of the lower significance pair. Besides, the "2PCM + 3T1C" unit combined long-term storage in phase change memory, near-linear and symmetric updates of volatile capacitors, and weight data transfer with "polarity inversion" to cancel out inherent device-to-device variations. Thus, this mixed hardware–software neural network made software-equivalent training accuracies for MNIST, CIFAR-10, and even CIFAR-100 by a simulated MLP model (involve up to 204 900 synapses). Significantly, the computational energy efficiency of 28 065 $GOPs^{-1}$ W^{-1} and throughput per area of 3582 $GOPs^{-1}$ mm^{-2} for the PCM-based neural network implementation exceed those of today's GPU by 2 orders of magnitude, which strongly indicated a path toward hardware accelerators that are both fast and energy efficient. Not only the simple MLP model, Eryilmaz et al. experimentally employed an HNN consisting of 100 synaptic devices and 10 recurrently connected neurons for implementation of brain-like associative learning (Figure 4.10c) [50]. Experimentally, on-chip evaluations were performed on a 2.5 M analog PCM array for forward inference of LSTM networks (Figure 4.10d) [51]. Despite nonideality and variability in PCM arrays, inference results close to software baselines were achieved on LSTM networks of competitive size. By storing a large number of parameters and offering in-memory computing capability, PCM synapse array has also been proved strongly that the application on LSTM platform would be a promising low-power and low-latency hardware implementation.

Different from traditional ANNs that have mature mathematical algorithms and convincing benchmarks, the emerging SNN is inspired by the biological neural systems. It performs high computing efficiency in terms of real-time information processing capabilities with comparable energy consumption due to the local learning rules and asynchronously emitting spikes. Generally, the STDP learning rule is adopted commonly as the weight update rule for the training phase of SNN, which shows more biological resemble plausibility. Therefore, the PCM synaptic

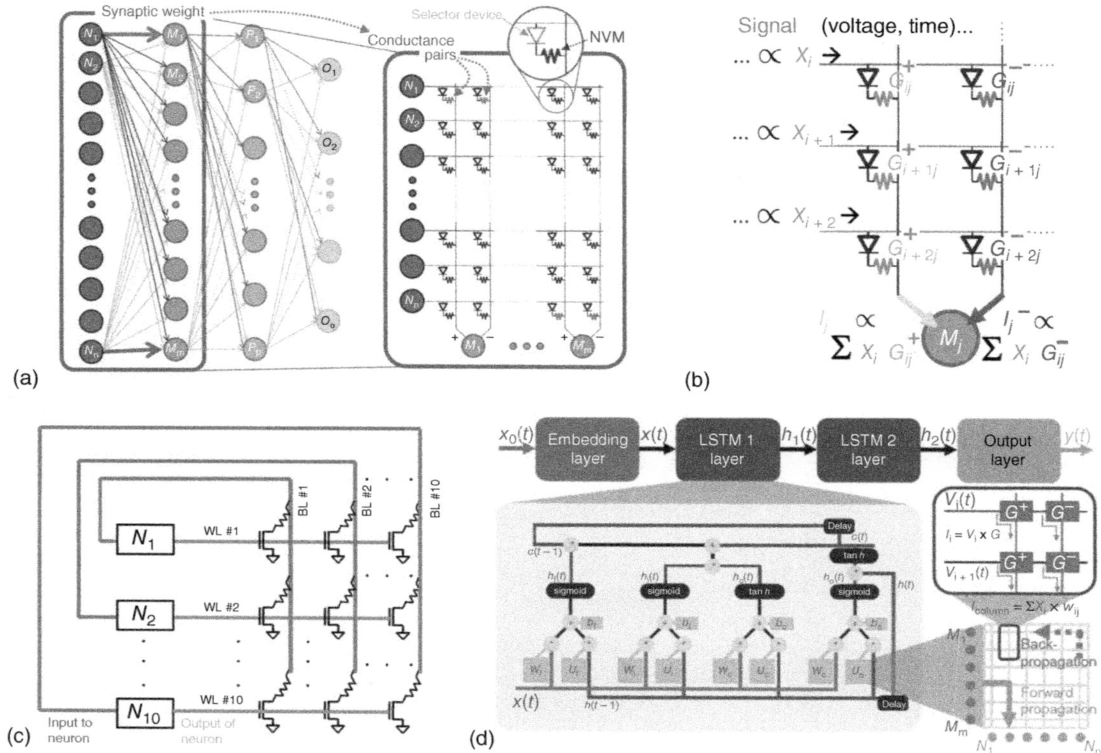

Figure 4.10 (a) Mapping a fully connected neural network onto PCM arrays. (b) Illustration of the 2-PCM synaptic unit. Source: (a, b) Burr et al. [49]. (c) PCM array schematic shown as a 10-neuron recurrent Hopfield neural network. Source: Eryilmaz et al. [50]. (d) Network with fully connected embedding and output layers and two LSTM layers and the eventual mapping of fully connected layers inside the LSTM block to analog PCM array. Source: Tsai et al. [51].

arrays are also flexible to build an in situ SNN platform with not only the matrix computing acceleration but also the biological synaptic plasticity.

A neuromorphic core with 64k-cell PCM synaptic array (256 axons by 256 dendrites) with in situ learning capability was developed by Kim et al. [53]. Two hundred fifty-six configurable on-chip neuron circuits perform leaky integrate and fire (LIF) and synaptic weight update based on STDP. The novel 2T1R (two transistors one resistor) structure, as shown in Figure 4.11a, enables both LIF and STDP learning to operate asynchronously within the array, avoiding additional circuit complications and power consumption associated with timing schemes. A set of patterns have been successfully learned using the STDP learning rule by feeding in a series of patterns. Output spike patterns also exhibited sparse coding with an increased representational capacity. Taking such a similar design principle, a more recent work from the same group designed and fabricated an ultralow-power SNN processor integrating a novel 6T2R PCM synaptic cell array and asynchronous stochastic LIF neurons (Figure 4.11b) [54]. The experimental results confirmed the capability of the processor to perform on-chip training and inference with the spiking restricted Boltzmann machine (RBM). STDP-based

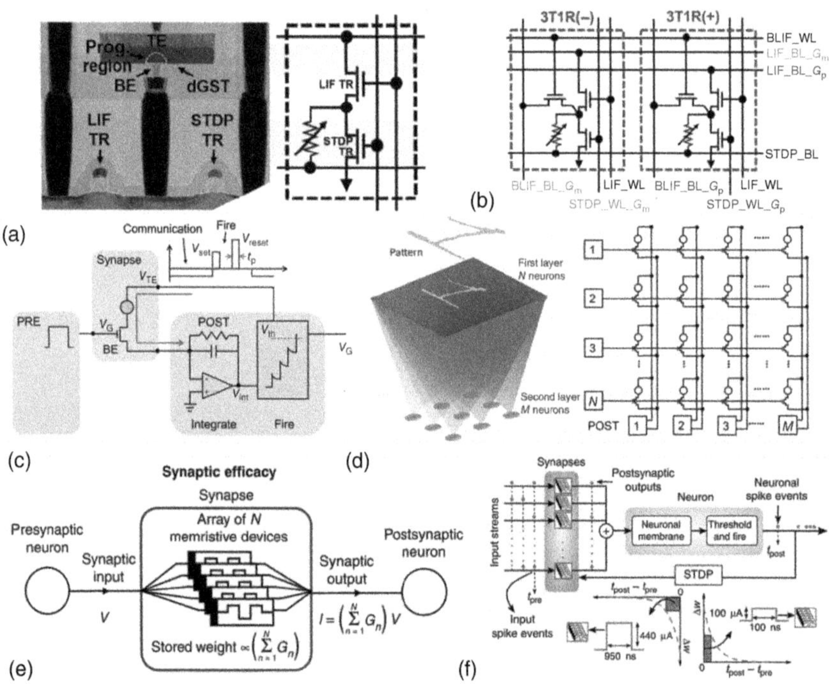

Figure 4.11 (a) TEM image of synaptic phase change memory cell in 2T-1R configuration. Source: Kim et al. [53]. (b) Circuit diagram of 6T2R PCM synapse cell. Source: Ishii et al. [54]. (c) Schematic illustration of the neuromorphic network with a 1T1R synapse. (d) Schematic illustration and corresponding circuit for simulated 1T1R synapse-based neuromorphic network. Source: (c, d) Ambrogio et al. [55]. (e) The multi-PCM synapse concept. (f) A multi-PCM synapse-based spiking neural network is trained to perform the task of temporal correlation detection through unsupervised learning. Source: (e, f) Boybat et al. [56].

unsupervised learning was first shown using 1T1R PCM arrays by Ambrogio et al. [55]. As shown in Figure 4.11c, the neuromorphic network with a 1T1R synapse was schematically illustrated, which could be replicated to achieve a generic multilayer neuromorphic array. STDP characteristics are demonstrated by experiments as a function of the initial resistance state and of the number of potentiating pulses. The ability to learn and recognize patterns in a fully connected neuromorphic network was demonstrated with the proposed scheme to utilize the input noise as a means to depress background synapses for the first time, thus enabling online pattern learning, forgetting, and updating (Figure 4.11d). However, precise modulation of the device conductance over a wide dynamic range, necessary to maintain high network accuracy, is still challenging. To address this, Boybat et al. presented a multi-PCM synaptic architecture with an efficient global counter-based arbitration scheme [56] (Figure 4.11e,f). In such a synapse, the synaptic weight is represented by the combined conductance of N devices. By using multiple devices to represent a synaptic weight, it is shown to overcome several significant challenges that are characteristic to nanoscale PCM devices proposed for synaptic implementation, such as the asymmetric conductance response, limitations in resolution and dynamic range, and device-level variability. The experimental hardware platform was built around a prototype PCM chip with 3 million GST devices with a four-bank inter-leaved architecture. Then, an experimental demonstration of the multi-PCM synapse architecture using the prototype PCM chip to train an SNN that detects temporal correlations in event-based data streams in an unsupervised way was presented. The proposed architecture and its experimental demonstration are a significant step toward the realization of highly efficient, large-scale neural networks based on nanoscale nonvolatile memory devices with typical experimentally observed nonideal characteristics.

4.6 Summary and Outlook

The research devoted to chalcogenide materials has made significant progress over the past decades, owing to their comparable electronic and interesting application prospects in storage class memory and neuromorphic computing. In this chapter, we discussed the feasible strategy to build artificial synapses and neurons by exploiting the memory switching and threshold switching behaviors of chalcogenides. We also present an overview of the progress in various neural network accelerators using chalcogenide memory devices. However, the optical properties of PCM materials are not discussed in detail in this chapter, which is also highly advanced for the investigation of neuromorphic photonic chips.

To construct energy-efficient and bio-plausible neuromorphic devices, there remain several unsolved problems. Firstly, despite all these years of intensive research, an in-depth understanding of the physical mechanisms of materials has the utmost impact on the development of high-performance devices. Especially, the electronic conduction behind the TS behaviors is still under debate. Secondly, to realize artificial synapses with ideal performances, material optimized is critical.

Besides, it also requires system-level co-design to develop scenarios that are better suited to PCM devices, which tolerate the inherent nonidealities and maximize the ability of in-memory computing. Last but not least, under the potential of biological plausibility on PCM devices, a fully PCM biologically relevant neuromorphic computing platform is greatly expected.

These challenges require an interdisciplinary approach for further breakthroughs to create chalcogenide-based neuromorphic systems. Combining the considerable advances already made by the community and the outstanding properties of chalcogenides, it is believed that ovonic devices offer unique advantages for future applications.

References

1 Tang, J., Yuan, F., Shen, X. et al. (2019). Bridging biological and artificial neural networks with emerging neuromorphic devices: fundamentals, progress, and challenges. *Adv. Mater.* 31 (49): 1902761.

2 Xia, Q. and Yang, J.J. (2019). Memristive crossbar arrays for brain-inspired computing. *Nat. Mater.* 18 (4): 309–323.

3 Sebastian, A., Le Gallo, M., Khaddam-Aljameh, R., and Eleftheriou, E. (2020). Memory devices and applications for in-memory computing. *Nat. Nanotechnol.* 15 (7): 529–544.

4 Wang, Z., Wu, H., Burr, G.W. et al. (2020). Resistive switching materials for information processing. *Nat. Rev. Mat.* 5 (3): 173–195.

5 Sangwan, V.K. and Hersam, M.C. (2020). Neuromorphic nanoelectronic materials. *Nat. Nanotechnol.* 15 (7): 517–528.

6 Tsai, H., Ambrogio, S., Narayanan, P. et al. (2018). Recent progress in analog memory-based accelerators for deep learning. *J. Phys. D: Appl. Phys.* 51 (28): 283001.

7 Burr, G.W., Shelby, R.M., Sebastian, A. et al. (2017). Neuromorphic computing using non-volatile memory. *Adv. Phys. X.* 2 (1): 89–124.

8 Sebastian, A., Le Gallo, M., Burr, G.W. et al. (2018). Tutorial: brain-inspired computing using phase-change memory devices. *J. Appl. Phys.* 124 (11): 111101.

9 Burr, G.W., Brightsky, M.J., Sebastian, A. et al. (2016). Recent progress in phase-change memory technology. *IEEE J. Emerg Select Topics Circuits Sys.* 6 (2): 146–162.

10 Ovshinsky, S.R. (1968). Reversible electrical switching phenomena in disordered structures. *Phys. Rev. Lett.* 21 (20): 1450.

11 Feinleib, J., DeNeufville, J., Moss, S.C., and Ovshinsky, S. (1971). Rapid reversible light-induced crystallization of amorphous semiconductors. *Appl. Phys. Lett.* 18 (6): 254–257.

12 Neale, R., Nelson, D., and Moore, G.E. (1970). Nonvolatile and reprogrammable, the read-mostly memory is here. *Electronics.* 43 (20): 56–60.

13 Shanks, R. and Davis, C. (eds.) (1978). A 1024-bit nonvolatile 15 ns bipolar read-write memory. In: *1978 IEEE International Solid-State Circuits Conference Digest of Technical Papers*. IEEE.

14 Yamada, N., Ohno, E., Akahira, N. et al. (1987). High speed overwritable phase change optical disk material. *Japan. J. Appl.Phys.* 26 (S4): 61.

15 Le Gallo, M. and Sebastian, A. (2020). An overview of phase-change memory device physics. *J. Phys. D: Appl. Phys.* 53 (21): 213002.

16 Raoux, S., Wełnic, W., and Ielmini, D. (2010). Phase change materials and their application to nonvolatile memories. *Chem. Rev.* 110 (1): 240–267.

17 Lencer, D., Salinga, M., Grabowski, B. et al. (2008). A map for phase-change materials. *Nat. Mater.* 7 (12): 972–977.

18 Kroll, D.M. (1974). Theory of electrical instabilities of mixed electronic and thermal origin. *Phys. Rev. B.* 9 (4): 1669.

19 Kroll, D. (1975). Theory of electrical instabilities of mixed electronic and thermal origin. II. Switching as a nucleation process. *Phy. Rev. B.* 11 (10): 3814.

20 Kaplan, T. and Adler, D. (1971). Thermal effects in amorphous-semiconductor switching. *Appl. Phys. Lett.* 19 (10): 418–420.

21 Karpov, V., Kryukov, Y., Savransky, S., and Karpov, I. (2007). Nucleation switching in phase change memory. *Appl. Phys. Lett.* 90 (12): 123504.

22 Karpov, V., Kryukov, Y., Karpov, I., and Mitra, M. (2008). Field-induced nucleation in phase change memory. *Phys. Rev. B.* 78 (5): 052201.

23 Adler, D., Henisch, H.K., and Mott, N. (1978). The mechanism of threshold switching in amorphous alloys. *Rev. Modern Phy.* 50 (2): 209.

24 Ielmini, D. and Zhang, Y. (2007). Analytical model for subthreshold conduction and threshold switching in chalcogenide-based memory devices. *J. Appl. Phys.* 102 (5): 054517.

25 Guo, Y., Li, H., Zhang, W., and Robertson, J. (2019). Structural changes during the switching transition of chalcogenide selector devices. *Appl. Phys. Lett.* 115 (16): 163503.

26 Li, H. and Robertson, J. (2019). Materials selection and mechanism of non-linear conduction in chalcogenide selector devices. *Sci. Rep* 9 (1): 1–9.

27 Velea, A., Opsomer, K., Devulder, W. et al. (2017). Te-based chalcogenide materials for selector applications. *Sci. Rep.* 7 (1): 1–12.

28 Huang, D., Miao, X., Li, Z. et al. (2011). Nonthermal phase transition in phase change memory cells induced by picosecond electric pulse. *Appl. Phys. Lett.* 98 (24): 242106.

29 Loke, D., Lee, T., Wang, W. et al. (2012). Breaking the speed limits of phase-change memory. *Science* 336 (6088): 1566–1569.

30 Rao, F., Ding, K., Zhou, Y. et al. (2017). Reducing the stochasticity of crystal nucleation to enable subnanosecond memory writing. *Science.* 358 (6369): 1423–1427.

31 Simpson, R., Fons, P., Kolobov, A. et al. (2011). Interfacial phase-change memory. *Nat. Nanotechnol.* 6 (8): 501–505.

32 Kim, S., Burr, G.W., Kim, W., and Nam, S.-W. (2019). Phase-change memory cycling endurance. *MRS Bull.* 44 (9): 710–714.

33 Kim, W., Bright Sky, M., Masuda, T. et al. (eds.) (2016). ALD-based confined PCM with a metallic liner toward unlimited endurance. In: *2016 IEEE International Electron Devices Meeting (IEDM)*. IEEE.

34 Fantini, P. (2020). Phase change memory applications: the history, the present and the future. *J. Phys. D: Appl Phys.* 53 (28): 283002.

35 Kuzum, D., Jeyasingh, R.G., Lee, B., and H-SP, W. (2012). Nanoelectronic programmable synapses based on phase change materials for brain-inspired computing. *Nano. Lett.* 12 (5): 2179–2186.

36 Suri, M., Sousa, V., Perniola, L. et al. (eds.) (2011). Phase change memory for synaptic plasticity application in neuromorphic systems. In: *2011 The International Joint Conference on Neural Networks*. IEEE.

37 Suri, M., Bichler, O., Querlioz, D. et al. (eds.) (2011). Phase change memory as synapse for ultra-dense neuromorphic systems: application to complex visual pattern extraction. In: *2011 International Electron Devices Meeting*. IEEE.

38 Li, Y., Zhong, Y., Xu, L. et al. (2013). Ultrafast synaptic events in a chalcogenide memristor. *Sci. Rep.* 3 (1): 1–7.

39 Li, Y., Zhong, Y., Zhang, J. et al. (2014). Activity-dependent synaptic plasticity of a chalcogenide electronic synapse for neuromorphic systems. *Sci. Rep.* 4 (1): 1–7.

40 Ding, K., Wang, J., Zhou, Y. et al. (2019). Phase-change heterostructure enables ultralow noise and drift for memory operation. *Science* 366 (6462): 210–215.

41 Indiveri, G., Chicca, E., and Douglas, R. (2006). A VLSI array of low-power spiking neurons and bistable synapses with spike-timing dependent plasticity. *IEEE Trans. Neu. Netw.* 17 (1): 211–221.

42 Tuma, T., Pantazi, A., Le Gallo, M. et al. (2016). Stochastic phase-change neurons. *Nat. Nanotechnol.* 11 (8): 693.

43 Indiveri, G., Linares-Barranco, B., Hamilton, T.J. et al. (2011). Neuromorphic silicon neuron circuits. *Front. Neurosci.* 5: 73.

44 Pickett, M.D., Medeiros-Ribeiro, G., and Williams, R.S. (2013). A scalable neuristor built with Mott memristors. *Nat. Mater.* 12 (2): 114–117.

45 Lee, M., Kim, Y., Cho, S.W. et al. (2020). A highly scalable and energy-efficient artificial neuron using an Ovonic Threshold Switch (OTS) featuring the spike-frequency adaptation and chaotic activity. *Phys. Rev. Appl.* 13 (6): 064056.

46 Wang, Z., Joshi, S., Savel'ev, S. et al. (2018). Fully memristive neural networks for pattern classification with unsupervised learning. *Nat. Electron.* 1 (2): 137–145.

47 Lim, H., Ahn, H.-W., Kornijcuk, V. et al. (2016). Relaxation oscillator-realized artificial electronic neurons, their responses, and noise. *Nanoscale* 8 (18): 9629–9640.

48 Lee, D., Kwak, M., Moon, K. et al. (2019). Various threshold switching devices for integrate and fire neuron applications. *Adv. Electron. Mater.* 5 (9): 1800866.

49 Burr, G.W., Shelby, R.M., Sidler, S. et al. (2015). Experimental demonstration and tolerancing of a large-scale neural network (165 000 synapses) using phase-change memory as the synaptic weight element. *IEEE Trans. Electron. Dev.* 62 (11): 3498–3507.

50 Eryilmaz, S.B., Kuzum, D., Jeyasingh, R.G. et al. (eds.) (2013). Experimental demonstration of array-level learning with phase change synaptic devices. In: *2013 IEEE International Electron Devices Meeting*. IEEE.

51 Tsai, H., Ambrogio, S., Mackin, C. et al. (eds.) (2019). Inference of long-short term memory networks at software-equivalent accuracy using 2.5 M analog phase change memory devices. In: *2019 Symposium on VLSI Technology*. IEEE.

52 Ambrogio, S., Narayanan, P., Tsai, H. et al. (2018). Equivalent-accuracy accelerated neural-network training using analogue memory. *Nature* 558 (7708): 60–67.

53 Kim, S., Ishii, M., Lewis, S. et al. (eds.) (2015). NVM neuromorphic core with 64k-cell (256-by-256) phase change memory synaptic array with on-chip neuron circuits for continuous in-situ learning. In: *2015 IEEE international electron devices meeting (IEDM)*. IEEE.

54 Ishii, M., Kim, S., Lewis, S. et al. (eds.) (2019). On-chip trainable 1.4 M 6T2R PCM synaptic array with 1.6 K stochastic LIF neurons for spiking RBM. In: *2019 IEEE International Electron Devices Meeting (IEDM)*. IEEE.

55 Ambrogio, S., Ciocchini, N., Laudato, M. et al. (2016). Unsupervised learning by spike timing dependent plasticity in phase change memory (PCM) synapses. *Front. Neurosci.* 10: 56.

56 Boybat, I., Le Gallo, M., Nandakumar, S. et al. (2018). Neuromorphic computing with multi-memristive synapses. *Nat. Commun.* 9 (1): 1–12.

5

Neuromorphic Devices Based on Organic Materials

Junyao Zhang and Jia Huang

Tongji University, Interdisciplinary Materials Research Center, School of Materials Science and Engineering, Frontiers Science Center for Intelligent Autonomous Systems, 4800 Caoan Road, Shanghai, 201804, P. R. China

5.1 Introduction

During the progress of neuromorphic electronics, enormous efforts have been made in the design and manufacture of electronic devices that simulate biological synapses. Among different kinds of materials, organic materials are regarded to be the most suited candidates for the design of artificial synaptic devices, owing to their typical characteristics of low cost, outstanding ductility, low energy consumption, prominent mechanical flexibility, and high compatibility with complementary metal–oxide–semiconductor (CMOS) technology [1–3]. In addition, organic materials can be tailored through functional groups to obtain particular electronic, mechanical, chemical, and photonic properties for desired applications by chemical synthesis [4]. Organic neuromorphic devices can be manufactured by simple solution processes, such as spin coating and printing, which are extraordinarily suited for mass production. Thus, organic neuromorphic devices will be the strong basis for the development of future energy-efficient brain-inspired computing and bionic perception, and might launch a new era for next-generation soft electronics.

Here, the recent studies in organic neuromorphic devices, such as both two-terminal memristors and three-terminal transistors, are reviewed and discussed, followed by some innovative applications of organic neuromorphic devices for bionic perception systems. At the end of the chapter, we analyze the main current challenges and future research directions.

5.2 Two-Terminal Organic Neuromorphic Devices

Memristors are one of the fundamental two-terminal devices with the nonvolatile characteristics, which are different from the features of conventional electronics [5–7]. Memristors composed of metal/insulator/metal structure have drawn immerse interests because of their unique properties, including excellent scalability,

Neuromorphic Devices for Brain-Inspired Computing: Artificial Intelligence, Perception, and Robotics, First Edition. Edited by Qing Wan and Yi Shi.

low energy consumption, high connectivity, and perfect compatibility with CMOS process. In most organisms, biological synapses are mainly two-terminal structures. Thus, memristors are the closest devices to biological synapses, because they have the capability to record the changes of voltage and current at any time on the basis of their resistance states, and then some basic synaptic behaviors have been successfully emulated.

In this section, the working principles and major advances of four kinds of two-terminal organic neuromorphic devices are reviewed and discussed, including metal filament conducting-based memristors, redox reaction-based memristors, ion migration-based memristors, and charge trapping-based memristors.

5.2.1 Metal Filament Conducting-Based Memristors

Metal filament is often formed when an electric spike is added on the top electrode, resulting in the reduction and electrocrystallization of the metal ions [8]. Among the polymers, poly(3,4-ethylenedioxythiophene):poly(styrene sulfonate) (PEDOT:PSS) is extensively researched due to the ease to modify the conductivity through various methods. Li et al. demonstrated the Ag/PEDOT:PSS/Ta memristor, where the conductive capacity of PEDOT:PSS layer could provide a conducting path for the Ag movement (Figure 5.1a) [9]. Compared with the initial state in the transmission electron microscopy (TEM) images, the Ag filament was clearly formed when a sequence of positive bias was exerted to the electrodes, and the energy-dispersive X-ray spectroscopy (EDX) profile also illustrated the increase of Ag peak around the Ta electrode (Figure 5.1b,c). Some important synaptic performance, including short-term plasticity (STP), long-term plasticity (LTP), spike timing dependent plasticity (STDP), and spike-rate-dependent plasticity (SRDP), were successfully simulated.

Bioinspired materials have great advantages for application to resistive switching memories and artificial synaptic devices, thanks to their simple manufacturing process, flexibility, and biocompatibility [7]. In general, biopolymers mean the materials extracted by organisms in nature, and diverse biopolymers have been studied. Kim et al. demonstrated a novel biopolymer called ι-carrageenan (ι-car), which was extracted from seaweed and created an Ag/ι-car/Pt structure (Figure 5.1d) [10]. If a positive voltage bias was exerted to the top electrode, the Ag electrode was dissolved, bringing about the generation of Ag cations. These cations diffused into the ι-car layer, leading to their movements to the Pt bottom electrode. Then, a reduction process was caused close to the interface between ι-car layer and bottom electrode with the growth of an Ag filament.

Halide perovskites have attracted tremendous interests owing to their tunable bandgap, excellent carrier mobility, and solution process availability [13–16]. Benefitting from these unique characteristics, halide perovskites have the potential to be potential materials for resistive switching memories. Recently, researchers have greatly investigated the features of halide perovskite memristors. Yoo et al. demonstrated an Ag/$CH_3NH_3PbI_{(3-x)}Cl_x$/fluorine-doped tin oxide (FTO) structure (Figure 5.1e) and explored its performance as an artificial synaptic device [11]. The I–V sweep characteristics of the memristor was exhibited by scanning voltages with

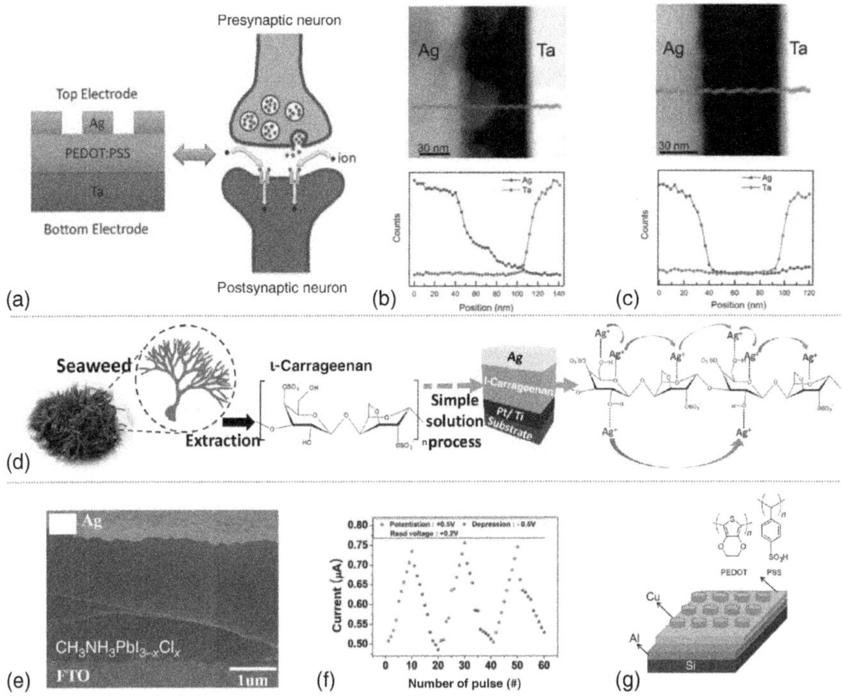

Figure 5.1 (a) Structural analogy between Ag/PEDOT:PSS/Ta memristor and a typical biological synapse. Cross-sectional TEM images and the related EDX spectra for Ag and Ta element profiles of (b) a device stimulated using 10 consecutive 2 V stimulations and (c) a pristine device. Source: Li et al. [9]. (d) Typical seaweed image and structure of ι-car and Ag/ι-car/Pt device structure with the proposed mechanism of Ag^+ ion migration in ι-car biopolymer matrix. Source: Kim and Lee [10]. (e) Cross-sectional SEM image of the Ag/$CH_3NH_3PbI_{3-x}Cl_x$/FTO structure. (f) Response to the pulses with +0.5 V for potentiation and −0.5 V for depression of the Ag/$CH_3NH_3PbI_{3x}Cl_x$/FTO device. Source: Yoo et al. [11]. (g) Schematic of the memory device with the Si/Al/PEDOT: PSS/Cu structure and the chemical structures of PEDOT and PSS. Source: Wang et al. [12].

0 V → 3 V → −3 V → 0 V. Thus, positive voltages promoted the state of the memristor to low resistive state (LRS) because of the gradual generation of Ag filament, and negative voltages switched the state of the memristor to high resistive state (HRS) due to the rupture of Ag filament. As depicted in Figure 5.1f, the potentiation and depression process were realized by applying successive pulse voltages of +0.5 and −0.5 V, respectively. As 10 positive pulses of +0.5 V were induced repeatedly, the current of the memristor had a linear increase from 0.5 to 0.75 μA, simulating a synaptic function of potentiation. Then, next 10 pulses of −0.5 V resulted in the current decreasing linearly from ~0.7 to 0.5 μA, which emulated the process of depression. In addition, the neuromorphic device demonstrated three sets of repetitions of potentiation and depression.

Moreover, Wang et al. reported the metal filaments formed from a resistive switching memory based on PEDOT:PSS with the Cu electrode (Figure 5.1g) [12]. Similar to the above Ag filaments, the resistivity of Cu electrode device was also adjusted through voltage. Nevertheless, owing to the characteristics of high crystallinity and unsatisfactory ionic conductivity of this device, the injected Cu cations could only migrate a short distance. As a result, the devices might slowly deteriorate in synaptic behaviors because of low ion mobility and the gradual damage to electrodes.

5.2.2 Redox Reaction-Based Memristors

Chemical redox reaction is usually accompanied by the variation of chemical state of reactants. When a substance presents different chemical states, its resistance is usually different. It is possible to realize memristive effects in organic synaptic devices by redox reaction, so recently, the application of voltage-induced redox reaction in organic memristors has made a great effort.

Liu et al. demonstrated an organic memristor based on ethyl viologen diperchlorate [$EV(ClO_4)_2$]/triphenylamine-containing polymer (BTPA-F) redox combination, with the structure of Ta/$EV(ClO_4)_2$/*N*,*N*′-diphenyl-4-bromoaniline dimer (BTPA)-F/Pt (Figure 5.2a) [17]. The BTPA-F with a closed ring and a steric tightly packed triphenylamine (TPA) unit has been utilized primarily because these groups show rich electrochemical redox reactions, while the $EV(ClO_4)_2$ group acts as the counter electrode and stabilizes the state of charge. Once a positive bias was

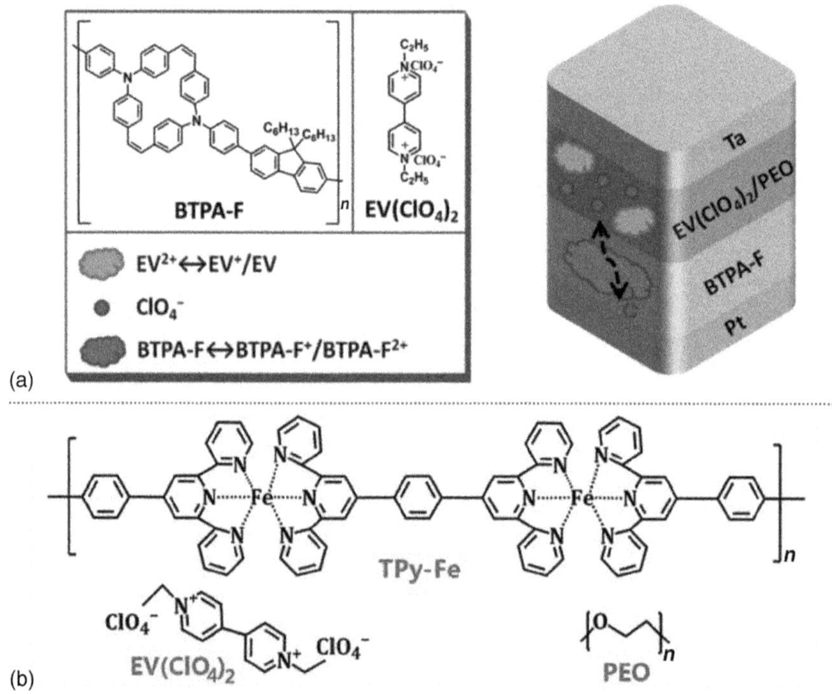

Figure 5.2 (a) Chemical structures of BTPA-F and $EV(ClO_4)_2$ and the electrochemical redox reaction of the $EV(ClO_4)_2$/BTPA-F bilayer structure and schematic illustration of the Ta/$EV(ClO_4)_2$/BTPA-F/Pt memristor. Source: Liu et al. [17]. (b) Chemical structures of TPy-Fe, $EV(ClO_4)_2$ and PEO. Source: Yang et al. [18].

induced to the top electrode, BTPA-F group was oxidized by separating unpaired electrons from N atoms, followed by ClO_4^- entering the oxidized N atoms of BTPA to neutralize the whole redox system. At the same time, the conductance of device was continuously and precisely enhanced through the intra/interchain hopping from the oxidized BTPA. If a negative voltage was applied to the top electrode, BTPA undergone a reduction process. As a result, ClO_4^- returned to its original value, leading to the low conduction of the memristor. Some basic synaptic behaviors, including STP, LTP, SRDP, and "learning experience" behavior, were successfully emulated. This study broadens the organic redox reaction-based memristors in the application of emulating artificial synapses.

Moreover, Yang et al. demonstrated memristive and synaptic behaviors in a Ta/EV$(ClO_4)_2$ and polyethylene oxide (PEO)/terpyridyl-iron polymer (TPy-Fe)/indium tin oxide (ITO) structure [18]. The chemical structures of EV$(ClO_4)_2$, TPy-Fe, and PEO were illustrated in Figure 5.2b. EV$(ClO_4)_2$ was used as the reaction material to form a coordination bond between iron ion and pyridine ring for the oxidation of TPy-Fe. PEO mixed in EV$(ClO_4)_2$ acted as a solid electrolyte, which could enhance the mobility of perchlorate ions in EV$(ClO_4)_2$. The potentiation or depression process in the memristor was achieved by exerting positive or negative voltage pulse to the device. Furthermore, two-terminal organic synaptic devices were also illustrated on PEO-EV$(ClO_4)_2$/triphenylamine-based polyimide (TPA-PI) system and PEO-$LiCF_3SO_3$/poly[2-methoxy-5-(2-ethylhexyloxy)-1,4-phenylenevinylene] (MEH-PPV) system. The above results show that it is possible to build neuromorphic devices for storing and processing information through chemical redox reaction. However, the anions in this device are expected to neutralize positive charges, which greatly affects synaptic functions. Thus, this type of organic neuromorphic devices needs to be further designed and optimized.

5.2.3 Ion Migration-Based Memristors

Poly(methylsilsesquioxane) (pMSSQ), as one promising organic material, has been highlighted for neuromorphic devices due to its electrochemical stability, three-dimensional stacking capability, and flexibility. Wu et al. demonstrated an organic memristor based on the migration of Cu cations, with the structure of Al/copper-ion-doped pMSSQ/Al (Figure 5.3a) [19]. The pMSSQ acted as the hole injection layer. The doped Cu cations controlled the conductance of pMSSQ, which had a concentration gradient that was higher around the top electrode while lower in the vicinity of the bottom electrode. A high concentration of Cu ions was dispersed to decrease the energy gap. As a result, as a positive voltage was exerted to the top electrode, numerous carriers could increase the conductivity of the memristor through low barrier between the Cu-pMSSQ layer and Al electrode. On the contrast, once an opposite bias was induced to the top electrode, the memristor conductance declined because the barrier between Cu-pMSSQ layer and Al electrode was higher, owing to the slight copper doping near the bottom electrode. Thus, they simulated several synaptic properties, including LTP, long-term depression (LTD), and STDP (Figure 5.3b,c).

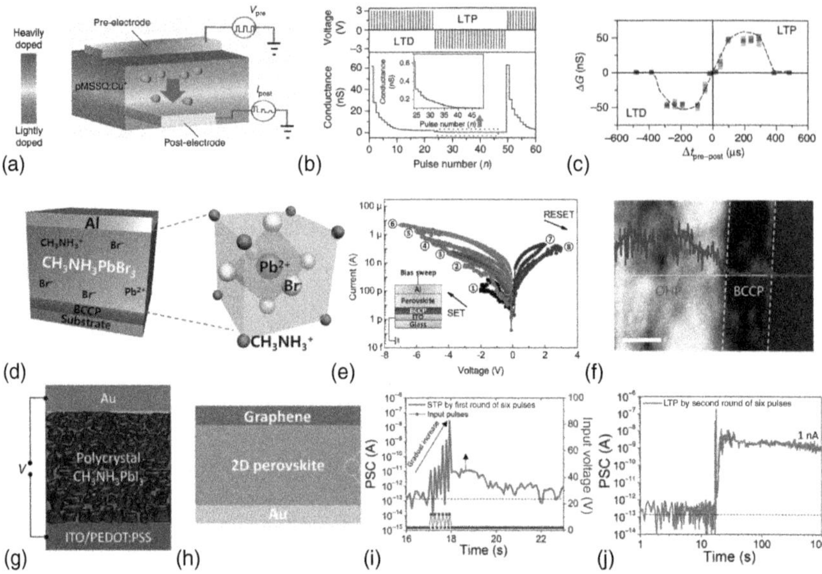

Figure 5.3 (a) Schematic diagram and band structure of the e-synapse consisting of a sandwiched structure of Al (bottom electrode)/lightly doped layer/highly doped layer/Al (top electrode). (b) Switching behaviors under a sequence of electric pulses. The inset presents enlarged view in the range from 25 to 50 pulses. (c) ΔG of the e-synapse stimulated by using a pair of spikes with different values of $\Delta t_{\text{pre-post}}$ ranging from −500 to 500 μs. Source: Wu et al. [19]. (d) Geometry of an OHP perovskite artificial synapse, and crystal structure of $CH_3NH_3PbBr_3$. (e) DC $I-V$ sweeps with consecutively increased ranges that induced gradual setting and resetting processes. Inset: Schematic of the architecture and measurement of the OHP thin-film-based artificial synapse. (f) TEM images and Br EDS profiles of the cross-section of OHP/BCCP structure after applying a series of pulses to form LTP (light grey line, position EDS line profile; dark grey line, intensity of Br element; scale bar of 20 nm). Source: Xu et al. [20]. (g) Schematic diagram of the resistive switching memory structure based on PEDOT:PSS. Source: Xiao et al. [21]. (h) Schematic diagram of a vertical device with graphene/2D perovskite/Au structure. (i) Short-term potentiation by a first round of six pulses. (j) Long-term potentiation by a second round of six pulses. Source: Tian et al. [22].

Organometallic halide perovskites have also attracted great attention in ion migration-based memristors due to their unique chemical formula of ABX_3, in which A refers to an organic cation (typically $CH_3NH_3^+$), B is a metal cation (typically Pb^{2+}), and X is a halide anion (I^-, Br^-, or Cl^-). The ion migration of organometallic halide perovskites enables memristive behaviors for the achievement of artificial synapses. Xu et al. fabricated a structure of Al/$CH_3NH_3PbBr_3$/buffer-capped conducting polymer (BCCP)/ITO, as shown in Figure 5.3d [20]. BCCP consisted of PEDOT:PSS, perfluorinated ionomer, and tetrafluoroethylene-perfluoro-3,6-dioxa-4-methyl-7-octenesulfonic acid copolyme (PFI). The current level increased gradually by continuously exerting step voltages ranging from −2 to −6 V. (Figure 5.3e). In addition, as 2 and 3 V amplitude positive step voltages were successively exerted, current level of the device decreased gradually. Thus, they mimicked excitatory postsynaptic current (EPSC), paired-pulse facilitation (PPF), SRDP, and STDP. When the intensity and number of pulses were not enough, some ions moved a short distance and returned to their original positions quickly, which contributed to the STP behavior. While a high intensity or numerous pulses were applied, the ions moved a long distance, followed by trapped into the $CH_3NH_3PbBr_3$/BCCP interface. The mobile ions of organometal halide perovskite (OHP) layer were captured by BCCP layer. Under an external electric field, the relatively low activation energy (E_a) of Br^- (~0.2 eV) enabled it more readily to migrate than $CH_3NH_3^+$ (E_a = ~0.8 eV), so the memristive mechanism could be originated from Br^- migration, and conductivity could be dominated by vacancy defects. Energy dispersive spectroscopy (EDS) profile of the BCCP layer was illustrated to explore the device working mechanism by revealing the Br^- peak (Figure 5.3f). Xiao et al. reported a two-terminal synaptic device with the structure of Au/$CH_3NH_3PbI_3$/PEDOT:PSS/ITO (Figure 5.3g) [21]. Under the action of applied voltage, the memristor could convert from p–i–n structure to n–i–p structure so as to realize the gradual change of conductivity. The working mechanism of this memristor was through the $CH_3NH_3^+$ interstitial ion or I^- ion migration. The device could emulate some biological synaptic behaviors, including STDP, SRDP, learning experience behavior, and forgetting behavior. Four different kinds of STDP behaviors, including symmetric Hebbian, symmetric anti-Hebbian, asymmetric Hebbian, and asymmetric anti-Hebbian learning rules, were obtained by applying different programmed pulse sequences. In addition to $CH_3NH_3PbI_3$ and $CH_3NH_3PbBr_3$, other organometallic halide perovskites also have the property of ionic conductivity, so they may also be utilized as ion migration-based neuromorphic devices.

$(PEA)_2PbBr_4$ with a wide band gap, as one of the promising 2D halide perovskites, has shown its potential in neuromorphic devices. Tian et al. explored a memristor with the structure of graphene/$(PEA)_2PbBr_4$/Au for artificial synaptic device (Figure 5.3h) [22]. The working mechanism of the memristor was also through the Br^- migration, which caused the increasing and decreasing of device conductance. When six electric pulses were applied to the electrode, this memristor simulated a gradual STP behavior (Figure 5.3i). While by continuously applying another six electric pulses, the transition from STP to LTP was obtained, and the EPSC maintained for over 10^3 seconds at ~1 nA (Figure 5.3j).

5.2.4 Charge Trapping-Based Memristors

Charge trapping in organic active materials enables to produce space electric field and therefore regulate the transition between high resistance and low resistance states of devices. Wang et al. fabricated a memristor based on an organic semiconducting small-molecule CuPc, which was fabricated through thermal evaporation (Figure 5.4a) [23]. Because of the polycrystalline characteristics of organic semiconductors, the domain of CuPc contained numerous traps in its interface area that acted as charge reservoirs to limit current. The working mechanism was mainly attributed to the charge trapping and detrapping processes caused by distinct filling ratios and the energy levels in CuPc micro-domains on the basis of space

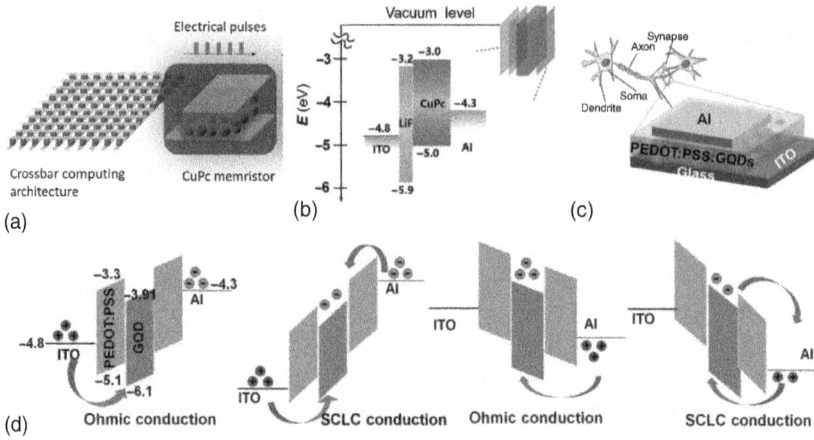

Figure 5.4 (a) Schematic diagram of the CuPc-based crossbar architecture and one typical CuPc element. Source: Wang et al. [23]. (b) Schematic of the energy levels of different layers in ITO/LiF/CuPc/Al device. Source: Wang et al. [24]. (c) Structure of neuron and schematic diagram of the Al/PEDOT: PSS: GQDs/ITO/glass device. (d) Schematic diagrams of the carrier transport behaviors for the devices under small positive bias voltages, under larger positive bias voltages, under small negative bias voltages, and under larger negative bias voltages. Source: Choi et al. [25].

charge-limited conduction (SCLC). Besides, Wang et al. also reported another organic rectification-regulated memristor with the structure of ITO/LiF/CuPc/Al, which possessed excitatory unidirectional and bidirectional electrical synapse-like behaviors [24]. Owing to the uniform interface effect of LiF anode buffer, the device exhibited stable, slow-changing memristor response. The LiF buffer layer and CuPc active layer had deterministic effects to reconfigure device energy level according to the energy level diagram of ITO/LiF/CuPc/Al (Figure 5.4b). The reconfiguration of energy levels had a significant influence in redistributing the processes of charge trapping and detrapping. In addition, associated functions of synaptic plasticity were successfully simulated in the memristor. On the basis of the mentioned mechanism, other types of metal phthalocyanine (MPc) might also own similar memristive characteristics due to their homologous molecular and electronic structures.

Moreover, introducing quantum dots or nanoparticles into organic active materials can also induce charge trapping effect. Choi et al. fabricated a memristor based on graphene quantum dots (GQDs)/organic hybrid systems, with the structure of Al/PEDOT:PSS:GQDs/ITO, which was illustrated in Figure 5.4c [25]. GQDs own prominent properties, including superior charge storage capability, which is considered to be connected with the existence of charge capture states. The memristive behavior of this device was regarded to be associated with the charge trapping and detrapping characteristics of GQDs. The working principle was depicted in Figure 5.4d. Under bias voltage, the processes of trapping and detrapping of charge carriers were primarily dominated by the SCLC mechanism, which was beneficial for regulating the conductance of the memristor. On the basis of the comprehension of the operating mechanism, the potentiation and depression behaviors were mimicked. However, it was expected to realize more synaptic functions, such as STDP and SRDP in further investigation.

5.3 Three-Terminal Organic Neuromorphic Devices

In terms of the abovementioned two-terminal organic neuromorphic devices, only one signal path can be used to transmit signals. More importantly, as we integrate two-terminal devices in the system, additional circuitry components are also required in application of selecting the target cell. Compared with two-terminal devices, three-terminal organic neuromorphic devices have the advantages of excellent stability, well-defined operation mechanism, and relatively controllable testing parameters. In addition, they can also be prepared from different kinds of materials, with the cost of more complicated structure [26]. Through reasonable structural design and material selection, three-terminal transistors enable to transform external stimuli (such as pressure, light, temperature, etc.) to the electrical signal, which makes it possible to prepare artificial synaptic devices that react to the external environment directly. Furthermore, synergistic control of one device is readily achieved in a three-terminal transistor-based artificial synaptic devices, which provides the possibility of exploiting robust neuron networks with remarkably fewer neural components. Significantly, three-terminal transistor-based artificial synaptic devices can implement signal transmission and self-learning simultaneously. Thus, compared with other kinds of devices, three-terminal transistors may be more suited to emulate synaptic performance, especially for mimicking dendrite integration and concurrent learning that demand multiterminal operation. In recent years, more and more attention has been paid to developing artificial synapses based on transistors while this field is still in its infancy.

In this section, the working mechanisms and major advances of four types of three-terminal organic neuromorphic devices are reviewed and discussed, including floating-gate transistors, ferroelectric-gate transistors, electrolyte-gate transistors (EGTs), and optoelectronic transistors.

5.3.1 Floating-Gate Transistors

Generally, floating-gate transistors have similar device structures with traditional field-effect transistors (FETs), in addition to adding the control gate that is usually embedded in the dielectric layer. The charges (electrons or holes) accumulated by applying gate bias in the channel of traditional FETs are dissipated completely and rapidly as the gate bias is removed. Thus, charges accumulated in the channel of traditional FETs normally show a "volatile" behavior. With regard to floating-gate transistors, electronic charges are readily injected into the floating gate based on thermal emission or quantum tunneling during the gate programming process. Owing to the presence of the strong charge blocking layer and tunneling layer, the captured charges are stored nonvolatile. The trapped charges shielded the vertical electric field between the gate electrode and the channel layer, leading to the change of threshold voltage and modulation of channel conductance. In addition, the number of trapped charges in the floating gate can be efficiently adjusted by the gate voltage pulse. Recent studies have shown that floating-gate transistors can be regarded as a potential platform for simulating biological synapses because the gate-adjustable nonvolatile channel conductance can be used to record synaptic weight.

Ren et al. reported a flexible floating-gate synaptic transistor with both learning and signal transmission functions (Figure 5.5a) [27]. Through a simple solution process, C_{60} nanoparticles were well dispersed in the polymethyl methacrylate (PMMA) layer and were served to capture the charges. The as-prepared synaptic transistor demonstrated a memory window of 2.9 V, a high current on/off ratio over 10^3, and more than 500 times of program/erase endurance. Furthermore, the synaptic device showed steady channel conductance modulating characteristic that could be used to simulate synapse-like potentiation and depression (Figure 5.5b). Basic synaptic functions, such as STP, LTP, and repetitive learning processes were also well mimicked. This study is an important step to future development of artificial intelligence utilizing nanoparticle floating-gate transistors.

The abovementioned work is to embed the floating gate into the dielectric. Alibart et al. demonstrated an organic synapse by embedding Au nanoparticles into the interface of organic semiconductors/dielectric (Figure 5.5c) [28]. The prepared synaptic transistors are also known as nanoparticle organic memory field-effect transistors (NOMFET). Nanoparticles are nanosized objects, so they are considered to be suited to manufacture nanosized devices. Moreover, utilizing a low-cost, bottom-up preparation method can manipulate and assemble these nanoparticles. Both potentiating and depressing synaptic behaviors were simulated by taking advantage of NOMFET (Figure 5.5d). In addition, the synaptic plasticity of NOMFET for real-time computing was demonstrated and described by a biological synaptic model. These significant synaptic functions were achieved by integrating two vital features of NOMFETs: transconductance gain of the synaptic transistors and nanoparticle electrical charge trapping effect. The transconductance of NOMFET could be easily modulated by adjusting the number of charges on the nanoparticles, and the charges on the nanoparticle showed a "leaky" behavior. These characteristics of NOMFET enabled synaptic weight to be worked in a dynamic range. More importantly, the dynamic behavior of NOMFETs could be modulated by regulating the size of nanoparticles and the size of the synaptic transistor (Figure 5.5e). Thus, it is possible that NOMFETs can be utilized in next generation of dynamical neuromorphic computing circuits.

5.3.2 Electrolyte-Gate Transistors

EGTs can effectively use ions in the electrolyte dielectric layer to regulate the channel conductivity of neuromorphic device. The operating mechanisms of EGTs are mainly classified to two forms: electrostatic modulation and electrochemical doping. When electrostatic modulation is used as the operating mechanism, semiconductors are impermeable into the ions in the electrolyte. Under the applied gate electric

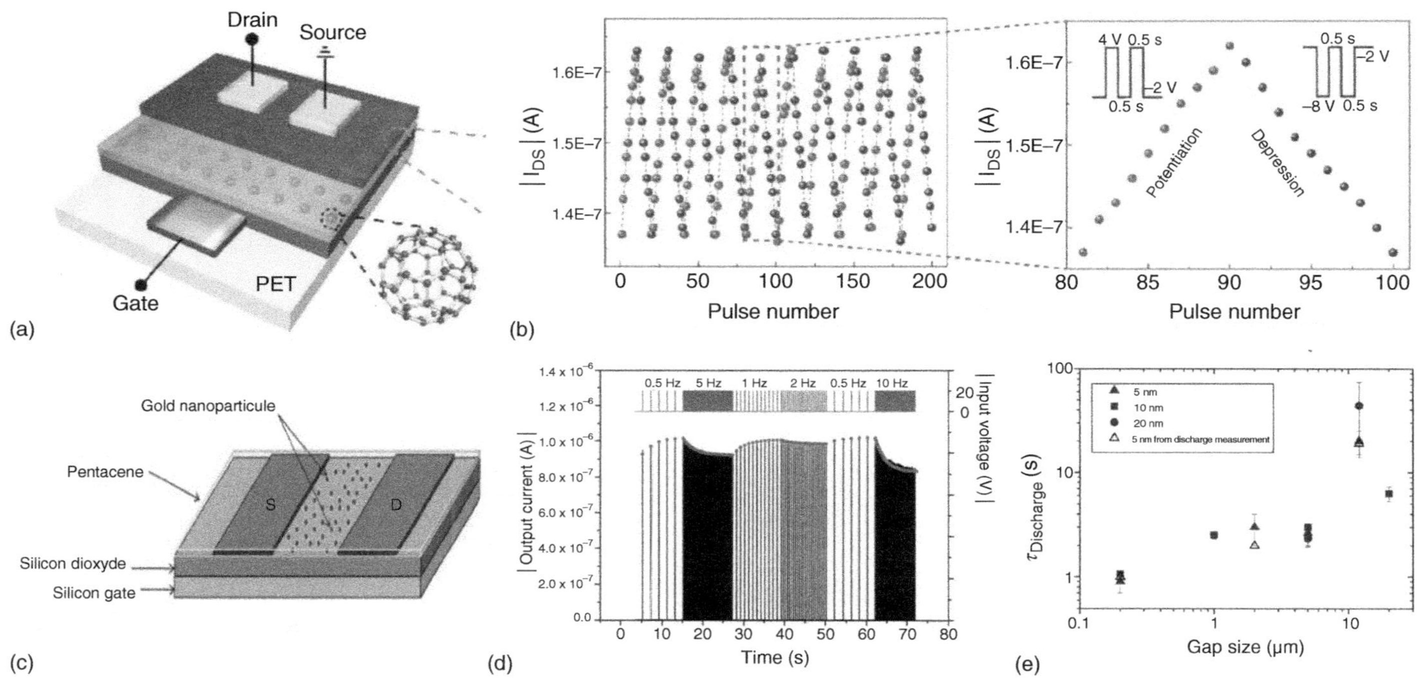

Figure 5.5 (a) Schematic illustration of the 3D bottom-gate top-contact flexible transistors. (b) Repetitive channel conductance modulation by applying repeated positive (4 V for 0.5 s spaced 0.5 s apart, $V_{DS} = -3$ V) and negative (−8 V for 0.5 s spaced 0.5 s apart, $V_{DS} = -3$ V) gate spikes. The picture on the right is an enlarge image of one waveform in the picture on the left. Source: Ren et al. [27]. (c) Schematic illustration of the organic nanoparticle transistor. (d) Response of the organic nanoparticle transistor. (e) Evolution of the nanoparticle discharge time constant as function of the organic nanoparticle transistor channel length and for the different sizes of nanoparticle. Source: Alibart et al. [28].

field, the ions in the electrolyte are able to migrate and accumulate into the interface between dielectric and semiconductor, contributing to the formation of a large capacity electric double layer (EDL) locally near the interface. When electrochemical doping is used as the operating mechanism, EDL is also formed, while semiconductors are permeable to the ions in electrolyte, which will further dope and regulate the device channel conductance in a nonvolatile pattern. Recently, using EGTs to simulate synaptic behaviors has become a new research direction. When the electric field is applied, the migration of ions in electrolyte is extremely similar to the behavior of releasing neurotransmitters in response to an action potential. As EGTs are worked in electrostatic regime, the channel conductance induced by gate voltage can realize reversible transformation, which provides a basis for simulating STP. When EGTs work in the electrochemical doping mode, due to the doping of ions into the channel, the change of channel conductance is partially irreversible, which provides a foundation for emulating the LTP. In biological systems, a single synapse event consumes about 10^{-15} J. The properties of EGTs that can operate in low voltage also provide the opportunity for the realization of ultralow energy consumption neuromorphic devices.

Malliaras et al. revealed a very systematic study in the implementation of biological synapse-like behaviors with PEDOT:PSS [29–32]. Figure 5.6a shows the schematic diagram of the synaptic transistors based on PEDOT:PSS, which were laterally gated by a KCl electrolyte [30]. Cations in the electrolyte diffused into the PEDOT:PSS layer by applying positive gate voltage, which was helpful for de-doping of the PEDOT:PSS film. Once the pulse voltage was removed, the de-doped PEDOT:PSS would come back to its initial state, because the cations injected previously would drift back into the electrolyte. Important synaptic behaviors, including paired-pulse depression (PPD), adaptation, and dynamic filter characteristics, were successfully mimicked by the PEDOT:PSS-based synaptic transistors. Furthermore, Malliaras et al. also demonstrated a synaptic transistor based on PEDOT:PTHF (poly(tetrahydrofuran)) with a function of nonvolatile memory (Figure 5.6b) [33]. By exerting large gate voltage pulses, the long-term plasticity could be achieved by modulating the conformation of PEDOT:PTHF. The conformational change of PEDOT:PTHF was reversed by exerting opposite gate voltage pulses.

The above-reviewed liquid electrolyte-gated transistors are able to realize various neuromorphic functions. Nevertheless, it is hard to process liquid electrolytes and integrate them into progressive neuromorphic devices and circuits. A three-dimensional network structure can be formed by polymers so that liquid electrolytes solidify in the gel network, so the macroscopic fluidity will lose. Thus, compared with liquid electrolyte variety, solid electrolyte-gated transistors have the advantages of better safety and less leakage, and are closer to practical application. Xu et al. revealed a synaptic transistor based on electrolyte-gate organic core–sheath nanowire to mimic the morphology and operating mechanism of nerve fibers [36]. The nanowire consisted of a poly(3-hexylthiophene) (P3HT) core and a PEO sheath. The prepared neuromorphic device consumed an ultralow energy of 1.23 fJ per synaptic pulse, which was 1 order of magnitude lower than that of biological synapse. A negative presynaptic pulse was applied to the gate electrode to realize

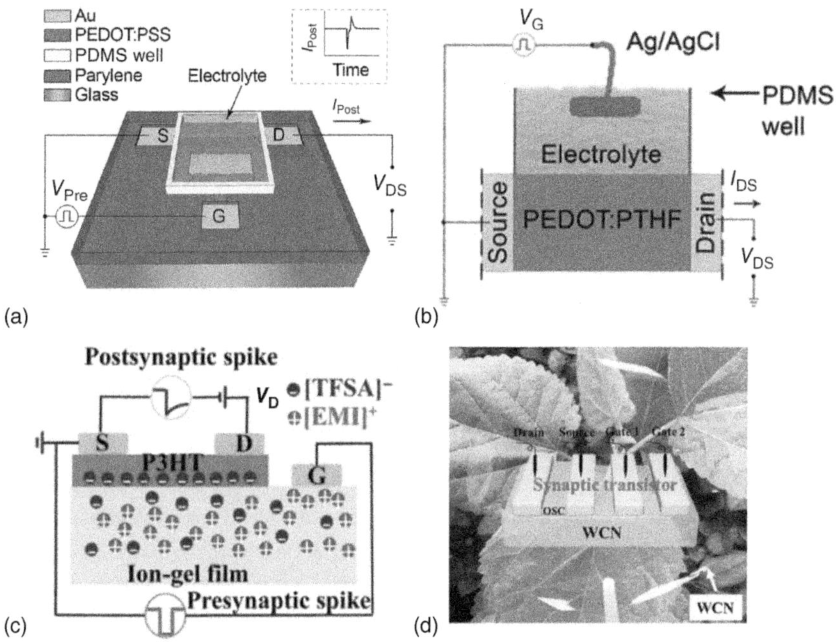

Figure 5.6 (a) The schematic diagram of the synaptic transistors based on PEDOT:PSS, which were laterally gated by a KCl electrolyte. Source: Gkoupidenis et al. [30]. (b) Simplified schematic of the synaptic transistor based on PEDOT:PTHF. Source: Gkoupidenis et al.[33]. (c) Detailed structure of a single self-supported multiterminal organic electrolyte-gate synaptic transistor. Source: Fu et al. [34]. (d) Schematic of a wood-derived cellulose nanopaper-based organic synaptic transistor. Source: Dai et al. [35].

short-term potentiation. When a negative presynaptic pulse was exerted, the anions in the electrolyte migrated to the interface of electrolyte/nanowire, which resulted in the increase of the device channel current. Once the presynaptic spike was eliminated, the anions rapidly returned to their balance place, contributing to the quick decay of the device channel conductance. Long-term potentiation was also achieved by inducing a series of continuous presynaptic spikes. After the operation, the PEO layer would trap some anions, leading to a nonvolatile variation in channel conductance. Moreover, the spatiotemporally correlated dynamic logic function was also achieved through two laterally coupled gates. In addition, Fu et al. exploited a highly interconnected neuromorphic architecture (HINA) based on self-supported multiterminal organic electrolyte-gate synaptic transistors (Figure 5.6c) [34]. Even after 1000 times of bending, the synaptic transistor showed extraordinary mechanical flexibility and electrical stability. Typical neuromorphic behaviors, such as four types of STDP, synaptic potentiation, and depression, were successfully simulated. Furthermore, dynamic memory, Pavlov's learning rule, and forgetting process were achieved by utilizing advantages of HINA. Recently, the development of "green" synaptic devices has attracted much attention. Dai et al. demonstrated an organic synaptic transistor based on 2,7-dioctyl[1]benzothieno[3,2-*b*]-[1]-benzothiophene (C8-BTBT), which was laterally gated through biocompatible and biodegradable wood-derived cellulose nanopaper with ionic conductivity (Figure 5.6d) [35]. Important synaptic behaviors, including dendritic integration and high-pass filter characteristics, were well emulated. This study provides a new direction for the future realization of green neuromorphic electronics.

5.3.3 Ferroelectric-Gate Transistors

Ferroelectric field-effect transistors (FeFETs) have been widely studied in applications of nonvolatile memory, thanks to the advantages of nondestructive readout, high operating speed, and low energy consumption [37–39]. The critical element of FeFETs in application of memory is the ferroelectric insulator with spontaneous polarization states. Because the carriers in the channel and the polarization of the ferroelectric insulator have Coulomb interaction, the carrier concentration of FeFETs can be accurately and gradually regulated by adjusting the polarization state of ferroelectric materials applying the gate voltage. Recently, FeFETs, as a promising platform, have attracted considerable attention for emulating biological synapses due to the extraordinary multidomain polarization switching capability of ferroelectric materials, which are utilized to achieve multilevel FeFET channel conductance. Multilevel channel conductance of FeFETs can be used to record synaptic weight.

In general, oxide ferroelectrics require a high crystallization temperature that limits their applications in large-scale electronic devices on plastic substrates. Besides, due to their inherent rigidity, oxide ferroelectrics are difficult to use in the field of wearable and flexible electronics. These limitations of oxide ferroelectrics can be overcome by using organic ferroelectrics. At recent, Jang et al. demonstrated ultrathin conformable organic synaptic transistors by using a freestanding poly(vinylidene fluoride-*co*-trifluoroethylene) [P(VDF-TrFE)] film as the dielectric layer (Figure 5.7a) [40]. The total thickness of the synaptic devices was ~500 nm. By precisely regulating the residual polarization of P(VDF-TrFE), typical synaptic functions, such as postsynaptic current (PSC), synaptic facilitation/depression process, and STDP, were successfully simulated through these synaptic transistors. Besides, a facile dry peeling-off and stick-on technique was used to steadily transfer the devices onto a variety of nonuniform substrates, such as thermal shrink plastic film, plastic film, jelly, textile, candy, etc. To investigate that the ultrathin conformable organic synaptic transistors capable of maintaining their excellent synaptic behaviors under severe bending conditions, the researchers measured the synaptic performance of the transistors on a brain-shaped PDMS mode and under an extraordinarily bending condition with a bending radius of 50 μm (Figure 5.7b). The prepared synaptic transistors demonstrated reliable long-term potentiation and long-term depression both under severe bending conditions (Figure 5.7c). This study can open up a new research orientation for the development of wearable artificial intelligent systems.

An organic synaptic transistor adjusted by both ferroelectric and electrochemical processes was demonstrated by Wang et al. to tailor its synaptic plasticity [41]. As illustrated in Figure 5.7d, the gate dielectric contained poly[(1-vinylpyrrolidone)-*co*-(2-ethyldimethylammonioethyl methacrylate ethyl

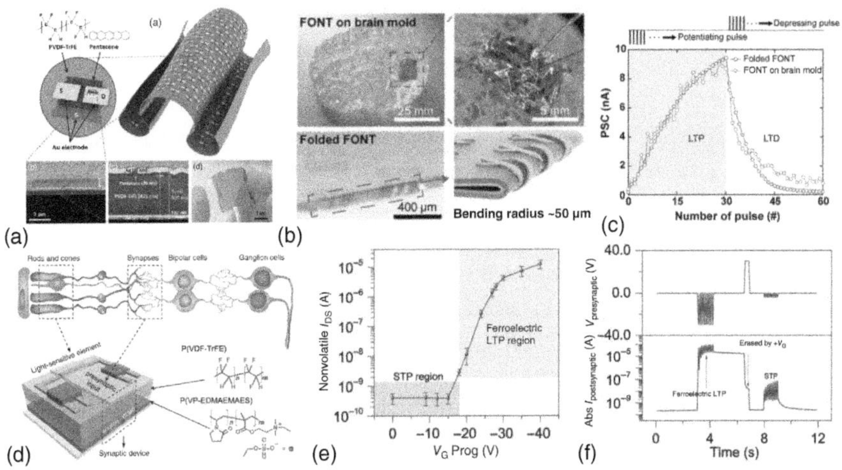

Figure 5.7 (a) Device structure of the ultrathin conformable organic synaptic transistors with a freestanding P(VDF-TrFE) film as the dielectric layer. (b) Photo images and (c) long-term potentiation and long-term depression characteristics of the FONT under two harsh bending conditions: on the surface of brain-shaped PDMS mold (upper panel) and under an extremely bending condition with a bending radius of 50 μm (lower panel). Source: Jang et al. [40]. (d) Schematic illustration of a retina's multilayer structure. (e) The synaptic transistor's nonvolatile current dependence on programming gate voltage (pulse width, 100 ms). (f) Triggering and erasing of ferroelectric LTP were performed by −30 and 30 V V_G pulses. Triggering of STP was conducted by −4 V V_G pulses. Source: Wang et al. [41].

sulfate)] [P(VP-EDMAEMAES)] acting as the electrochemical layer and P(VDF-TrFE) acting as the ferroelectric layer. Besides, poly(isoindigo-*co*-bithiophene) [P(IID-BT)], a p-type copolymer, was served as the channel material, and the photosensitive ingredients on the top of the device with the series connection consisted of an organic heterojunction and a phthalocyanine load resistor. By modulating the gate bias, both STP and ferroelectric LTP were revealed (Figure 5.7e). As 10 higher V_G pulses (−30 V) were applied, the synaptic transistor exhibited ferroelectric LTP. By adding lower V_G pulses (−4 V), STP was triggered. V_G pulse of 30 V was utilized to reset the device to off state between the above two operations (Figure 5.7f). Furthermore, applying distinct voltages over the ferroelectric LTP threshold voltage approximately from −18 to 40 V, nonvolatile currents showed a substantial increase of ~10^4 times. Interestingly, electrochemical LTP could also be achieved by applying low-amplitude but high-frequency electric pulses (−2 V, 32 Hz), which led to a postsynaptic current maintaining 10% higher than the residual current after 100 seconds. In addition, the researchers constructed an ultra-flexible light-stimulated organic neuromorphic conformal array through a water-assisted peeling-off process to recognize colors and to simulate the function of retina. This conformal device both mimicked important typical synaptic functions and emulated the simple properties of retina, which accelerated the further development in fully biological-compatible neuromorphic devices.

5.3.4 Optoelectronic Transistors

In spite of remarkable progresses, the latest electrical-stimulated artificial synaptic devices still face certain apparent limitations, such as large interconnection energy loss, limited bandwidth, etc. Recently, optical control of synaptic weight has been proposed that can give the synaptic devices the performance of ultrafast signal transmission, large bandwidth, and low interconnection energy loss, which is helpful to build novel artificial neural network architectures. Besides, optoelectronic synaptic devices supply a noncontact writing supply that is conducive to the development of optical wireless communication. More significantly, optoelectronic artificial synapses are suitable to simulate functions of retinal neurons, thus paving a way between brain computing and visual systems.

Dai et al. reported a facile processing method for building an organic phototransistor to mimic synaptic functions (Figure 5.8a) [42]. The researchers strictly chose polyacrylonitrile (PAN) film as the gate dielectric and 2,7-dioctyl[1]benzothieno [3,2-*b*][1]benzothiophene (C8-BTBT) as the channel material. Important synapse-like behaviors, including EPSC, PPF, and learning behaviors, have been well-emulated, contributed to the charge trapping effect around or between the PAN/C8-BTBT interfaces instead of using the distinctive photosensitive characteristic of the organic semiconductor. Finally, learning and forgetting behaviors were explored via a T-shaped array consisting of 20 transistors fabricated without the silicon wafer substrate. Hence, during the learning process, after 20 light pulses, synaptic weights of major devices increased to a higher level, and the memory levels decreased quickly in the first stage, followed by a little slow decrease. Therefore, the operation principle of simulating the synaptic behaviors was further identified to the charge trapping effect around or between the PAN/C8-BTBT interfaces.

Moreover, using the outstanding light absorption characteristic of perovskite materials to prepare optoelectronic synaptic transistors has also been attracted considerable attention. Wang et al. reported organic photonic synaptic transistors prepared by using inorganic halide perovskite quantum dots (QDs) (Figure 5.8b) [43]. The heterojunction formed between the $CsPbBr_3$ QDs layer and the organic semiconductor layer afforded the device the foundation for optically programmable and electrically erasable properties (Figure 5.8c). During the programming process, photoinduced holes in the $CsPbBr_3$ QDs were readily transferred to pentacene via a bending of energy band, and the electrons would be retained in the $CsPbBr_3$ QDs. The $CsPbBr_3$ QDs with trapped electrons could serve as an additional control gate that would further speed up the injection of holes into the semiconductor layer. Because of the presence of potential well, the capture electrons were left in $CsPbBr_3$ QDs for quite a long time, contributing to the photonic memory property. An enough negative gate bias pulse was required to completely erase the memory effect. Based on this photonic synaptic transistor, synaptic behaviors, including EPSC and PPF, were simulated. Besides, these synaptic functions could be effectively modulated by adjusting the parameters of light pulses, such as pulse duration, pulse intensity, and pulse wavelength. Furthermore, the transition from PPF to PPD was mimicked

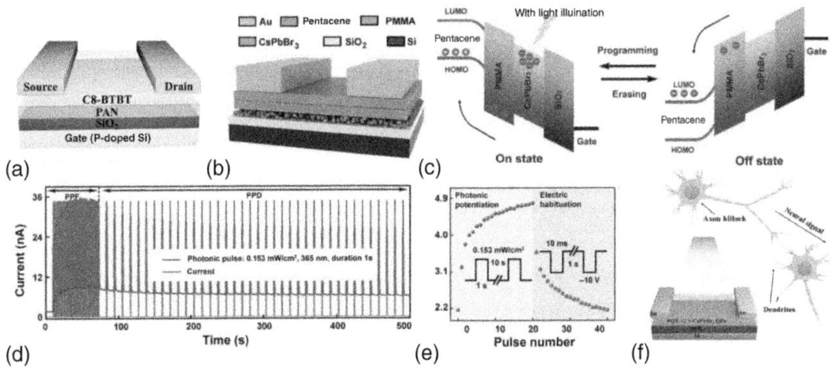

Figure 5.8 (a) Device structure of the light-stimulated organic synaptic transistors. Source: Dai et al. [42]. (b) Device structure of a photonic synapse based on inorganic halide perovskite quantum dots. (c) Schematic energy diagram of the device during light programming operation and during electrical erasing operation under dark condition. (d) Experimental demonstration of PPD following PPF in the memory device. (e) Gradual channel current modulation of the synaptic devices subjected to a train photonic pulses (365 nm, 0.153 mW/cm^2, one second duration with 10 second interval) and negative electrical pulses (−20 V, 10 ms duration with one second interval). Source: Wang et al. [43]. (f) Schematic image illustrating the neural signal transmission process through biological synapses in neuron and simplified device structure of the $CsPbBr_3$ QD/PQT-12 hybrid synaptic transistors. Source: Wang et al. [44].

by modulating the time intervals of light pulses (Figure 5.8d). The synaptic plasticity with nonvolatile performance was emulated through the synaptic transistor under optical/electric stimuli. The functions of stable synaptic potentiation and habituation were also achieved, as illustrated in Figure 5.8e. Recently, Wang et al. explored an effective and facile solution process to construct optoelectronic synaptic transistors by combining inorganic halide perovskite quantum dots (IHP QDs) with organic semiconductors (Figure 5.8f) [44]. This work demonstrated that organic semiconductor poly(3,3-didodecylquarterthiophene) (PQT-12) was mixed with $CsPbBr_3$ QDs. The energy band of PQT-12 and $CsPbBr_3$ QDs matched well, which was beneficial for the effective separation of photoinduced charges at the interface between PQT-12 and $CsPbBr_3$ QDs. Hence, the synaptic transistors showed a high photoresponsivity. More importantly, the photocurrent could be maintained for a long time after the light illumination was removed, which was caused by the charge trapped in the PQT-12 film. The high photosensitivity and slow decay of the photocurrent were the key to the achievement of high-efficiency optoelectronic synaptic transistors. On the basis of these devices, typical synapse-like behaviors, such as EPSC, PPF, and high-pass filter characteristic, have been successfully simulated.

5.4 Innovative Applications of Organic Neuromorphic Devices for Bionic Perception Systems

Recently, aside from imitating basic synaptic behaviors and adjusting synaptic plasticity, organic neuromorphic devices have also been utilized to construct bionic perception systems. Sensory nervous systems can detect external stimuli, such as pressure, light, chemicals, and temperature, from sensory receptors, transmitting sensory signals to the brain [45]. Bionic perception systems, which integrate bionic perception organs and artificial synapses, recognize stimuli and then produce postsynaptic responses through the same method like biological sensory receptors and synapses. Synapse-like responses are then transmitted to the brain or operative organs to develop learning and cognitive functions and sensory systems of future electronics. In this section, we will make an introduction to two main bionic perception systems: artificial visualization systems and tactile-perception systems. Bionic perception systems can be configured in certain methods, including direct stimulation of sensitive organic neuromorphic devices and connecting external sensors and electronic components to organic neuromorphic devices. In the later configuration, presynaptic electrical signals from the stimulated sensors are transferred to neuromorphic devices, which then contribute to postsynaptic responses.

5.4.1 Artificial Visualization Systems

In human retina, vision transforms intensity and λ information into nerve impulses, which are transmitted to the brain through the optic nerve for interpretation. Light cognition is one of the most important sensory characteristics for artificial visualization systems. Researchers have delicately used light-sensitive components combining with neuromorphic devices to explore a variety of optoelectronic synapses for bionic perception systems [46]. Therefore, the research on artificial visualization systems is also growing explosively. So far, optoelectronic synapses have two main innovative applications in artificial visualization systems: (i) simulating the characteristics of human retina, such as discerning numbers and letters in the database of the Modified National Institute of Standards and Technology (MNIST), recognizing various colors and color-mixed patterns, and achieving the image preprocessing function and (ii) integrating with motion systems to simulate the feeling and movement human functions.

To mimic the functions of human retina, Ham et al. used MNIST images to study the learning efficiency of organolead halide perovskite optoelectronic artificial synapses (Figure 5.9a) [47]. After 2000 times of training, the recognition accuracy of low energy consumption synaptic device had reached 82.7%, which paved the way for improving the learning and recognition ability of visualization systems. In addition, Lv et al. employed MNIST pattern to explore the functions of learning and recognition of a single-layer perceptron network composed of 785 input neurons and 10 output neurons, which was based on carbon dot/silk-based organic synaptic transistors (Figure 5.9b) [48]. Through 15 000 times of training, the recognition rates of the letters “A” and “I” reached about 73% and 65%, respectively.

In biomimetic electronics, it is of great significance to emulate perception and motion characteristics of human. Lee et al. demonstrated an organic optoelectronic sensorimotor system by combining an organic optoelectronic synaptic device with a stretchable organic nanowire synaptic transistor (s-ONWST) [49]. The key of this system was the s-ONWST having the ability to sense and transmit the signals of

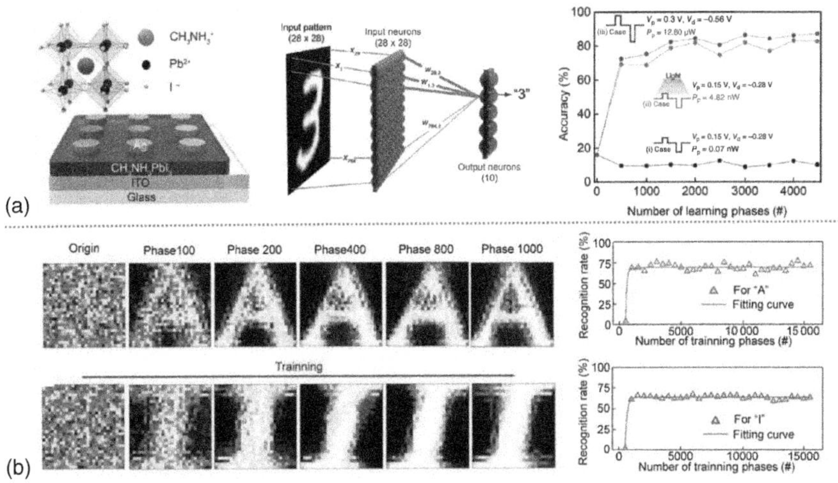

Figure 5.9 (a) Schematics of the organolead halide perovskite synaptic device and estimated recognition accuracy for the MNIST patterns. Source: Ham et al. [47]. (b) Mapping images and recognition rate of "A" and "I" with the evolution of training phases. Source: Lv et al. [48].

input light illumination and then generate the followed motor outputs. Using International Morse code as the light inputs, output information was generated via the optoelectronic synaptic device, which was a novel optical wireless communication for human–machine interfaces. "SOS," the typical emergency signal, was regarded as light input signal to trigger EPSC response on the condition of 100% strain, which was nearly similar to the EPSC response triggered under 0% strain. In addition to visible illumination, distinct wavelengths of light, including ultraviolet and infrared, were also utilized to be the input signals. The results showed that distinctive EPSC responses were obtained showing that optoelectronic artificial synapses could be used in broad-spectrum absorption. Furthermore, as light pulses were exerted continuously, EPSC responses obtained from the optoelectronic synaptic device were transferred to the artificial actuator. As the pulse increased, the displacement δ and the output voltage also increased. More significantly, the actuator can work stably even under 100% strain. The organic optoelectronic sensorimotor system provides a creative method for the further progress of soft robotics and neurorobotics.

5.4.2 Tactile-Perception Systems

In human bodies, touch can be sensed by receptors on sensory neurons in the skin, and then, the touch signals are sent to axons and synapses. Afterward, the postsynaptic neurons will further process them. Zang et al. simulated the function of tactile-perception systems by assembling pressure sensitive organic transistor and organic synaptic transistor (Figure 5.10a) [50]. In the tactile-perception system, the capacitance of the insulator in the pressure sensitive organic transistor varies with the change of the applied pressure, which also results in the change of the output current of the synaptic transistor. The response is consistent with the signal

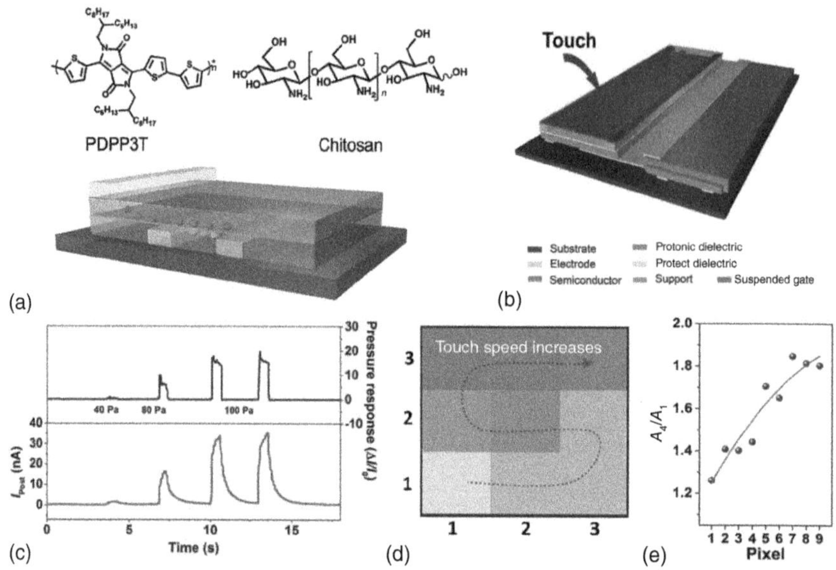

Figure 5.10 (a) The proposed organic synaptic transistor. (b) Schematics of the tactile-perception system. (c) The relative changes in current in the sensing device and the I_{post} responses of the synaptic transistor under different pressures. (d) Schematic illustrations of the tactile-perception functionality of the dual-organic-transistor-based tactile-perception array. (e) The gain A_4/A_1 for each pixel. Source: Zang et al. [50].

perception, transduction, and processing functions of human touch. With the increase of the tactile stimuli (contact duration, intensity, and number of the stress stimuli), the synaptic weight was enhanced (Figure 5.10b,c). In addition, a 3×3 pixel array of synaptic transistors was exploited to emulate the touch perception of position-resolved dynamic mechanical contact (Figure 5.10d). With the increase of repetitive touch frequency, synaptic potentiation led to the increase of EPSC gain (A_4/A_1) (Figure 5.10e). The tactile-perception system could utilize pressure to encode and detect information.

A hybrid reflex arc gained by linking the flexible artificial afferent nerve (AAN) to the efferent nerve is able to control the biological motor nerve. Kim et al. mimicked the characteristic of slowly fitting type-I sensory nerves by designing an AAN system including organic ring oscillators, resistive pressure sensors, and synaptic transistors [51]. This system could gather the information of pressure signal via a variety of tactile sensors and then transfer it through motor nerves. A pyramid elastomer by using carbon nanotubes formed a pressure sensor that was capable of biological sensitivity and operating range. The frequency of the ring oscillator matched that of the action potential of the biological sensory. By linking the ring oscillators to the synaptic transistors, the output voltages of the ring oscillators stimulated the synaptic transistors and then produced synaptic currents. By the movement of a sticklike object in different orientations on the pressure sensors of the system, the moving of recognition was emulated. The response analysis of the output current of synaptic transistors in the AAN demonstrated the moving orientation of the object over the pressure sensor.

Braille characters were recognized by building a pixel array on the AAN system. Due to the integration of synaptic signals, the minimum Victor–Purpura distance between various characters increased, which improved the resolution of Braille letters. A hybrid reflex arc showed that the information introduced from the pressure sensor transmitted a postsynaptic signal to the efferent nerve of a cockroach leg, thus activating leg muscles. In this system, when the frequency, duration, and intensity of stimuli of the AAN changed, there would be remarkable different muscle responses. The hybrid reflex system demonstrated its enormous potential applications in neuroprosthetics and neurorobotics.

5.5 Summary and Outlook

In this chapter, we reviewed the recent studies in organic neuromorphic devices, including both two-terminal memristors and three-terminal transistors, and bionic perception systems based on organic synapses. Although great advances have been made in organic neuromorphic devices, certain main problems and challenges remain unsolved. Many efforts have been done on the design of physical organic synaptic devices, which can emulate biological synapses at similar energy levels. At present, scientists have a relatively simple comprehending of brain functions and mechanisms, so it is still an enormous challenge for the implementation of "brain-like computing." With the joint efforts of scientists in materials science, biochemistry, physics, electronics, and other related fields, new applications of organic neuromorphic devices can be further developed and investigated. The system-level research and learning mechanism of neuromorphic system based on organic synaptic devices with STDP/SRDP rules have not been sufficiently studied and still need to be paid great attention to by the scientists. The utilization of organic materials needs to be improved in many aspects, including transmission, potential switching, and conducting path formation mechanisms of organic neuromorphic devices, which have not been expounded completely. In addition, the organic neuromorphic devices also suffer from aging-related performance problems, and the synaptic performance may not be reliable. These organic devices usually have a relatively high-power consumption. Most of previously reported researches in the field of organic neuromorphic devices have focused on the implementation of basic synapse-like functions, regardless of these performance metrics. Future studies should further evaluate and optimize the performance of organic neuromorphic devices.

Bionic perception system based on organic neuromorphic devices is a relatively new multidisciplinary research field. The rise of bionic perception system may also be suitable for the construction of coexisting–cooperative–cognitive (tri-co) neuromorphic systems. Presently, bionic perception system is in the primary stage of imitating peripheral nervous system. Efforts should be made to develop versatile and sophisticated bionic perception systems in the future. It is believed that the application of organic neuromorphic devices in neuromorphology and bioelectronics will continue to be valued. The enhancement of the organic synapses will contribute to the future development of next-generation soft electronics.

References

1 Lee, Y. and Lee, T.W. (2019). Organic synapses for neuromorphic electronics: from brain-inspired computing to sensorimotor nervetronics. *Acc. Chem. Res.* 52 (4): 964–974.

2 van de Burgt, Y., Melianas, A., Keene, S.T. et al. (2018). Organic electronics for neuromorphic computing. *Nat. Electron.* 1 (7): 386–397.

3 Sun, J., Fu, Y., and Wan, Q. (2018). Organic synaptic devices for neuromorphic systems. *J. Phys. D: Appl. Phys.* 51 (31): 314004.

4 Chen, Y., Liu, G., Wang, C. et al. (2014). Polymer memristor for information storage and neuromorphic applications. *Mater. Horizons.* 1 (5): 489.

5 Kim, K.M., Zhang, J.M., Graves, C. et al. (2016). Low-power, self-rectifying, and forming-free memristor with an asymmetric programing voltage for a high-density crossbar application. *Nano. Lett.* 16 (11): 6724–6732.

6 Kim, S.G., Han, J.S., Kim, H. et al. (2018). Recent advances in memristive materials for artificial synapses. *Adv. Mater. Technol.* 3 (12): 1800457.

7 Lv, Z.Y., Zhou, Y., Han, S.T., and Roy, V.A.L. (2018). From biomaterial-based data storage to bio-inspired artificial synapse. *Mater. Today.* 21 (5): 537–552.

8 Waser, R., Dittmann, R., Staikov, G., and Szot, K. (2009). Redox-based resistive switching memories - nanoionic mechanisms, prospects, and challenges. *Adv. Mater.* 21 (25–26): 2632–2663.

9 Li, S., Zeng, F., Chen, C. et al. (2013). Synaptic plasticity and learning behaviours mimicked through Ag interface movement in an Ag/conducting polymer/Ta memristive system. *J. Mater. Chem. C.* 1 (34): 5292.

10 Kim, M.K. and Lee, J.S. (2018). Short-term plasticity and long-term potentiation in artificial biosynapses with diffusive dynamics. *ACS Nano.* 12 (2): 1680–1687.

11 Yoo, E., Lyu, M., Yun, J.-H. et al. (2016). Bifunctional resistive switching behavior in an organolead halide perovskite based Ag/$CH_3NH_3PbI_3$–xClx/FTO structure. *J. Mater. Chem. C.* 4 (33): 7824–7830.

12 Wang, Z., Zeng, F., Yang, J. et al. (2011). Resistive switching induced by metallic filaments formation through poly(3,4-ethylene-dioxythiophene): poly(styrenesulfonate). *ACS Appl. Mater. Interfaces.* 4 (1): 447–453.

13 Yuan, Y. and Huang, J. (2016). Ion migration in organometal trihalide perovskite and its impact on photovoltaic efficiency and stability. *Acc. Chem. Res.* 49 (2): 286–293.

14 Choi, J., Le, Q.V., Hong, K. et al. (2017). Enhanced endurance organolead halide perovskite resistive switching memories operable under an extremely low bending radius. *ACS Appl. Mater. Interf.* 9 (36): 30764–30771.

15 Walsh, A. (2015). Principles of chemical bonding and band gap engineering in hybrid organic-inorganic halide perovskites. *J. Phys. Chem. C.* 119 (11): 5755–5760.

16 Xing, G.C., Mathews, N., Sun, S.Y. et al. (2013). Long-range balanced electron- and hole-transport lengths in organic-inorganic $CH_3NH_3PbI_3$. *Science* 342 (6156): 344–347.

17 Liu, G., Wang, C., Zhang, W. et al. (2016). Organic biomimicking memristor for information storage and processing applications. *Adv. Electron. Mater.* 2 (2): 1500298.

18 Yang, X., Wang, C., Shang, J. et al. (2016). An organic terpyridyl-iron polymer based memristor for synaptic plasticity and learning behavior simulation. *RSC Adv.* 6 (30): 25179–25184.

19 Wu, C., Kim, T.W., Choi, H.Y. et al. (2017). Flexible three-dimensional artificial synapse networks with correlated learning and trainable memory capability. *Nat. Commun.* 8 (1): 752.

20 Xu, W., Cho, H., Kim, Y.H. et al. (2016). Organometal halide perovskite artificial synapses. *Adv. Mater.* 28 (28): 5916–5922.

21 Xiao, Z.G. and Huang, J.S. (2016). Energy-efficient hybrid perovskite memristors and synaptic devices. *Adv. Electron. Mater.* 2 (7): 8.

22 Tian, H., Zhao, L., Wang, X. et al. (2017). Extremely low operating current resistive memory based on exfoliated 2D perovskite single crystals for neuromorphic computing. *ACS Nano.* 11 (12): 12247–12256.

23 Wang, L., Wang, Z., Lin, J. et al. (2016). Long-term homeostatic properties complementary to Hebbian rules in CuPc-based multifunctional memristor. *Sci. Rep.* 6: 35273.

24 Wang, L., Yang, J., Zhu, Y. et al. (2017). Rectification-regulated memristive characteristics in electron-type CuPc-based element for electrical synapse. *Adv. Electron. Mater.* 3 (7): 1700063.

25 Choi, H.Y., Wu, C., Bok, C.H., and Kim, T.W. (2017). Organic electronic synapses with pinched hystereses based on graphene quantum-dot nanocomposites. *NPG Asia Mater.* 9 (7): e413.

26 Dai, S., Zhao, Y., Wang, Y. et al. (2019). Recent advances in transistor-based artificial synapses. *Adv. Funct. Mater.* 29: 1903700.

27 Ren, Y., Yang, J.Q., Zhou, L. et al. (2018). Gate-tunable synaptic plasticity through controlled polarity of charge trapping in fullerene composites. *Adv. Funct. Mater.* 28 (50): 1805599.

28 Alibart, F., Pleutin, S., Guérin, D. et al. (2010). An organic nanoparticle transistor behaving as a biological spiking synapse. *Adv. Funct. Mater.* 20 (2): 330–337.

29 Gkoupidenis, P., Koutsouras, D.A., and Malliaras, G.G. (2017). Neuromorphic device architectures with global connectivity through electrolyte gating. *Nat. Commun.* 8: 15448.

30 Gkoupidenis, P., Schaefer, N., Garlan, B., and Malliaras, G.G. (2015). Neuromorphic functions in PEDOT:PSS organic electrochemical transistors. *Adv. Mater.* 27 (44): 7176–7180.

31 Koutsouras, D.A., Malliaras, G.G., and Gkoupidenis, P. (2018). Emulating homeoplasticity phenomena with organic electrochemical devices. *MRS Commun.* 8 (02): 493–497.

32 Gkoupidenis, P., Rezaei-Mazinani, S., Proctor, C.M. et al. (2016). Orientation selectivity with organic photodetectors and an organic electrochemical transistor. *Aip Adv.* 6 (11): 6.

33 Gkoupidenis, P., Schaefer, N., Strakosas, X. et al. (2015). Synaptic plasticity functions in an organic electrochemical transistor. *Appl. Phys. Lett.* 107 (26): 5.

34 Fu, Y., Kong, L.-a., Chen, Y. et al. (2018). Flexible neuromorphic architectures based on self-supported multiterminal organic transistors. *ACS Appl. Mater. Interfaces.* 10 (31): 26443–26450.

35 Dai, S., Wang, Y., Zhang, J. et al. (2018). Wood-derived nanopaper dielectrics for organic synaptic transistors. *ACS Appl. Mater. Interfaces.* 10 (46): 39983–39991.

36 Xu, W., Min, S.Y., Hwang, H., and Lee, T.W. (2016). Organic core-sheath nanowire artificial synapses with femtojoule energy consumption. *Sci. Adv.* 2 (6): 7.

37 Hoffman, J., Pan, X., Reiner, J.W. et al. (2010). Ferroelectric field effect transistors for memory applications. *Adv. Mater.* 22 (26–27): 2957–2961.

38 Naber, R.C., Asadi, K., Blom, P.W. et al. (2010). Organic nonvolatile memory devices based on ferroelectricity. *Adv. Mater.* 22 (9): 933–945.

39 Miller, S.L. and McWhorter, P.J. (1992). Physics of the ferroelectric nonvolatile memory field-effect transistor. *J. Appl. Phys.* 72 (12): 5999–6010.

40 Jang, S., Jang, S., Lee, E.-H. et al. (2018). Ultrathin conformable organic artificial synapse for wearable intelligent device applications. *ACS Appl. Mater. Interfaces.* 11 (1): 1071–1080.

41 Wang, H., Zhao, Q., Ni, Z. et al. (2018). A ferroelectric/electrochemical modulated organic synapse for ultraflexible artificial visual-perception system. *Adv. Mater.* 30 (46): e1803961.

42 Dai, S., Wu, X., Liu, D. et al. (2018). Light-stimulated synaptic devices utilizing interfacial effect of organic field-effect transistors. *ACS Appl. Mater. Interfaces.* 10 (25): 21472–21480.

43 Wang, Y., Lv, Z., Chen, J. et al. (2018). Photonic synapses based on inorganic perovskite quantum dots for neuromorphic computing. *Adv. Mater.* 30 (38): e1802883.

44 Wang, K., Dai, S., Zhao, Y. et al. (2019). Light-stimulated synaptic transistors fabricated by a facile solution process based on inorganic perovskite quantum dots and organic semiconductors. *Small* 15 (11): e1900010.

45 Han, H., Yu, H., Wei, H. et al. (2019). Recent progress in three-terminal artificial synapses: from device to system. *Small* 15: e1900695.

46 Zhang, J., Dai, S., Zhao, Y. et al. (2020). Recent progress in photonic synapses for neuromorphic systems. *Adv. Intell. Syst.* 2: 1900136.

47 Ham, S., Choi, S., Cho, H. et al. (2019). Photonic organolead halide perovskite artificial synapse capable of accelerated learning at low power inspired by dopamine-facilitated synaptic activity. *Adv. Funct. Mater.* 29 (5): 1806646.

48 Lv, Z., Chen, M., Qian, F. et al. (2019). Mimicking neuroplasticity in a hybrid biopolymer transistor by dual modes modulation. *Adv. Funct. Mater.* 29: 1902374.

49 Lee, Y., Oh, J.Y., Xu, W. et al. (2018). Stretchable organic optoelectronic sensorimotor synapse. *Sci. Adv.* 4 (11): eaat7387.

50 Zang, Y., Shen, H., Huang, D. et al. (2019). A Dual-organic-transistor-based tactile-perception system with signal-processing functionality. *Adv. Mater.* 29 (18): 1606088.

51 Kim, Y., Chortos, A., Xu, W.T. et al. (2018). A bioinspired flexible organic artificial afferent nerve. *Science* 360 (6392): 998–1003.

6

Neuromorphic Computing Systems with Emerging Devices

Qiumeng Wei[1], Jianshi Tang[1,2], Bin Gao[1,2], Xinyi Li[1], He Qian[1,2], and Huaqiang Wu[1,2]

[1] *Tsinghua University, School of Integrated Circuits, Beijing National Research Center for Information Science and Technology (BNRist), Beijing, 100084, P. R. China*
[2] *Tsinghua University, Beijing Innovation Center for Future Chips (ICFC), Beijing, 100084, P. R. China*

6.1 Introduction

6.1.1 Background Introduction

The origins of artificial neural networks derive from the efforts to model information processing in biological systems by abstracting the behavioral functions of biological neurons and synapses and mimicking the hierarchical network structures in the biological nervous system [1]. Early in the development of the neural network concept, attempts were made to replicate the learning abilities of the human brain by directly applying certain behaviors of neural units observed in neuroscience researches to networks [2–4]. With Werbos's backpropagation algorithm succeeding in the training of multilayer networks [5, 6], more practical methods based on statistics and information theory have been explored to train different neural networks. To cope with more complex tasks, diverse network structures and training algorithms have been developed, such as fully connected networks (FCNs), convolutional neural networks (CNNs), recurrent neural networks (RNNs), and deep belief networks (DBNs). These networks are collectively referred to as deep neural networks (DNNs) [7] in which synapses are simplified as connections with tunable strengths (i.e. trainable weights), while neurons integrate weighted synaptic inputs and then output activation results. Spiking neural networks (SNNs) [8, 9], as another type of network that more closely resembles biological neural networks (BioNNs), are expected to further improve the energy efficiency due to the spike-based information encoding and event-driven activities. The introduction of bionic features is expected to provide novel solutions for the implementation of artificial intelligence.

Figure 6.1 illustrates several representative network architectures of DNNs and SNNs, along with their manners of information processing. As the size of neural networks scales up, the drastically increasing computation burden on the existing digital computing platforms leads to the significant latency and power

Neuromorphic Devices for Brain-Inspired Computing: Artificial Intelligence, Perception, and Robotics, First Edition. Edited by Qing Wan and Yi Shi.

consumption. Therefore, it is urgent to develop new hardware systems with high computing power and energy efficiency for the implementation of neural networks, especially in resource-limited application scenarios, such as automated driving.

6.1.2 The Motivation for Neuromorphic Computing

Nowadays, graphics processing units (GPUs), tensor processing units (TPUs), and application-specific integrated circuits (ASICs) have been widely used for the hardware acceleration of neural networks. However, their energy efficiency still falls behind that of the human brain, which operates on about 20 W [10]. The concept of neuromorphic computing was first developed by Carver Mead [11], referring to an integration system composed of analog circuit modules to emulate the architecture and behaviors of BioNNs. It is desired that by drawing on the mechanisms of the biological nervous system, energy efficiency and performance could be improved.

There are many distinct features of the human brain different from digital computers. For example, the memory and computing regions in the brain are fully entangled, where information is stored in the form of synaptic connections and processed through hierarchical propagation and neuronal activities. It should be noted that this is fundamentally different from the von Neumann architecture adopted by modern computers, where memory and computing units are physically separated. Inspired by the human brain, computing-in-memory (CIM) hardware has been developed to accelerate the neural network computing by eliminating the frequent data shuffling between memory and computing units. In addition, resulting from various plastic behaviors, the brain features continuous self-adaptation in response to external stimuli. There is no clear boundary between training and inference phases in the brain. Therefore, integrating bionic functions at the circuit level may contribute to the replication of intelligent behaviors and achieve a high power efficiency comparable with the brain.

6.1.3 Progress and Challenges of CMOS-Based Neuromorphic Computing

So far, a variety of CMOS-based neuromorphic computing hardware has been developed and evaluated comprehensively from different perspectives. For example, Zhang et al. reviewed in detail four key metrics for neuro-inspired computing hardware [12], including the computing density, energy efficiency, computing accuracy, and learning capability. Here, the specific metrics that will be used for neuromorphic hardware benchmarks are described as follows:

(1) *Energy efficiency*: it is usually evaluated by the number of operations per second under the power of 1 W (OPS/W). Another widely used metric for energy efficiency in neural network systems is MACS/W, describing the number of multiply-and-accumulate (MAC) operations per watt (1MAC = 2 OP s). Higher OPS/W means that the system is more efficient in terms of energy utilization to perform computations.

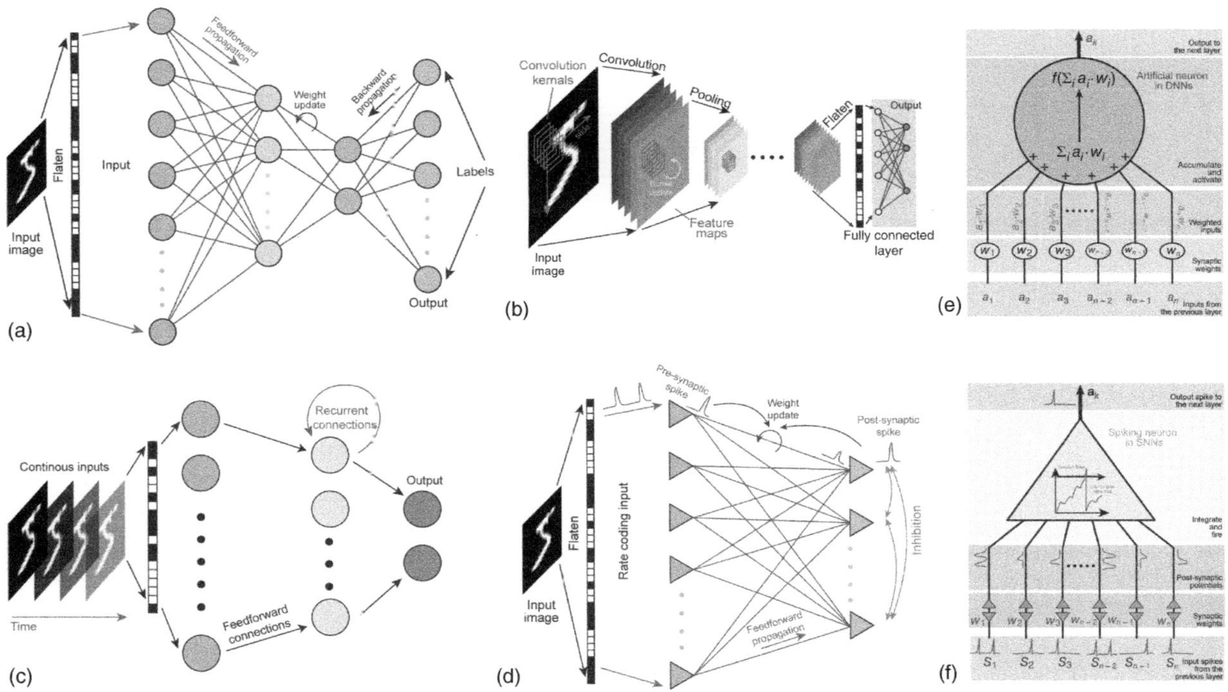

Figure 6.1 Structures of typical neural networks. (a) The structure of an FCN, in which the activation result of each neuron in one layer is transmitted to all the neurons in the next layer. Synaptic weights are updated according to the backpropagated errors and inputs in each layer. (b) A CNN usually consists of convolutional layers, pooling layers, and fully connected layer as an output layer. Convolutional kernels slide through the whole image to generate feature maps. (c) The structure of RNN, in which the outputs of hidden neurons are recurrently connected to serve as new inputs. (d) A spike-based FCN, where synaptic weight update is determined by the pre- and post-synaptic spikes. A lateral inhibition layer is added to induce competition between output neurons. (e) Illustration of an artificial neuron in DNNs, which accumulates the weighted synaptic inputs and outputs, the activated results as the input to the next neuron. (f) A spiking neuron in SNNs, which integrates post-synaptic potentials that are modulated by the connected synapses and generates a spike once the neuron is activated.

(2) *Computing power*: tera/giga operations per second (T/GOPS) is a standard performance metric to evaluate the peak computing power. Some works also use ns/MAC as the metric to evaluate the latency caused by an MAC operation.
(3) *Computing precision*: to improve energy efficiency, computations in neuromorphic hardware are usually completed with fixed-point arithmetic, rather than float-point arithmetic in conventional computers. Thus, the precision is expressed by the number of bits required to represent input/output data and store weights.
(4) *Accuracy*: it evaluates how well the system performs in a certain task. The accuracy is, on the one hand, affected by the adopted algorithms and datasets. On the other hand, it depends on the computing precision of hardware.

In the CMOS implementation of neuromorphic systems, conventional memories like static random-access memory (SRAM), dynamic random-access memory (DRAM), and flash memory have been used. They are viewed as charge-based memory [13] because 0 or 1 is represented by the number of charges stored in a memory cell. By controlling the charge redistributions among memory cells, matrix–vector multiplications (MVMs), the most intensive computing operations in DNNs, could be accelerated based on memory arrays. Besides DNNs, CMOS-based SNNs have also been implemented for neuromorphic computing, which replicate the synaptic plasticity behaviors, dynamics of neurons, and information transmission in networks. Many of those computing platforms have adopted mixed-signal circuit designs by utilizing the exponential I–V characteristic of transistors working in the sub-threshold regime, such as Stanford's Neurogrid [14], Braindrop [15], and UZH's ROLLS [16]. Given that sub-threshold circuits are susceptible to voltage fluctuations, fully digital neuromorphic hardware of SNNs has also been developed with higher robustness to noise and signal distortions, like IBM's TrueNorth [17], Intel's Loihi [18], and ZJU's Darwin [19]. Furthermore, digital hardware could adopt more advanced transistor technology to decrease leaky currents and improve integration density. However, a large number of transistors are usually required whether in digital and analog neuromorphic hardware. In the post-Moore era, adverse effects caused by the transistor scaling challenge the further scalability of CMOS-based neuromorphic computing.

6.1.4 Principles of Memristors

To address the above challenges faced by CMOS-based hardware implementations, emerging devices could provide a new pathway to realize high-efficiency and scalable neuromorphic computing. More recently, resistive switching memory devices, such as magnetic random-access memory (MRAM), phase-change memory (PCM), resistive random-access memory (RRAM), ferroelectric field-effect transistor (FeFET), and electrochemical random-access memory (ECRAM) [20–23], have been extensively explored to implement neuromorphic computing. The structures and working mechanisms of several types of neuromorphic devices are schematically illustrated in Table 6.1. As to be reviewed in Sections 6.2–6.4, by integrating synaptic crossbar arrays and artificial neurons based on these devices, energy-efficient neuromorphic systems are capable of fulfilling various tasks.

Table 6.1 Typical neuromorphic devices.

Devices	Filamentary memristor[a)]	Interfacial memristor	PCM	MRAM	ECRAM	FeFET
Device structures	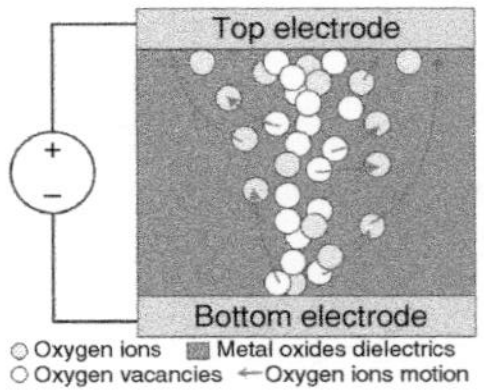	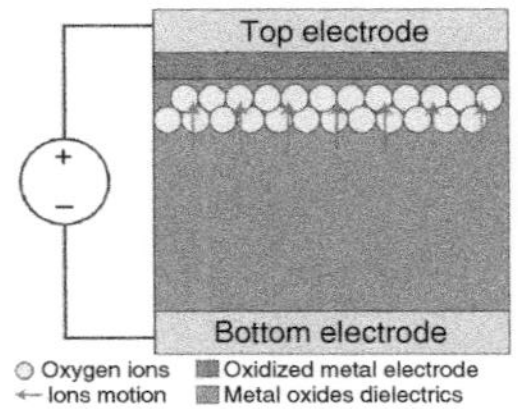	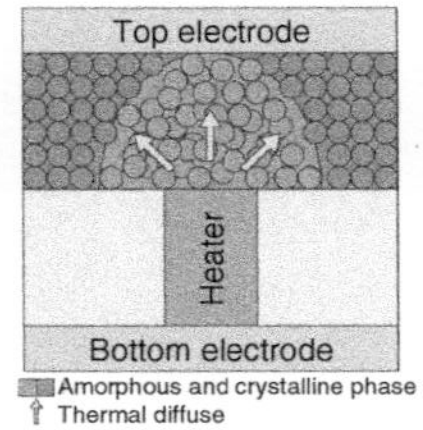	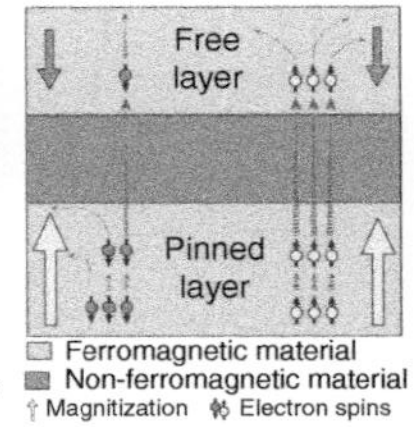	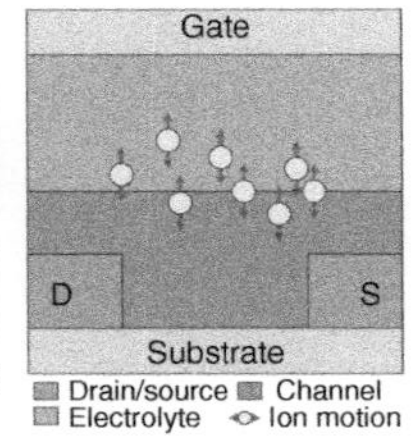	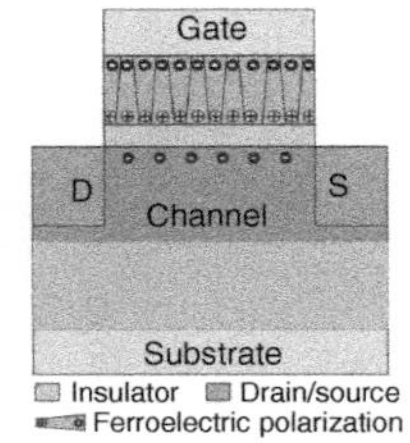
Resistive switching mecha-nisms	Redox reaction-induced conductive filaments	Schottky barrier modulated by oxygen ion motion	Switching between the crystalline phase and the amorphous phase induced by Joule heating	Change in the relative orientation of the magnetization between the free and pinned layers	Gate-driven ion intercalation leading to the electrochemical reaction and channel conductance change	Voltage-induced polarization switching modulates the channel conductivity of the transistor
Materials	TiN/TEL/HfO_x/TiN [24]	PCMO based [25]	$Ge_2Sb_2Te_5$ [26]	CoFeB/MgO/CoFeB [27]	LiPON-WO_3 [23] Metal oxide-WO_3 [28]	TiN/Si : HfO_2/SiON/Si gate stack [29]

a) Take the anionic conductive filament as an example here.

6.2 DNNs Based on Synaptic Devices

By leveraging the resistive switching behavior, memristor crossbar-based CIM hardware is applied in neuromorphic computing for DNNs. As shown in Figure 6.2a, each memory cell located at the intersection of source lines (SLs) and bit lines (BLs) corresponds to a synapse, whose conductance state represents the synaptic weights. With the MVM time complexity of O(1), crossbar arrays mapped with synaptic weights can substantially accelerate the computing of neural networks. However, there are still challenges hindering the performance boost of the synaptic device-based CIM systems and large-scale integration. Here, we will first discuss the performance requirements of the resistive switching devices. Then, the generic structure of a crossbar-based system is reviewed. Array demonstrations and chip implementations will be summarized in Sections 6.2.2 and 6.2.3. Finally, algorithm–architecture co-optimizations concerning the nonidealities of devices and arrays will be discussed.

6.2.1 Device Performance Requirements

To implement high-efficiency neuromorphic computing hardware for DNNs, there are many requirements in terms of device characteristics, such as retention, variations, yield, level numbers, dynamic range, linearity and symmetry of programming, programming voltage, writing speed, endurance, etc. There have been many review articles discussing the performance requirements of devices. Recently, Xi et al. [30] developed a device-to-application simulation system to thoroughly evaluate the influence of different device characteristics on the in situ learning of a multilayer perceptron of $784 \times 200 \times 10$. Moreover, a comprehensive analysis of the reliability of analog resistive switching memory in neuromorphic computing was presented in [31]. Techniques for device optimizations have been summarized in multiple review articles [32–34]. The definitions of several key characteristics are summarized in Table 6.2. Moreover, their effects on the hardware implementations are briefly discussed as follows.

For inference-only hardware, the cost caused by closed-loop programming with the iterative read-verify-write scheme is usually inexpensive. In this case, the requirement on the symmetry and linearity of programming can be loosened. By contrast, device retention is significant and needs to be long enough to support the long-term service of equipment. Level numbers should be as large as possible to alleviate the accuracy loss induced by the quantization of weights. Furthermore, device variations and defects need to be minimized because these nonidealities lead to inaccurate computing. In comparison, hardware systems that support in situ training could have a higher tolerance to device variations and defects because synaptic weights could be continuously readjusted to compensate for the accuracy loss. On the other hand, during the in situ training, synaptic cells are updated frequently, which raises strict requirements for device endurance and programming energy. Consequently, a linear and symmetric relation between the conductance change and the applied stimuli is preferred to facilitate efficient and accurate

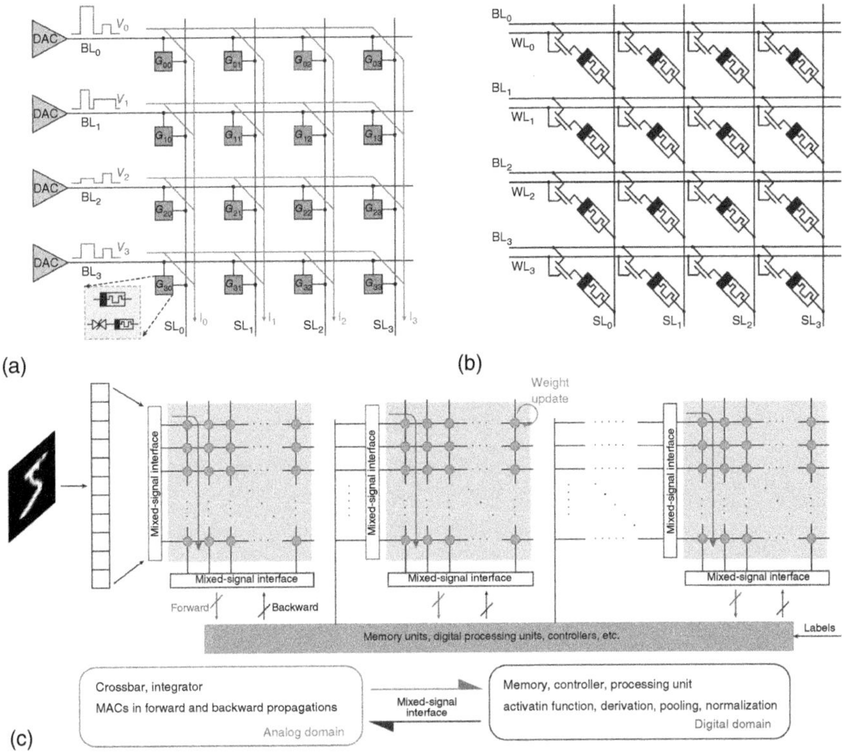

Figure 6.2 Schematic of memristor-based crossbar array for neural network hardware implementations. (a) Crossbar based on 0T1R or 1S1R. Waveforms that encode inputs are applied on BLs, and the sensed summation currents at the end of SLs represent the result of MACs. (b) Crossbar in which the synaptic cells are configured as 1T1R. By controlling the voltages applied on the WLs, transistors are set in different regimes to facilitate the synaptic cell reading and programming. (c) The schematic of a hardware system, which integrates several crossbar arrays, mixed-signal interfaces, and digital modules.

programming. For both in situ training or inference-only hardware, a large dynamic range of devices could enlarge the sense margin of peripheral modules and improve the resilience to device variations, read noise, and mismatch of transistors.

6.2.2 Array Demonstrations

In crossbar arrays, synaptic cells could be configured as one-transistor-one-resistor (1T1R), one-selector-one-resistor (1S1R), or only consist of a passive memristive device (0T1R), as shown in Figure 6.2a,b. In 1T1R or 1S1R, the serially connected transistor or selector device helps suppress the sneak path problem. Mixed-signal peripheral modules, including digital-to-analog converters (DACs), trans-impedance amplifiers (TIAs), and analog-to-digital converters (ADCs), are usually required for array operations such as waveform generation and signal conversion. When programming a 1T1R cell, setting proper compliance currents by controlling the gate voltage facilitates accurate programming. In a passive

Table 6.2 Analog resistive switching device characteristics and their definitions.

Types	Definitions and quantitative indicators
Programming linearity	The linearity of the conductance change under consecutive programming pulses. It is common to model the weight update behavior using an exponential function of the number of applied pules: $G_{SET} = G_0(1 - e^{-\nu p}) + G_{min}$, $G_{RESET} = -G_0(1 - e^{\nu(p-1)}) + G_{max}$ [23], where the level of nonlinearity is indicated by the fitting parameter ν and p is the normalized number of applied pulses (relative to the maximum number of pulses)
Programming symmetry	The resemblance of the conductance change curves in the SET and RESET processes
Level numbers	The number of distinguishable conductance states between maximum and minimum conductance's
Operating voltage and speed	The required amplitude and duration of an applied programming pulse to effectively tune the conductance state
Dynamic range	The ON/OFF ratio between maximum and minimum conductances
Retention	The stability of the device conductance state over time to ensure that the conductance states do not overlap mutually and retain sufficient sense margins
Endurance	In conventional memory applications, the endurance usually refers to the maximum number of full-range SET or RESET operations, where the device is switched between HRS and LRS. While for neuromorphic computing, the endurance concerns the switching cycles between arbitrary two accessible conductance states
Variations	The fluctuations of device parameters, including device-to-device and cycle-to-cycle variations
Yield	The percentage of memory cells in a finished array that function properly

crossbar or 1S1R array, the synaptic cells are usually programmed using by the half or one-third selection scheme [35]. Synaptic cell programming could be optimized on specific occasions. For example, by utilizing the stochastic computing and gradient descent algorithm, Gokmen et al. simplified the weight update at each cross-point to an AND operation and achieved the O(1) time complexity for the update operation in each layer [36]. During the feedforward propagation, information can be encoded as voltage amplitudes [37–40], pulse duration [41–43], or pulse numbers [44, 45], where the first two are preferred to alleviate sensing result deviations caused by the nonlinear I–V characteristic of the synaptic devices. For inference-only systems, A/D interfaces can be replaced by analog activation units to realize the direct transmission of analog signals between adjacent layers [46]. Besides A/D interfaces, digital processors and extra memory units are required for in situ training to calculate activation functions, temporarily store activation results, propagate errors backward, and calculate gradients. An illustration of the complete system integrating crossbars with peripherals is shown in Figure 6.2c.

So far, several experimental demonstrations of DNNs based on synaptic device arrays have been reported [37–40, 43, 44, 46–50], as summarized in Table 6.3. Here, we highlight some of the representative works. At the early stage of research, simple recognition tasks were demonstrated on small-scale memristor crossbars. The first array-level demonstration was made on a 12 × 12 passive memristor array to recognize three simple patterns in 3 × 3 binary images, where the network was trained in situ with the Manhattan update rule [47]. Based on a 128 × 8 1T1R memristor array, a single-layer FCN was trained with in situ weight update and achieved the recognition accuracy of 85% on the Yale Face dataset [44]. Thanks to the technology improvement in device fabrication and integration, larger-size crossbars were adopted to enlarge the scale of neuromorphic computing. By incorporating an external computer, communication modules, and mixed-signal peripherals, in situ training of a two-layer FCN, LeNet-5 network, LSTM network, and reinforcement learning [37–40] have all been demonstrated based on a 128 × 64 1T1R array, where MVMs were computed on the crossbar, and then analog results were sensed, converted, and transmitted to a computer to complete other calculation-expensive operations. Because in situ training can partially alleviate the accuracy degradation caused by the device nonidealities, the in situ trained two-layer FCN based on the 128 × 641T1R crossbar array achieved an accuracy of 91.71% on the widely used Modified National Institute of Standards and Technology (MNIST) dataset even if 11% of devices were stuck during the training.

Introducing robust training algorithms and compensation schemes could enable DNN implementations on larger-scale crossbar arrays without obvious performance degradations. For example, Joshi et al. [50] improved the resilience of the hardware system to the device nonidealities by injecting Gaussian noise to the synaptic weights in the training phase and designed a drift compensation technique to extend the accuracy retention. A hardware system including 1 million PCM devices was constructed to validate the proposed optimization techniques, in which a synaptic cell with a 4-bit precision was composed of two passive PCM devices. After adopting the proposed methodology and compensation strategy, a mapped ResNet achieved an accuracy of 93.7% on CIFAR-10 and 71.6% on ImageNet, with negligible degradation of accuracy after one day.

As discussed above, the variations and stochasticity of memristors are usually viewed as an undesired factor; however, they can also be used to enhance the performance of DNNs. For example, Lin et al. [49] utilized the inherent stochastic characteristics of memristors to generate a Gaussian distribution for each synaptic weight, rather than a deterministic value, to implement a four-layer Bayesian neural network based on a 160 kb 1T1R memristor array. The network showed the anti-attack capability of detecting adversarial images successfully and achieved an accuracy of 97% on the MNIST dataset. Besides, the intrinsic randomness in memristors has also been used in the simulated annealing of systems, which will be discussed in Section 6.4.5.

Table 6.3 Array demonstrations for DNNs.

References	Materials	Array and synapse configurations	Mixed-signal interfaces	Input encoding	Training	Task and accuracy
[50]	$Ge_2Sb_2Te_5$	512 × 512 >16 level PCM passive (0T1R)	8-bit ADC	—	Ex situ	ResNet-32 CIFAR-10: 93.7%
[38]	Pd/HfO_2/Ta	128 × 64 RRAM 1T1R	16-bit DAC 14-bit ADC	Pulse amplitude	In situ	LSTM-RNN
[39]	Pt/TaO_x/Ta	128 × 64 RRAM 1T1R	16-bit DAC 16-bit ADC	Pulse amplitude	In situ	LeNet-5-like MNIST: 92.13%
[40]	Pd/HfO_2/Ta	128 × 64 >16 level RRAM 1T1R	16-bit DAC 16-bit ADC	Pulse amplitude	In situ	Deep-Q Reinforcement Learning
[49]	TiN/TaO_x/HfO_x/TiN	160Kb RRAM 1T1R	—	—	Ex situ	Bayesian Neural Network MNIST: 97%
[37]	Ta/HfO_2/Pt	128 × 64 RRAM 1T1R	12-bit DAC 14-bit ADC	Pulse amplitude	In situ	Two-layer FCN MNIST: 91.71%
[46]	Pt/Al_2O_3/TiO_x/Ti/Pt	Two 20 × 20 RRAM passive (0T1R)	Analog activation	Pulse amplitude	Ex situ	Two-layer FCN Pattern recognition
[48]	TiN/TaO_x/HfO_2/TiN	1 kb RRAM 1T1R	NXOR logic	Binary inputs	Ex situ	Five-layer BNN MNIST: 97.4%
[44]	TiN/TaO_x/$HfAl_yO_x$/TiN	128 × 8 RRAM 1T1R	—	Pulse number	Ex situ	Single-layer FCN Yale face database : 85.04%
[43]	WO_x	32 × 32 RRAM passive (0T1R)	DAC ADC	Pulse width	In situ	Image reconstructions based on sparse coding
[47]	Pt/Al_2O_3/TiO_{2-x}/Pt	12 × 12 RRAM passive (0T1R)	—	Pulse amplitude (binary image inputs)	In situ	Recognition of 3 × 3 binary images

6.2.3 Chip and System Implementations

To further exploit the potential of memristor devices on the hardware implementation of DNNs, neuromorphic computing chips (including macros) and systems have also been developed [41, 45, 51–57], as to be reviewed in this Section 6.2.3. These implementations published in recent years are summarized in Table 6.4, including their synaptic cell configurations, technology nodes, computing precision of MACs, etc.

Memristor-based CIM macros have been demonstrated to implement high-performance and energy-efficient DNN computing. In ISSCC'20, Xue et al. recently reported a CIM macro [56] integrating 2 Mb 1T1R binary RRAMs, fabricated with a 22 nm CMOS logic process. The proposed scrambled 2's complement weight mapping effectively decreased the area overhead for high-precision weights. Incorporating a BL-IN-OUT multi-bit computing scheme and a dual-bit small-offset current-mode sense amplifier, the reduced BL currents and stabilized clamp voltage on BLs enhanced the computing precision and lowered the latency. With a precision configuration of 4-bit inputs, 4-bit weights, and 11-bit outputs, the system achieved an accuracy of 66.46% on CIFAR-100, with an energy efficiency of 28.93 TOPS/W.

Another work published on ISSCC'20 [45] demonstrated a fully integrated CIM chip with 158.8 kb analog RRAMs. To alleviate the IR drop, SL currents were reduced by configuring the synaptic units in a 2T2R structure. Also, a low-power resolution-adjustable ADC was designed to accommodate various requirements of power dissipation and accuracy. A $784 \times 100 \times 10$ FCN was implemented based on the fabricated chip and achieved the recognition accuracy of 94.4% on MNIST, the energy efficiency of 78.4 TOPS/W, and the inference speed of 77 μs/image.

In addition, a hybrid training method has been developed to mitigate the accuracy loss caused by inaccurate mapping, which is caused by device noideal characteristics. Yao et al. [57] demonstrated a fully hardware-implemented CNN using multiple memristor arrays that supported the retraining of the last fully connected layer to enhance the resilience to device nonidealities. Eight memristor processing element (PE) units were fabricated with 130 nm CMOS technology, integrated with an ARM core, DRAM blocks, calculation blocks, and other peripherals on a printed circuit board (PCB) broad. Each PE included a 2048-cell TiN/TaO_x/HfO_2/TiN memristor array, 8-bit ADCs, and other modules. A synaptic unit was composed of a differential pair of 1T1R cells, obtaining at least 15-level conductance states with a closed-loop programming scheme. A five-layer CNN based on the constructed system achieved an accuracy of more than 96% for the classification of the MNIST dataset and more than 2 orders of magnitude higher energy efficiency than state-of-the-art GPU.

Reprogrammable neuromorphic computing hardware of DNNs is required to facilitate the implementation of different network architectures and algorithms. By integrating a 54×108 WO_x-based passive memristor crossbar, D/A interfaces, a reduced instruction set computing (RISC) processor, and on-chip memory (SRAM), Cai et al. [41] firstly designed a reprogrammable CMOS-RRAM CIM system to maximize flexibility and programmability. Using a developed instruction set, this system supported bidirectional information propagation, online training, and both

Table 6.4 Chips based on memristive devices for DNNs.

References	Technology (nm)	Synapse configurations	A/D interface[a)]	Precision[b)]	Computing power[c)]	Energy efficiency	Tests
[56]	22	2 Mb 1T1R RRAM	DbSO-CSA	4–4–11 bit	18.3 ns/MAC	28.9 TOPS/W	ResNet-20 CIFAR-100: 66.46%
				1–2–6 bit	9.8 ns/MAC	121.3 TOPS/W	
[45]	130	158.8 kb SW-2T2R/1T1R RRAM	LPAR-ADC	1–3–1–8 bit	51.1 ns/MAC	78.4 TOPS/W	784 × 100 × 10 FCN MNIST: 94.4%
[57]	130	Eight 128 × 16 1T1R RRAM	8-bit ADC	8–4–8 bit	81.92 GOPS	11.01 TOPS/W	Five-layer CNN MNIST: 96.19%
[41]	180	54 × 108 1T1R RRAM	13-bit ADC	4-Analog-13 bit	57.5 GOPS	187.62 GOPS/W	Sparse coding, Two-layer network with PCA analysis
[54]	55	1 Mb 1T1R RRAM	TMCSA	1–3-4 bit	11.75 ns/MAC	53.17 TOPS/W	ResNet-18 CIFAR-10: 88.52%
				2-3-4 bit	14.6 ns/MAC	21.9 TOPS/W	
[52]	65	1 Mb 1T1R RRAM	DR-CSA	1-Ternary-3 bit	14.8 ns/MAC	19.2 TOPS/W	—
[53]	180	2 Mb 1T1R RRAM	SA	1-Analog-1 bit	330 GOPS	20.7 TOPS/W	LeNet MNIST: 98.1% CIFAR-10: 95.9%
	40	4 Mb 1T1R RRAM			660 GOPS	66.5 TOPS/W	
[55]	130	32 kb 1T1R RRAM	SA	1–1–1 bit	—	60.9 TOPS/W	—
[51]	150	4 × 1 kb 1T1R RRAM	SA	1–1–1–3 bit	—	462 GOPS/W	—

a) Dual-bit small-offset current-mode sense amplifier (DbSO-CSA), low-power resolution-adjustable ADC (LPAR-ADC), triple-margin small-offset current-mode sense amplifier (TMCSA), and distance-racing current-mode sense amplifier (DR-CSA).
b) Precision is represented as input precision – synaptic weight precision – output precision.
c) In the crossbar-based analog computing, the number of operations included in an MAC operation is dependent on the size of the crossbar array.

spiking and non-spiking networks, with a computing power of 57.5 GOPS and an energy efficiency of 187.62 GOPS/W when adopting 4-bit inputs and analog weights. It is noted that this energy efficiency was much lower than those reported for memristor-based CIM macros, which is mainly due to the utilization of passive crossbars with relatively low device resistance and the integration of massive digital modules.

Multi-core architectures were adopted in several CMOS-based neuromorphic hardware designs to increase the scale of systems. Similarly, enhanced multi-core architectures [58–60] for crossbar-based DNN hardware systems have been proposed, which integrate distributed CIM subcores and reconfigurable interconnection to accommodate various sizes and flexible dataflow paths of networks. Nevertheless, the hardware implementations of such a system are still lacking. In addition to hardware architectures, a compiler [61] compatible with the adopted hardware system is indispensable to flexibly map a pretrained network and make necessary adjustments according to the actual properties of hardware.

6.2.4 Architecture and Algorithm Optimization

In comparison with the floating-point computing in the conventional digital processors, several factors could limit the accuracy of synaptic device crossbar-based analog computing, including a limited number of conductance levels, errors caused by nonidealities of devices, signal distortions in arrays, etc. Here, we discuss their effects on the system performance and also how to mitigate them in the architecture and algorithm levels.

6.2.4.1 Quantization

Accuracy degradation occurs with the low-precision quantization of weights. However, there exists a trade-off between precision and computing efficiency; higher-precision programming usually requires higher-resolution A/D interfaces and a more redundant verification process. In this case, to achieve high energy efficiency, it is sensible to maintain moderate accuracy without introducing too much overhead on the peripheral circuits.

A special class of networks, called binarized neural networks (BNNs) [62], adopt 1-bit synaptic weights in the inference. Even memristive devices possessing only high-resistance state (HRS) and low-resistance state (LRS) meet the demands in BNN implementations [48, 63, 64]. Besides, the activation functions in BNNs are also binarized, so sense amplifiers (SAs) and simple logic units can serve as the activation units. However, the outcome of high energy efficiency is observable performance degradation. Therefore, increasing the precision and enlarging the available dynamic range of synaptic cells is inevitable. Bit slicing scheme [65] is widely adopted to implement multiple-bit synaptic weights with binary memristive device arrays. By adding current mirrors or post-processing in digital domains (like shift adder), different weights are assigned to multiple devices to constitute one synaptic cell with the desired multi-bit precision.

6.2.4.2 Nonideal Analog Switching Characteristics

Performance degradation also results from nonideal device characteristics, including the analog switching nonlinearity, asymmetry, conductance fluctuation or drift, finite endurance and retention, nonlinear I–V relation, and variations. Due to these factors, the mapped device conductance states could deviate from the expected ones. Therefore, fault-tolerant algorithms have been proposed to enhance the neural network robustness to device variations without introducing extra overhead [66, 67]. Besides, aiming to solve the conductance drift problems of PCMs, correction schemes [50, 68] were proposed to compensate for the adverse effects caused by the conductance drift, such as gradually increasing the slope of the activation function over time according to the empirical law and extracted device parameters.

In addition, in situ training involving self-adaptation could compensate for nonidealities of devices, which updates synaptic weights frequently. However, the device conductance may saturate during the training. Agarwal et al. [69] proposed a periodic carry strategy to avoid the saturation problem, which designated extra levels to store the carry information and determined whether to update bits with higher significance by periodically reading the carry information. To lower the cost of programming, open-loop programming is preferred for online training. Methods like paired pulses with opposite polarities and heating pulses could improve the linearity and lead to gradual switching [70, 71].

Besides nonlinearity in resistive switching, asymmetry is another challenge for online training. Innovations about synaptic cell structures and hardware architectures such as a zero-shifting technique along with the "Tiki-Taka" training algorithm and 2T1R+3T1C synaptic configuration [72, 73] have been proposed. Replacing the gradient descent with optimized training algorithms for crossbar-based hardware could ease the burden of weight update on synaptic cells. For example, the sign-based backpropagation (SBP) training method [74] has been developed for the training of FCNs with binary activation functions. In SBP, the partial derivatives of the loss function to the synaptic weights are only valued as −1, 0, or 1. In this case, the synaptic cell programming during the training merely involves the directions of weight update, rather than accurate values, which greatly simplifies the weight update process. Furthermore, simple logic units can serve as mixed-signal interfaces for the feedforward and backward information propagations, significantly reducing the peripheral circuit cost and improving robustness to device variations.

6.2.4.3 Synaptic Array Size

According to the analysis in [75, 76], systems with large synaptic arrays can fully exploit the advantages of parallelism and the high efficiency of the in-memory process. However, in a large crossbar array, the increasing IR drop could cause inaccurate synaptic weight programming and deviations of sensed currents and write disturbances caused by sneak paths (for passive arrays only). Besides, the shot noise also grows as a square root function of the array size, leading to the inevitable increment of the integration period [36]. Finally, owing to the currently immature fabrication technology of emerging memories, the yield of crossbar array may worsen with increasing array size. The common defect where cells are stuck in HRS or LRS and do not respond to the applied programming pulses is referred to as stuck-at-fault (SAF).

Replacing large memory arrays with small sizes may avoid those problems [76, 77]. However, hardware improvements in alleviating the performance degradation are necessary to take full advantage of parallelism on large arrays. Introducing selector devices with high nonlinearity can suppress the IR drop and sneak currents [78, 79]. To reduce summed word line (WL) currents, split weight mapping and 2T2R structures were adopted [45, 80]. Besides optimizations on the hardware, compensation methods, and defect-aware algorithms have also been developed. Generally, incorporating the model of array nonidealities in the training phase can improve resilience [81]. Moreover, compensation strategies can be applied in the inference phase to correct the distorted sensed currents [82].

In a summary, the above sections have reviewed the DNN implementations based on synaptic devices mainly from two perspectives: the construction of resistive switching device-based hardware systems and the system optimizations concerning nonidealities. For a complete hardware system, design parameters such as the array size, the precision of ADCs, etc. should be carefully configured. Optimizations of peripheral circuits and hardware architectures are necessary to fully exploit the parallelism and high energy efficiency. To accommodate various network structures and algorithms, reprogrammable single-chip systems and multi-core architectures have also been proposed. Furthermore, compilers are required to satisfy the requirement to map versatile network topologies. Regarding nonidealities, fault-tolerant algorithms, defect-aware training, and compensation strategies are needed. Therefore, to further exploit the full potential of memristive devices for implementing high-efficiency neuromorphic computing hardware, more device architecture–algorithm co-optimizations are in demand.

6.3 SNNs Based on Neuromorphic Devices

Compared with the above-discussed DNNs, SNNs attempt to adopt more neuroscience findings on BioNNs to build a bio-plausible intelligent system and pay more attention to the replication of the neural behaviors and learning mechanisms of BioNNs. Firstly, the information in SNNs carried in a train of spikes is represented not only by the firing rate but also in precise timing order or phase relations, empowering SNNs with extraordinary abilities in processing temporal information. In addition, the event-driven activities suggest that most neural units are at rest until they receive spikes or are ready to fire spikes. Based on the above two aspects, neuromorphic computing hardware for SNNs has great potentials for high energy efficiency and temporal information processing. It is worth noting that various emerging devices could emulate neural units in BioNNs, including synapses, somas, and dendrites. These behavioral similarities can be fully exploited to implement an emerging device-based complete SNN hardware system. Therefore, in this section, we mainly focus our discussion on the following aspects. At first, learning mechanisms in SNNs and relevant properties of neural units are introduced. Then, implementations of these properties based on neuromorphic devices are discussed. Finally, we reviewed the hardware demonstrations of SNNs capable of learning.

6.3.1 Learning Rules and Memory Principles

In BioNNs, neurotransmitters released by pre-synaptic neurons bind with receptors distributed among post-synaptic neurons and induce the post-synaptic potentials (PSPs) to complete information transmission. Meanwhile, abundant ion channels and complex morphology endow neurons with the powerful ability of information processing. According to the behavioral features of synapses and neurons, various models are proposed and applied in SNNs to empower the networks with learning abilities similar to BioNNs.

6.3.1.1 Synapses and Neurons in SNNs

The strength of the induced PSPs depends on the number of receptors and released neurotransmitters that are influenced by post-synaptic calcium ions. The phenomenon that synaptic connections are strengthened or weakened according to the activities of pre- and post-synaptic neurons is called synaptic plasticity, including short-term facilitation (STF)/short-term depression (STD) or long-term potentiation (LTP)/long-term depression (LTD). One of the prevalent long-term plasticity is Hebb's rule [4], which emphasizes the causality between spikes of pre- and post-synaptic neurons. Deriving from Hebb's rule, spike-timing-dependent plasticity (STDP) is a widely used learning rule [83], in which the synaptic weight update is related to the relative timing between pre- and post-synaptic spikes. Moreover, other types of plasticity, such as heterosynaptic plasticity [84], are also involved in maintaining the normal function of networks. In contrast to long-term plasticity, short-term plasticity lasts for on the order of hundreds to thousands of milliseconds in biological neural systems [85].

Synapses in SNNs are responsible for information propagation while neurons possess powerful information processing capabilities. Various neuron models have been proposed to describe how different neurons respond to different types of stimuli [86]. Among them, the most prevalent is the leaky integrate-and-fire (LIF) model, where the neuron generates an action potential when the integration of synaptic inputs reaches the firing threshold. However, the Hodgkin–Huxley neuron model is the most complete one that is capable of replicating most of the firing modes observed in BioNNs, at the cost of high computing complexities. The common neuronal models used in SNNs today remain the simplified point-neuron models, without no internal morphology. However, increasing attention has been focused on multi-compartment neuron models [87], especially those incorporating dendritic features, which will be discussed in Section 6.4.4. There is no doubt that neurons in SNNs need further refinements to replicate the complex dynamics of BioNNs.

6.3.1.2 Learning Algorithms and Benchmarks for SNNs

By assuming that information propagated in SNNs is encoded in the firing rate, the synaptic weights in SNNs could be obtained from a pretrained DNN with the same architecture, which is called the conversion-based training algorithm. The advantage of this method is that the software framework and training algorithms of DNNs are relatively mature. However, to fit the characteristics of SNNs, there are constraints on the DNN training [88]. Methods like weight scaling and normalization are used to set the firing rate of neurons in each layer in a proper range [89]. In other words, DNNs whose activation results are represented with spike trains could be viewed as conversion-based SNNs.

In addition, spike-based training algorithms have also been developed. Approximate derivatives are widely adopted to facilitate gradient descent-based backpropagation [90]. Nevertheless, the computation is expensive and not bio-plausive. As described in the previous part, STDP is a type of local synaptic plasticity that plays a significant role in BioNNs. It has been demonstrated that a fully connected unsupervised SNN, trained with the STDP-based algorithm, achieved an accuracy of 95% on the MNIST dataset [91]. Furthermore, supervised training algorithms incorporating STDP have been widely explored to train larger, deeper SNNs and handle more complex tasks, such as the remote supervised method (ReSuMe) and backpropagation-spike-timing-dependent plasticity (BP-STDP) [92, 93].

The above-described algorithms are based on firing rate coding. However, rate coding requires a long observation window and redundant neurons to ensure precise information transmission, which contradicts the characteristics of event-driven sparseness and high energy efficiency in BioNNs [94]. Moreover, several experiments have revealed that the precise relative timing between spikes is the underlying mechanism of specific functions in some areas of the brain [95]. In this case, temporal coding [94, 96] has been proposed as an alternative encoding method to fully exploit the capability of spatial–temporal information processing in SNNs. For example, with the rank-order code [97], the neuron only fires once, leading to the enhanced sparseness of the networks.

Nevertheless, for both SNNs using rate coding or temporal coding, the achieved performances are still inferior to that in conventional DNNs. On the one hand, optimizations of algorithms are necessary. On the other hand, finding proper datasets and application scenarios is also critical [88, 98]. Deng et al. [99] simulated ANNs and SNNs on both ANN- and SNN-oriented benchmark workloads to comprehensively compare their performances, concluding that good neuromorphic datasets, suitable benchmark tasks, and evaluation metrics beyond the recognition accuracy are desired to be developed.

6.3.2 SNNs with Synaptic Devices and Neuronal Devices

Aiming for replicating the learning abilities of the BioNN, firstly, it is critical to imitate the behaviors of its components, especially the synapse and neuron. Comparing with ANNs, one of the essential differences in the synaptic units in SNNs is the activity-dependent synaptic plasticity.

6.3.2.1 Synaptic Devices with Plasticity

About STDP, various programming schemes have been implemented by applying designed voltage waveforms to the terminals of a synaptic cell when the pre- and post-synaptic spikes occur [100, 101]. Zamarreño-Ramos et al. [102] designed a set of pre- and post-synaptic waveforms and simulated the layer V_1 in the visual cortex based on passive memristor crossbars and CMOS neurons. Perceptive fields emerged after receiving several patches of visual data, showing the ability to extract input features under the control of STDP. Simulations about STDP-based SNN training employing the abovementioned synaptic cell designs have been illustrated in many papers with unsupervised [103, 104] or supervised manners [105], from the simple pattern recognition to the handwritten digit classification of the MNIST dataset. Concerning the temporal information processing, Wang et al. [106] adopted 1T1R memristor-based synapses to simulate brain-like spatiotemporal pattern recognition, as shown in Figure 6.3a,b. With the STDP-based supervised training, a 16 × 1 network shown in Figure 6.3c learned and recognized specific spike sequences. Besides, Boybat et al. [110] considered the performance degradation caused by the variation of devices and proposed multi-memristive synapse design to improve the accuracy of weight update and lower the number of programming cycles for each device.

The above programming schemes that integrate extra waveform generators would introduce extra design complexities. In fact, the intrinsic properties and switching mechanisms of memristive devices could yield more efficient hardware implementations. For example, the concept of *n*th-order memristor refers to the memristive devices whose state is controlled by external stimuli and a set of *n* internal state variables [111]. An experimental demonstration of TaO_x-based bilayer memristors [107] showed that, by considering the local temperature as the second-order variable, the device could mimic calcium ion-like internal dynamics under the paired programming pulse, as shown in Figure 6.3d. Due to the similarity between the motion of Ag nanoparticles in the gap region and calcium ions transporting through the ion channel, a serial connection of SiO_xN_y : Ag diffusive memristor and nonvolatile drift memristor also exhibited the second-order memristive behavior [112]. The diffusive memristor mentioned here, whose conductance was modulated by the morphologic change of conducting filaments formed with metallic nanoparticles, featured both threshold switching behavior and fast relaxation. Similar diffusive dynamics also found in devices fabricated with inorganic materials like AgS_2 [113], SiO_xN_y [112], or organic biopolymer materials [114].

Other kinds of synaptic plasticity have also been explored based on emerging memristive devices. For example, by introducing a third terminal to modulate the growth of Ag nanoclusters between the source and drain terminals, the fabricated physically evolving network device [115] was capable of emulating heterosynaptic plasticity. A multi-terminal memtransistor design based on polycrystalline monolayer molybdenum disulfide (MoS_2) has been demonstrated to implement flexible gate-tunable resistive switching [116]. By contrast, a bio-voltage diffusive memristor [117] with a programming voltage ranging from 40 to 100 mV was experimentally realized by introducing protein nanowires to facilitate the reduction in the diffusive memristors. The device showed potentials for direct interactions with biological samples and the sensed bio-signal processing.

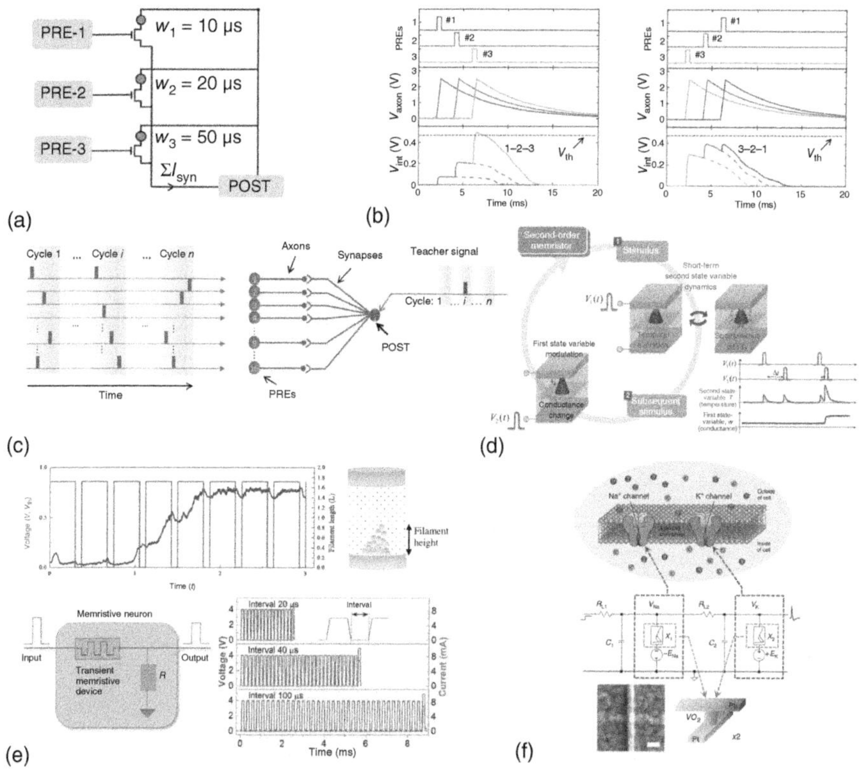

Figure 6.3 Synaptic behaviors based on memristive devices. (a, b) A 3×1 network could recognize different spatiotemporal patterns determined by the connection strengths of synapses. The simulated waveforms are presented in (b), where the accumulated PSP depended on the different orders of pre-synaptic spikes. (c) With the 1T1R-based STDP, the network could learn to recognize specific patterns with the supervision of a teacher signal. Source: (a–c) Wang et al. [106]. Reproduced with permission of AAAS. (d) The dynamic evolution of the second-order (temperature) variable allows the memristor to natively replicate the STDP behavior that the conductance changes are modulated by pulse intervals. Source: Kim et al. [107]. (e) A capacitor-less neuron design. The top waveform shows that conductive filaments grow with the applied voltage but diffuses away once the bias is removed. Source: Zhang et al. [108]. (f) The circuit design of Hodgkin–Huxley neuron based on VO_2 Mott devices. The threshold switching behavior of Mott devices is similar to the voltage-triggered activation of ion channels in neuron membranes. Source: Yi et al. [109].

6.3.2.2 Neuronal Devices

Since the very beginning of neuromorphic computing, CMOS circuits [118] have been widely used in neuromorphic systems to model neurons. Alternatively, emerging devices [119] are potential candidates for artificial neuron designs by taking advantage of their various intrinsic properties. For example, integrating a threshold switching memristor (TSM) with a capacitor in a neuronal cell can emulate the integrate-and-fire behavior [120], where the abrupt set and spontaneously reset resemble the neuron firing and resting procedure. Similar behaviors were found in NbO_x- and VO_x-based Mott devices [121] and metal atom-based filamentary memristor with diffusive dynamics [108, 112]. Other than the LIF neuron model, the group in HP Labs reported an artificial neuron emulating the Hodgkin–Huxley model based on Nb_2O_5-based Mott devices connected with several capacitors by utilizing the similarity between Mott devices and voltage-controlled ion channels in neuron membranes [122]. It has been verified that this unit could reproduce most of the known biological neuronal dynamics by tuning the design parameters [109]. Figure 6.3e,f depicts the applications of TSM devices in several artificial neurons.

In addition, applying three-terminal devices, like FeFETs, in the artificial neuron design could enable more flexibility due to the separate reading and writing paths. Capacitor-less neuron [123, 124] designs have been developed based on leaky FeFETs. Finally, it is worth mentioning that besides the behavior of neurons, researchers also need to take into consideration other aspects that have received little attention yet, such as the signal driving ability of artificial neurons, the impact of frequent firing on the device reliability, etc.

6.3.3 SNN Implementations with Synaptic Arrays

In this section, we would like to highlight several hardware implementations of SNNs with neuromorphic device-based synaptic arrays and artificial neurons [125–131]. A general overview of these works is summarized in Table 6.5.

To remove extra conversion modules required for generating voltage pulses to drive the next layer, Wang et al. [126] developed neuro-transistors as neurons by integrating dynamic pseudo-memcapacitors (DPMs) with transistors to construct a capacitively coupled SNN. A DPM was composed of a serial connection of a $Pt/Ag/SiO_x:Ag/Ag/Pt$ diffusive memristor with a $Pt/Ta_2O_5/TaO_x/Pt$ capacitor. When the accumulated external stimuli caused the resistive switching of the diffusive memristor from HRS to LRS, the equivalent capacitance of the neuron unit abruptly increased, leading to an increase in the current flowing through the transistor, which resembles the neuronal LIF behavior. Similarly, a synapse was constructed by replacing the diffusive memristor in DPM with a nonvolatile memristor whose conductance determined the connection strength. Based on these two neural devices, a 4×4 memcapacitive array was fabricated to implement associative learning with the Hebbian-like synaptic plasticity. To realize larger networks, Zhang et al. [129] designed a hybrid memristor-CMOS LIF neuron, capable of generating "all-or-none" spikes to drive the next layer and update synaptic strengths according to Hebb's rule. The proposed capacitor-less neuron integrated a $Ti/Au/SiO_2:Ag/Ag/Au$ device to emulate the integrate-and-fire behavior and CMOS logic circuits to detect and generate spikes. Furthermore, a global clock was adopted to coordinate the device operations. Based on these, they further demonstrated a fully hardware-implemented $30\times10\times10$ SNN with hybrid in situ learning, where synaptic cells were configured as 1T1R in $Pd/HfO_2/Ta$ memristor arrays. The first layer was trained with unsupervised Hebb's rule, with an additional 10×10 1T1R array to implement the lateral inhibition, while the second layer was trained in a supervised way. The fabricated hardware successfully realized the recognition of 0–9 digital images.

Table 6.5 Hardware demonstrations of SNNs based on synaptic arrays and neuronal devices.

References	Synapses	Neurons	Scales	Demonstrations	Highlights
[125]	1T1R Pd/HfO_2/Ta drift memristor	Pt/Ag/SiO_x : Ag/Ag/Pt diffusive memristor, LIF	8 × 8 array	Unsupervised learning of a fully connected layer based on simplified STDP	—
[126]	HfO_2-based drift memristor + Pt/Ta_2O_5/TaO_x/Pt capacitor	Pt/Ag/SiO_x : Ag/Ag/Pt diffusive memristor + Pt/Ta_2O_5/TaO_x/Pt capacitor, LIF	4 × 4 array	Associative learning based on the Hebbian-like mechanism; signal classifications	Low static power dissipation in the capacitive neural network. Signal amplification of the neuro-transistor
[129]	1T1R Ta/HfO_2/Pd drift memristor	Hybrid CMOS-Ti/Au/SiO_2:Ag/Ag/Au diffusive memristor-based circuit module, LIF	30 × 10 – 10 × 10 – 11 × 1 array	A two-layer fully connected SNN with unsupervised STDP learning in the first layer, supervised learning in the second layer, realizing the number image classification	The hybrid neuron supported in situ synaptic weight updating and can generate "all-or-none" spikes with the driving capacity.
[128]	1T1R memristor	1T1R Si/NbO_x/TiN Mott device, ReLU	128 × 64 array	A 320 × 10 single-layer SNN achieved an accuracy of 85.7% on the MNIST dataset, with the conversion-based training	—

Table 6.5 (Continued)

References	Synapses	Neurons	Scales	Demonstrations	Highlights
[127]	1T1R memristor	CMOS-based spiking activation function	64 kb array	FCN on MNIST and LeNet on CIFAR-10 with the conversion-based training	Tunable precision of the spiking activation function. Comprehensive evaluations of energy efficiency, performance, and computing power
[124]	1T1R $TiN/TaO_x/HfO_x/TiN$ memristor	CMOS-based integrate-and-fire neuron	64 kb array, including 64 subcores, and each subcore contained 16 × 16 synapses	Generative restricted Boltzmann machine to reconstruct MNIST digit images	Reconfigurable bidirectional dataflow. Comprehensive evaluations of energy efficiency, performance, and computing power
[131])	PCM devices	Software SNU	52 800 devices	Image recognition using the MNIST dataset, language modeling using the Penn Treebank dataset, and music prediction using the Johann Sebastian Bach chorales dataset, with in situ training	Exploit the dynamics of spiking neuron, and incorporate it into a recurrent ANN unit

It is noted that the scale of the hardware demonstrations of unsupervised SNNs so far is quite small. One reason would be that the performance of unsupervised SNNs trained with simple Hebb's rule or STDP still lags behind that of conventional supervised DNNs, even compared with shallow FCNs. Another reason would be that similar optimizations to alleviate the adverse effects of device nonidealities as DNNs, especially for emerging neuronal devices, are still lacking, resulting in the possible serious degradation of the performance of in situ training. Consequently, the reported neuromorphic hardware with larger-scale arrays mainly focused on the conversion-based SNNs. In fact, neuromorphic computing hardware for conversion-based SNNs could be viewed as hardware implementations of DNNs whose ReLU activation results are represented by spike frequency. For example, Yan et al. [127] integrated an array of 64 kb 1T1R binary memristors with CMOS-based in situ nonlinear activation (ISNA) modules to construct a CIM processing engine, whose synaptic weights could be configured as binary or ternary. The spike frequency of an ISNA represented the activation results, with configurable precision up to 8 bits. The system was validated with four-layer LeNet, exhibiting the energy efficiency of 16.9 TOPS/W and only 2.5% accuracy drop compared with the software inference with binarized weights in the recognition of MNIST.

The demonstrations mentioned above were only limited to conventional classification tasks with static input images. However, SNNs are expected to be favorable in temporal information processing. Therefore, Woźniak et al. proposed a concept of spiking neural unit (SNU) that abstracted an LIF spiking neuron into a recurrent ANN unit [131]. In this case, the training of the network could be realized with backpropagation through time (BPTT). Based on the simulation results, 52 800 PCM devices were used to implement in situ training for the music prediction task, where each synaptic weight was mapped to a differential 2-PCM pair, with 4-bit precision. The in situ training showed a competitive performance (the negative log-likelihood loss of 9.51) comparing with the software-simulated model using the same hyperparameters.

In this section, we have reviewed the neuromorphic devices and systems for SNNs. Compared with DNNs, the intrinsic behaviors of neural units are the underlying mechanisms for memory and learning abilities of the brain. Thus, by mimicking the neural dynamics, device-based artificial synapses and neurons provide fundamentals to implement large-scale SNN hardware systems. In terms of network algorithms, temporal coding and local STDP were exploited to take advantage of the sparseness and event-driven features. Finally, this section summarizes SNN hardware implementations based on synaptic arrays and neuronal devices. Currently, emerging device-based hardware implementations for SNNs have still been mostly limited to small-sized arrays and static recognition tasks. Therefore, besides neuromorphic devices, SNN-oriented benchmarks and training algorithms that leverage the event-driven property and sparseness are critical to implementing efficient spike-based neuromorphic computing systems.

6.4 Other Neuromorphic Systems

In addition to DNNs and SNNs, there are several other systems inspired by the behavioral characteristics of the brain. In this section, we will mainly focus on hyperdimensional computing, dendritic computing, reservoir computing, oscillatory neural network, and Hopfield neural network. The fundamentals and corresponding hardware demonstrations based on neuromorphic devices will be discussed. For each type of brain-inspired computing, a representative work with key achievements is summarized in Table 6.6.

6.4.1 Hyperdimensional Computing

Hyperdimensional (HD) computing is a type of one-shot learning proposed by Kanerva [137]. By encoding basic entities with HD random vectors, cognitive functions can be realized through a serial of operations on vectors and matrices, such as multiplication–addition–permutation (MAP). Represented by HD vectors, the involved entities are viewed as points in the HD space, in which highly associated items are close, while the distances between unrelated items in the HD space are almost half of the dimension. Accordingly, by calculating the Hamming distance between cognitive codes stored in item memories and query vectors, associative memory can be implemented. The encoding procedure and mechanisms are presented in Figure 6.4a.

Logic-in-memory circuits based on binary memristive devices show vast potentials in accelerating HD computing [139], where inherent device variations provide valid random sources to initialize HD representations of independent entities. A fabricated four-layer three-dimensional (3D) vertical RRAM (VRAM)/FinFET kernel demonstrated the implementation of in-memory MAP operation. Furthermore, an architecture-level simulation achieved an accuracy of 90.4% in pairwise language classification in the presence of device variations [140]. A chip prototype [132] was fabricated with memristors and carbon nanotube FETs using 3D integration, utilizing inherent variations in devices and the gradual RESET property of memristors, and exhibited 98% accuracy in language recognition despite 78% stuck bits, which verified the tremendous error resilience of HD computing. G. Karunaratne et al. [141] reported a programmable in-memory HD computing system composed of PCM crossbars and CMOS-based peripherals. Tasks including language classification, news classification, and hand gesture recognition were demonstrated. The implementations on more complicated dynamic perception tasks and large-scale integration remain to be explored.

6.4.2 Dendritic Computing

As mentioned in Section 6.3.1, neurons, especially pyramidal neurons, have lots of dendrites with complex morphology and abundant ion channels, endowing neurons with complicated dynamics [138], as shown in Figure 6.4b. Their active and passive properties greatly enhance the neurons' ability to process information. For example,

Table 6.6 Hardware demonstrations for other neuromorphic computing systems.

Types and references	Technology, scales, and configurations	Task demonstrations	Illustrations for representative works
HD Computing [132]	A system fabricated with the 3D monolithic integration with CNFETs and memristors	98% accuracy for the language classification on the European Parallel Corpus	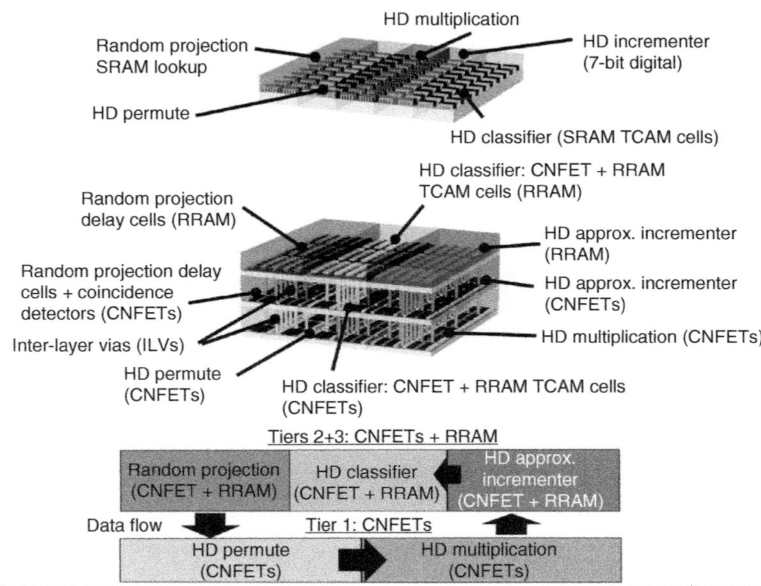

Table 6.6 (Continued)

Types and references	Technology, scales, and configurations	Task demonstrations	Illustrations for representative works
Dendritic Computing [133]	Synapse device: TiN/TaO_y/HfO_x/TiN Soma device: Pd/NbO_x/Pt Dendrite device: Pt/TaO_x/AlO_δ/Al A 256 × 4 FCN consisted of 1K synapses (2K memristors), 12 dendrites, and 4 somas and was hardware implemented	87% accuracy on the SVHN dataset, with an energy efficiency of 4.97 TOPS/W estimated by multilayer FCN simulation	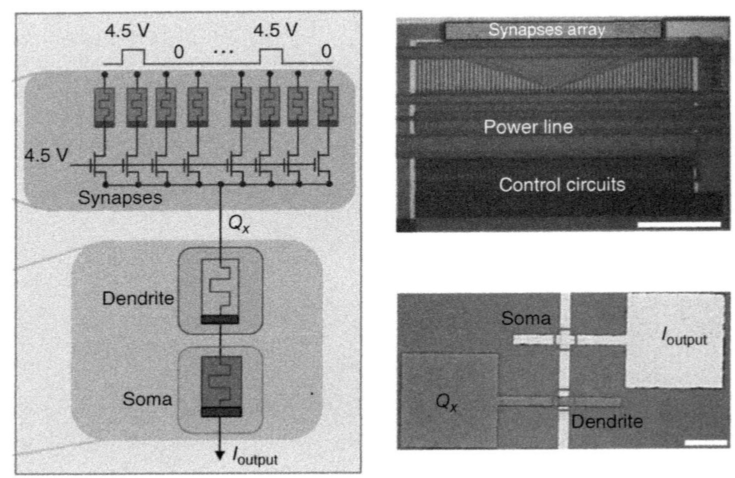

<table>
<tr><td>RC
[134]</td><td>Neuronal node: WO_x-based memristor
A RC system was demonstrated based on a 32 × 32 memristor array</td><td>The 99.2% accuracy of the speech recognition is based on the NIST TI46 database.
Time series forecasting with the Mackey–Glass time series</td><td>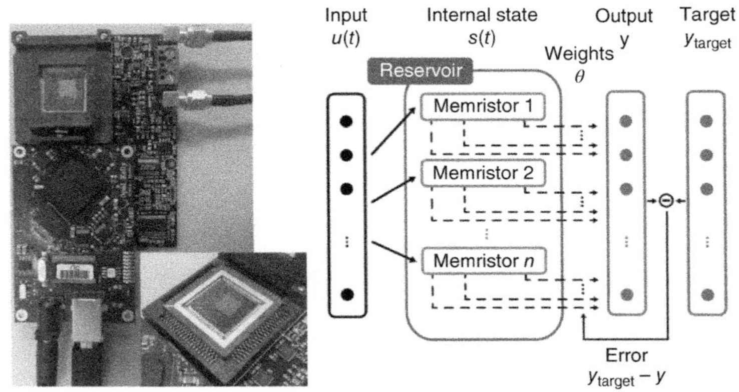</td></tr>
<tr><td>ONN
[135]</td><td>Four spin-torque nano-oscillators were connected through an millimeter-long wire, where microwave currents influenced the dynamic states of the oscillators</td><td>88% accuracy of the vowel recognition based on the Hillenbrand database</td><td>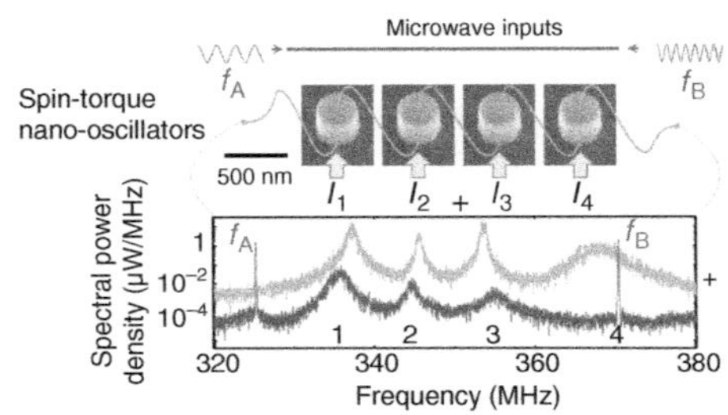</td></tr>
</table>

Table 6.6 (Continued)

Types and references	Technology, scales, and configurations	Task demonstrations	Illustrations for representative works
HNN [136]	The HNN was demonstrated on a 4K passive TiO_{2-x}/Al_2O_3-based memristor array. By utilizing the readout current noise, stochastic simulated annealing, chaotic simulated annealing, and exponential annealing were implemented	Five-node weighted maximum-clique problem Weighted vertex cover problem 10-node independent set problem Graph partitioning optimization problem	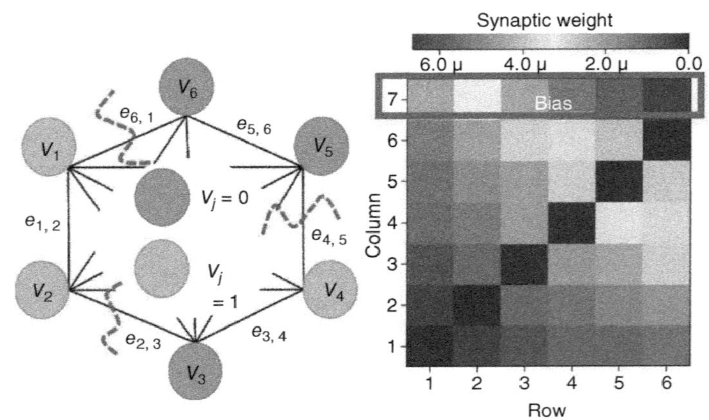

Images are respectively reproduced with permissions of IEEE, Wu et al. 2018 [132]; Springer Nature, Li et al. 2020 [133]; Springer Nature, Moon et al. 2019 [134]; Springer Nature, Romera et al. 2018 [135]; IEEE, Mahmoodi et al. 2019 [136].

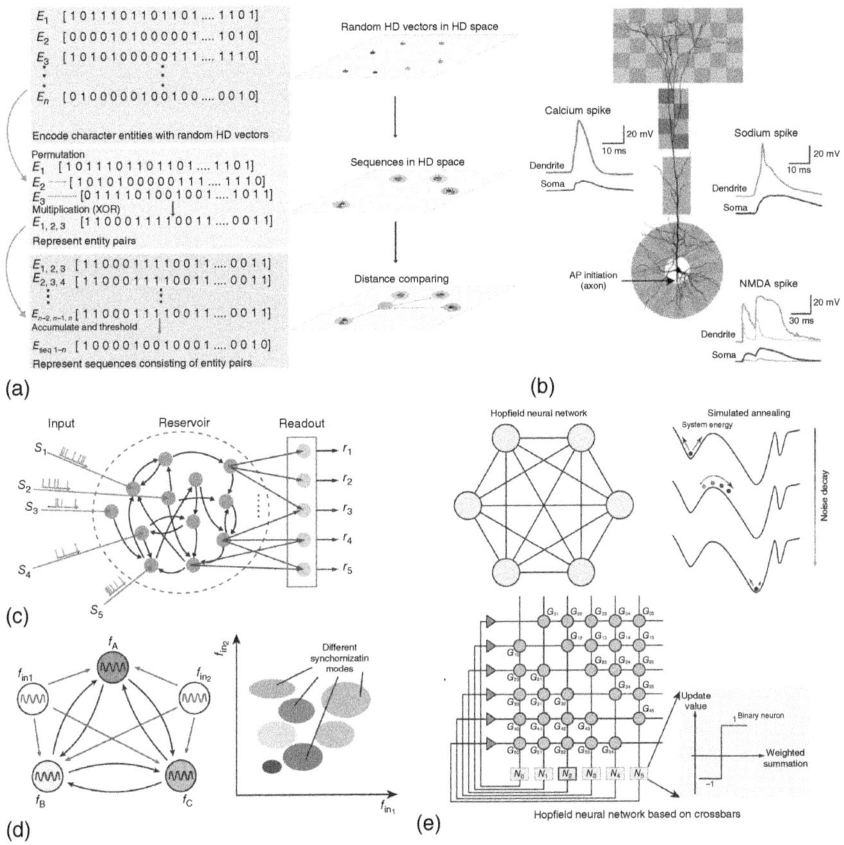

Figure 6.4 Illustrations of other forms of neuromorphic computing. (a) The mechanism of HD computing. Take the language classification as an example. Every character entity is represented by a random HD binary vector E_k that is scattered in a HD space. Then by permutating and multiplying, these entities are assembled into new vectors that represent sequentially grouped trigrams in the text. Finally, these vectors of trigrams are accumulated and binarized to generate the HD vector that represents the whole text (a sequence of character trigrams). In this case, vectors that encode texts in the same language are similar and distributed closely in the HD space. (b) Ca^{2+}, Na^{+}, and NMDA channels are distributed in different regions of the neuronal dendrites. Source: (b) Stuart and Spruston [138]. (c) The structure of an RC system. Randomly connected neuronal nodes constitute a reservoir. Input pulse trains are transmitted to several nodes. A readout layer takes the states of neuronal nodes as inputs and outputs, the result of a linear classification. (d) The schematic of an ONN system, where three oscillators with different intrinsic frequencies are mutually coupled and affected by the states of two input oscillators. Inputs are encoded as different frequencies of the two input oscillators. Different combinations of input frequencies induce different synchronization modes, which could be utilized in classification tasks. (e) HNN is a fully connected recurrent network, where each neuronal node updates its state according to whether the summation of synaptic inputs from other nodes exceeds the threshold. Simulated annealing helps the system energy successfully evolve to the global minimum. Initially, with a large noise disturbance, the energy state could jump out from local energy minimum states. With more iterations, noise decays and the system energy stably converges to the global minimum.

dendrites can perform logic operations, detect the coincidence of synaptic inputs, and amplify signals selectively [142]. Furthermore, the multi-compartment neuron structure is believed to contribute to the basic behavior of biological neural systems that the brain associates perceived sensory inputs with the internal world model to make predictions [143]. Moreover, the interactions between dendritic and somatic compartments provide a rational perspective to explain how BioNNs possibly implement information backpropagation. In conclusion, by incorporating dendritic features, it is expected to develop more biologically plausive training algorithms to enhance the learning ability and energy efficiency of neuromorphic computing hardware.

However, CMOS-based implementations of the dendrite function generally involve a large number of transistors [144]. It is therefore highly desired to emulate the critical functions of dendrites, such as nonlinear integration, signal regeneration, etc. using emerging devices in a more compact fashion. For example, Li et al. fabricated a Pt/TaO_x/AlO_y/Al dynamic memristor device to emulate the dendritic nonlinear integration and filtering functions [133]. A 1 kb HfO_x-based 1T1R artificial synapse array was integrated with the developed artificial dendrites and NbO_x-based Mott memristors as artificial somas to implement a complete neural network, in which all the three key computing components were realized by three different types of memristors. To evaluate the performance of such a hardware-implemented neural network, it was used to recognize digits in the Street View House Numbers (SVHN) dataset by extracting critical information from a noisy background. Experimental results indicate that the incorporation of artificial dendrites in the network could significantly improve both classification accuracy and power efficiency.

It should be pointed out that dendritic computing using emerging neuromorphic devices is still at the early stage of research, and the filtering and integration function implemented in previous works are just part of many important functions of biological dendrites. New designs of artificial dendrites with richer internal dynamics and also co-optimizations in devices, algorithms, circuits, and architectures are needed to build a more intelligent neural network that can handle sophisticated tasks.

6.4.3 Reservoir Computing

By introducing recurrent connections, RNNs are capable of processing continuous inputs with arbitrary lengths. Previous output states are fed back and combined with the current external inputs to serve as the input to the neural processing units. With this short-term memory property, RNNs are applicable to handle temporal tasks. However, the gradient descent-based training of RNN is prone to either vanishing or exploding gradient problem [145]. A new computing paradigm, called reservoir computing (RC), provides an efficient solution to the RNN training [146]. A typical RC network consists of two separate parts: a recurrent network called a reservoir and a fully connected part for readout. The reservoir is responsible for mapping the input data to a linear segmentable high-dimensional space so that only the readout part needs to be trained, as shown in Figure 6.4c.

Devices with short-term memory could be used in reservoirs to eliminate the requirement of complex feedback connections. For example, WO_x memristors were used as neuron nodes in the reservoir to receive the time series of the input data. Meanwhile, by adding a pretrained readout part, the system successfully implemented tasks like time series forecasting [134, 147]. Zhu et al. [117] demonstrated that an RC hardware system based on perovskite halide-based memristors could be directly driven by the detected neural signals, thanks to the low programming voltage (tens of millivolts), helping to construct an integrated system to sense and analyze electrophysiological data simultaneously.

Nevertheless, a fixed network initialized with random connections generally cannot satisfy the computing requirements on all occasions. To enrich the reservoir stats of a single memristor, Zhong et al. [148] applied a mask operation to implement the time multiplex and achieved an accuracy of 97.6% on the MNIST dataset. Moreover, theories suggest that inputting information encoded by networks at the edge of chaos into a simple linear classifier can emulate a broad range of complex nonlinear functions [149]. Therefore, edge-of-chaos found in spin-torque nano-oscillators and self-organized atomic switch networks [150, 151] could be utilized to optimize the reservoirs.

6.4.4 Oscillatory Neural Network

Inspired by the rhythmic behaviors in BioNNs, oscillatory neural networks (ONNs) [152] were proposed. ONNs are composed of mutually coupled oscillators, where information is encoded as the state of phases and oscillation frequencies. ONN systems tend to converge to a stable state due to the coupling between two arbitrary oscillators, which could serve as autocorrelative associative memory and implement pattern recognition tasks. Figure 6.4d depicts the schematic of a typical ONN system.

Many research works have revealed the potentials of emerging devices in constructing compact oscillators, such as nanoscale magnetic tunnel junctions (MTJs) [135] and Mott devices [153]. Compared with CMOS-based oscillators, these neuromorphic devices could facilitate more compact hardware implementations of ONNs. For example, Romera et al. [135] experimentally exhibited that a hardware network composed of spin-torque nano-oscillators, whose oscillation frequency could be precisely tuned in a wide range, was capable of realizing spoken vowel recognition. Four nano-oscillators were serially connected and mutually coupled through the propagating microwave currents. After the online training, the combinations of oscillation frequencies corresponding to different categories of input vowels were scattered, yielding the classification result. Although several demonstrations have been reported, hardware implementations of ONN based on emerging devices are limited to small-scale networks so far. To construct large-scale systems, mismatches and variabilities of devices should be suppressed [154]. Furthermore, efficient implementations of electrical coupling between oscillators are necessary to improve the scalability of ONN hardware.

6.4.5 Hopfield Neural Network and Simulated Annealing

Hopfield neural network (HNN) is a fully connected recurrent network without self-connections, where each neuron state switches between +1 and −1 according to the accumulation of weighted synaptic inputs [155]. Under the adaptation algorithm, the system could be attracted to a local stable state by iteratively decreasing the system energy. Thus, HNN can implement associative memory and solve optimization problems.

Memristive device crossbars have the potential in accelerating the update of HNN neuron states that involves a large amount of dot product operations [156, 157]. Furthermore, to avoid the system getting stuck in a local minimum, the simulated annealing algorithm introduces a noise factor, which decays with increasing iterations. With this random disturbance, the system could jump out from local minima and likely converge to a global minimum [158], as shown in Figure 6.4e. By utilizing the current-controlled stochastic switching behaviors, integrating devices like MTJs and conductive-bridge random-access memory (CBRAMs) [159, 160] with deterministic CMOS circuits can effectively implement the simulated annealing. In the work elucidated in [161], the decay of noise intensity was mimicked by introducing a hysteretic threshold to determine the next neuron state under the condition where the noise source intensity was constant. Besides adding noise, HNN hardware systems with chaotic dynamics based on emerging memristive devices also perform well in finding the energy minimum to solve optimization problems [136, 162].

6.5 Summary and Outlook

As reviewed above, emerging resistive switching devices, such as RRAM and PCM, have shown great promise in embedded memory applications. With artificial neural networks exhibiting superior performance on many intelligent tasks, highly efficient hardware implementations of neural networks, including both DNNs and SNNs, are attracting more and more research interests. Neuromorphic computing emerges as a new computing paradigm that could potentially eliminate a large amount of power consumption and latency induced by frequent data shuffling between the physically separated data storage and computing units in traditional von Neumann architecture. To achieve that, many custom hardware systems accelerate neuromorphic computing by performing intensive computations in memory and implementing artificial synapses and neurons to replicate the working mechanisms of BioNNs.

While the CMOS-based neuromorphic hardware has a potential scalability issue amid the slowdown of Moore's law scaling, analog resistive switching devices have shown great potentials to implement neuromorphic computing systems, where MVM operations could be conveniently performed on synaptic crossbars with extremely high efficiency. Cases of array demonstrations and chip implementations with both in situ and ex situ training for various networks and datasets are all reviewed to highlight the feasibility of building power-efficient neuromorphic computing hardware based on emerging memory devices. The limited precision and nonideal device characteristics, such as nonlinearity and asymmetry, would cause a certain degree of accuracy degradation compared with the software baseline. Nevertheless, there is plenty of room for algorithm–hardware co-optimizations to alleviate the effect of nonidealities and further enhance the system performance.

In addition to being used for MVM acceleration in neuromorphic computing, by leveraging the dynamic behaviors of emerging memory devices, it is possible to design compact neural modules that feature synaptic and neuronal behaviors without introducing too much overhead. Simulations and experimental demonstrations have been realized to exhibit brain-like learning capabilities. Furthermore, other brain-inspired computing deriving benefits from neuromorphic devices has also been reviewed, including HD computing, dendritic computing, RC, ONNs, and HNNs with simulated annealing.

Moving forward, we would like to give a brief perspective on the future development of neuromorphic computing hardware systems based on emerging neuromorphic devices.

For DNNs, firstly, synaptic devices need further optimizations to improve the programming linearity and symmetry to improve the training efficiency, lower the operating voltage to reduce the power consumption, enlarge the dynamic range and level numbers to enhance noise tolerance, etc. In terms of hardware systems, the accuracy, latency, and area overhead need to be carefully weighed to avoid overly redundant peripheral circuitry that may diminish the efficiency benefits of synaptic array-based CIM. System flexibility is another key point. A multi-core architecture consisting of reconfigurable dataflow routing and tunable computing precision, along with a matching compiler, is necessary to accommodate the requirements of different network structures and algorithms. Moreover, simulators [163] covering the software framework for algorithm development, complier for network mapping, and backend simulator for benchmark are required to facilitate system optimizations.

For SNNs, besides improving the reliability of devices, the main goal of device optimization is to precisely emulate various behaviors of neural units by leveraging intrinsic device properties. Moreover, one of the key challenges in fulfilling the potentials of SNNs is the lack of proper benchmarks and training algorithms. The intrinsic capability of temporal information processing and event-driven induced sparseness has not been well exploited. Consequently, developing SNN-efficient algorithms and suitable datasets are necessary. For example, bio-plausive training algorithms that integrate local synaptic plasticity and information backpropagation [164] are expected to realize high energy efficiency in neuromorphic computing hardware. Furthermore, the brain–machine interface [165] is a promising application scenario where SNNs could directly process recorded spike signals that have very rich temporal dynamics.

Finally, for techniques of brain-inspired computing like RC and HNNs, further optimizations for dealing with more complex tasks are required. The reported experimental works are mostly proof-of-concept demonstrations just to verify the feasibility while fully integrated hardware systems and comprehensive evaluations for the performance and efficiency are still lacking.

In summary, it is quite promising to apply various emerging devices in neuromorphic computing hardware systems to accelerate the computations, increase energy efficiency, and improve scalability. However, to take full advantage of the unique properties of these devices, device algorithm–architecture co-optimizations are indispensable.

References

1 McCulloch, W.S. and Pitts, W. (1943). A logical calculus of the ideas immanent in nervous activity. *Bull. Math. Biophys.* 5 (4): 115–133.

2 Farley, B. and Clark, W. (1954). Simulation of self-organizing systems by digital computer. *Trans. IRE Profession. Group Informat. Theory.* 4 (4): 76–84.

3 Rosenblatt, F. (1958). The perceptron: a probabilistic model for information storage and organization in the brain. *Psychol. Rev.* 65 (6): 386–408.

4 Hebb, D.O. (2005). *The Organization of Behavior: A Neuropsychological Theory.* Oxfordshire: Taylor & Francis.

5 Werbos, P. (1974). *Beyond Regression: New Tools for Prediction and Analysis in the Behavioral Sciences.* Massachusetts, USA: Harvard University.

6 Werbos, P.J. (1990). Backpropagation through time: what it does and how to do it. *Proc. IEEE.* 78 (10): 1550–1560.

7 LeCun, Y., Bengio, Y., and Hinton, G. (2015). Deep learning. *Nature* 521 (7553): 436–444.

8 Roy, K., Jaiswal, A., and Panda, P. (2019). Towards spike-based machine intelligence with neuromorphic computing. *Nature* 575 (7784): 607–617.

9 Tang, J., Yuan, F., Shen, X. et al. (2019). Bridging biological and artificial neural networks with emerging neuromorphic devices: fundamentals, progress and challenges. *Adv. Mater.* 31 (49): e1902761.

10 Kandel, E.R., Schwartz, J.H., Jessel, T.M. et al. (2000). *Principles of Neural Science.* New York: McGraw-Hill Professional Pub.

11 Mead, C. (1990). Neuromorphic electronic systems. *Proc. IEEE.* 78 (10): 1629–1636.

12 Zhang, W., Gao, B., Tang, J. et al. (2020). Neuro-inspired computing chips. *Nat. Electron.* 3 (7): 371–382.

13 Sebastian, A., Le Gallo, M., Khaddam-Aljameh, R., and Eleftheriou, E. (2020). Memory devices and applications for in-memory computing. *Nat. Nanotechnol.* 15 (7): 529–544.

14 Benjamin, B.V., Gao, P., McQuinn, E. et al. (2014). Neurogrid: a mixed-analog-digital multichip system for large-scale neural simulations. *Proc. IEEE* 102 (5): 699–716.

15 Neckar, A., Fok, S., Benjamin, B.V. et al. (2019). Braindrop: a mixed-signal neuromorphic architecture with a dynamical systems-based programming model. *Proc. IEEE.* 107 (1): 144–164.

16 Qiao, N., Mostafa, H., Corradi, F. et al. (2015). A reconfigurable on-line learning spiking neuromorphic processor comprising 256 neurons and 128K synapses. *Front. Neurosci.* 9: 141.

17 Merolla, P.A., Arthur, J.V., Alvarez-Icaza, R. et al. (2014). Artificial brains. A million spiking-neuron integrated circuit with a scalable communication network and interface. *Science* 345 (6197): 668–673.

18 Davies, M., Srinivasa, N., Lin, T.-H. et al. (2018). Loihi: a neuromorphic many-core processor with on-Chip learning. *IEEE Micro.* 38 (1): 82–99.

19 Ma, D., Shen, J., Gu, Z. et al. (2017). Darwin: a neuromorphic hardware co-processor based on spiking neural networks. *J. Sys. Architec.* 77: 43–51.

20 Yuasa, S., Nagahama, T., Fukushima, A. et al. (2004). Giant room-temperature magnetoresistance in single-crystal Fe/MgO/Fe magnetic tunnel junctions. *Nat. Mater.* 3 (12): 868–871.

21 Hoffman, J., Pan, X., Reiner, J.W. et al. (2010). Ferroelectric field effect transistors for memory applications. *Adv. Mater.* 22 (26–27): 2957–2961.

22 Wong, H.S.P., Raoux, S., Kim, S. et al. (2010). Phase change memory. *Proc. IEEE.* 98 (12): 2201–2227.

23 Tang, J., Bishop, D., Kim, S. et al. (eds.) (2018). ECRAM as scalable synaptic cell for high-speed, low-power neuromorphic computing. In: *2018 IEEE International Electron Devices Meeting (IEDM)*. San Francisco, CA: IEEE.

24 Wu, W., Wu, H., Gao, B. et al. (2017). Improving analog switching in HfO_x-based resistive memory with a thermal enhanced layer. *IEEE Elect. Dev. Letts.* 38 (8): 1019–1022.

25 Chouhan, S., Kumbhare, P., Khanna, A. et al. (eds.) (2017). Effect of thermal resistance and scaling on dc-IV characteristics of PCMO based RRAM devices. In: *2017 75th Annual Device Research Conference (DRC)*. South Bend, IN: IEEE.

26 Servalli, G. (ed.) (2009). A 45 nm generation phase change memory technology. In: *2009 IEEE International Electron Devices Meeting (IEDM)*. San Francisco, CA: IEEE.

27 Ikeda, S., Miura, K., Yamamoto, H. et al. (2010). A perpendicular-anisotropy CoFeB–MgO magnetic tunnel junction. *Nat. Mater.* 9 (9): 721–724.

28 Kim, S., Todorov, T., Onen, M. et al. (eds.) (2019). Metal-oxide based, CMOS-compatible ECRAM for deep learning accelerator. In: *2019 IEEE International Electron Devices Meeting (IEDM)*. San Francisco, CA: IEEE.

29 Mulaosmanovic, H., Ocker, J., Müller, S. et al. (eds.) (2017). Novel ferroelectric FET based synapse for neuromorphic systems. In: *2017 Symposium on VLSI Technology*. Kyoto, Japan: IEEE.

30 Xi, Y., Gao, B., Tang, J. et al. (2020). In-memory learning with analog resistive switching memory: a review and perspective. *Proc. IEEE.*: 1–29.

31 Zhao, M., Gao, B., Tang, J. et al. (2020). Reliability of analog resistive switching memory for neuromorphic computing. *Appl. Phys. Rev.* 7 (1): 1–2.

32 Wu, H., Wang, X.H., Gao, B. et al. (2017). Resistive random access memory for future information processing system. *Proc. IEEE.* 105 (9): 1770–1789.

33 Islam, R., Li, H., Chen, P.-Y. et al. (2019). Device and materials requirements for neuromorphic computing. *J. Phys. D: Appl. Phys.* 52 (11): 721–724.

34 Zhang, W., Gao, B., Tang, J. et al. (2019). Analog-type resistive switching devices for neuromorphic computing. *Phys. Status Solidi (RRL) – Rapid Res. Lett.* 13 (10): 35.7.1–35.7.4.

35 Yu, S. and Chen, P.-Y. (2016). Emerging memory technologies: recent trends and prospects. *IEEE Solid-State Circuits Mag.* 8 (2): 43–56.

36 Gokmen, T. and Vlasov, Y. (2016). Acceleration of deep neural network training with resistive cross-point devices: design considerations. *Front. Neurosci.* 10: 333.

37 Li, C., Belkin, D., Li, Y. et al. (2018). Efficient and self-adaptive in-situ learning in multilayer memristor neural networks. *Nat. Commun.* 9 (1): 2385.

38 Li, C., Wang, Z., Rao, M. et al. (2019). Long short-term memory networks in memristor crossbar arrays. *Nat. Mach. Intell.* 1 (1): 49–57.

39 Wang, Z., Li, C., Lin, P. et al. (2019). In situ training of feed-forward and recurrent convolutional memristor networks. *Nat. Mach. Intell.* 1 (9): 434–442.

40 Wang, Z., Li, C., Song, W. et al. (2019). Reinforcement learning with analogue memristor arrays. *Nat. Electron.* 2 (3): 115–124.

41 Cai, F., Correll, J.M., Lee, S.H. et al. (2019). A fully integrated reprogrammable memristor–CMOS system for efficient multiply–accumulate operations. *Nat. Electron.* 2 (7): 290–299.

42 Choi, S., Shin, J.H., Lee, J. et al. (2017). Experimental demonstration of feature extraction and dimensionality reduction using memristor networks. *Nano. Lett.* 17 (5): 3113–3118.

43 Sheridan, P.M., Cai, F., Du, C. et al. (2017). Sparse coding with memristor networks. *Nat. Nanotechnol.* 12 (8): 784–789.

44 Yao, P., Wu, H., Gao, B. et al. (2017). Face classification using electronic synapses. *Nat. Commun.* 8: 15199.

45 Liu, Q., Gao, B., Yao, P. et al. (eds.) (2020). 33.2 A fully integrated analog ReRAM based 78.4TOPS/W compute-in-memory chip with fully parallel MAC computing. In: *2020 IEEE International Solid-State Circuits Conference – (ISSCC)*. San Francisco, CA: IEEE.

46 Bayat, F.M., Prezioso, M., Chakrabarti, B. et al. (2018). Implementation of multilayer perceptron network with highly uniform passive memristive crossbar circuits. *Nat. Commun.* 9 (1): 2331.

47 Prezioso, M., Merrikh-Bayat, F., Hoskins, B.D. et al. (2015). Training and operation of an integrated neuromorphic network based on metal-oxide memristors. *Nature* 521 (7550): 61–64.

48 Zhou, Z., Huang, P., Xiang, Y. et al. (eds.) (2018). A new hardware implementation approach of BNNs based on nonlinear 2T2R synaptic cell. In: *2018 IEEE International Electron Devices Meeting (IEDM)*. San Francisco, CA: IEEE.

49 Lin, Y., Zhang, Q., Tang, J. et al. (eds.) (2019). Bayesian neural network realization by exploiting inherent stochastic characteristics of analog RRAM. In: *2019 IEEE International Electron Devices Meeting (IEDM)*. San Francisco, CA: IEEE.

50 Joshi, V., Le Gallo, M., Haefeli, S. et al. (2020). Accurate deep neural network inference using computational phase-change memory. *Nat. Commun.* 11 (1): 1–8.

51 Su, F., Chen, W., Xia, L. et al. (eds.) (2017). A 462GOPs/J RRAM-based nonvolatile intelligent processor for energy harvesting IoE system featuring nonvolatile logics and processing-in-memory. In: *2017 Symposium on VLSI Technology*. Kyoto, Japan: IEEE.

52 Chen, W., Li, K., Lin, W. et al. (eds.) (2018). A 65 nm 1 Mb nonvolatile computing-in-memory ReRAM macro with sub-16 ns multiply-and-accumulate for binary DNN AI edge processors. In: *2018 IEEE International Solid-State Circuits Conference – (ISSCC)*. San Francisco, CA: IEEE.

53 Mochida, R., Kouno, K., Hayata, Y. et al. (eds.) (2018). A 4 M synapses integrated analog ReRAM based 66.5 TOPS/W neural-network processor with cell current controlled writing and flexible network architecture. In: *2018 IEEE Symposium on VLSI Technology*. Honolulu, HI: IEEE.

54 Xue, C.-X., Chen, W.-H., Liu, J.-S. et al. (eds.) (2019). 24.1 A 1 Mb multibit ReRAM computing-in-memory macro with 14.6 ns parallel MAC computing time for CNN based AI edge processors. In: *2019 IEEE International Solid-State Circuits Conference – (ISSCC)*. San Francisco, CA: IEEE.

55 Zha, Y., Nowak, E., and Li, J. (eds.) (2019). Liquid silicon: a nonvolatile fully programmable processing-in-memory processor with monolithically integrated ReRAM for big data/machine learning applications. In: *2019 Symposium on VLSI Circuits*. Kyoto, Japan: IEEE.

56 Xue, C., Huang, T., Liu, J. et al. (eds.) (2020). 15.4 A 22 nm 2 Mb ReRAM compute-in-memory macro with 121-28TOPS/W for multibit MAC computing for tiny AI edge devices. In: *2020 IEEE International Solid-State Circuits Conference – (ISSCC)*. San Francisco, CA: IEEE.

57 Yao, P., Wu, H., Gao, B. et al. (2020). Fully hardware-implemented memristor convolutional neural network. *Nature* 577 (7792): 641–646.

58 Luo, Y. and Yu, S. (2020). Accelerating deep neural network in-situ training with non-volatile and volatile memory based hybrid precision synapses. *IEEE Trans. Comp.* 69 (8): 1113–1127.

59 Shafiee, A., Nag, A., Muralimanohar, N. et al. (eds.) (2016). ISAAC: a convolutional neural network accelerator with in-situ analog arithmetic in crossbars. In: *2016 ACM/IEEE 43rd Annual International Symposium on Computer Architecture (ISCA)*. Seoul, Korea: IEEE.

60 Nag, A., Balasubramonian, R., Srikumar, V. et al. (2018). Newton: gravitating towards the physical limits of crossbar acceleration. *IEEE Micro.* 38 (5): 41–49.

61 Ji, Y., Zhang, Y., Chen, W., and Xie, Y. (2018). Bridge the gap between neural networks and neuromorphic hardware with a neural network compiler. In: *Proceedings of the Twenty-Third International Conference on Architectural Support for Programming Languages and Operating Systems*, 448–460. Williamsburg, VA: Association for Computing Machinery.

62 Courbariaux, M., Hubara, I., Soudry, D. et al. (2016). Binarized neural networks: training deep neural networks with weights and activations constrained to +1 or −1. In: arXiv preprint arXiv: 160202830.

63 Sun, X., Peng, X., Chen, P.-Y. et al. (eds.) (2018, 2018). Fully parallel RRAM synaptic array for implementing binary neural network with (+1,−1) weights and (+1, 0) neurons. In: *23rd Asia and South Pacific Design Automation Conference (ASP-DAC)*. Jeju, Korea: IEEE.

64 Pan, Y., Ouyang, P., Zhao, Y. et al. (2018). A multilevel cell STT-MRAM-based computing in-memory accelerator for binary convolutional neural network. *IEEE Trans. Mag.* 54 (11): 1–5.

65 Feinberg, B., Wang, S., and Ipek, E. (2018). Making memristive neural network accelerators reliable. In: *IEEE International Symposium on High Performance Computer Architecture (HPCA)*. Vienna, Austria: IEEE.

66 Liu, T., Wen, W., Jiang, L. et al. (2019). A fault-tolerant neural network architecture. In: *56th ACM/IEEE Design Automation Conference (DAC)*. Las Vegas, NV: IEEE.

67 Kim, H., Kim, T., Kim, J., and Kim, J.-J. (2018). Deep neural network optimized to resistive memory with nonlinear current–voltage characteristics. *ACM J. Emerg. Technol. Comput. Syst.* 14 (2): 1–17.

68 Ambrogio, S., Gallot, M., Spoon, K. et al. (eds.) (2019). Reducing the impact of phase-change memory conductance drift on the inference of large-scale hardware neural networks. In: *2019 IEEE International Electron Devices Meeting (IEDM)*. San Francisco, CA: IEEE.

69 Agarwal, S., Gedrim, R.B.J., Hsia, A.H. et al. (eds.) (2017). Achieving ideal accuracies in analog neuromorphic computing using periodic carry. In: *2017 Symposium on VLSI Technology*. Kyoto, Japan: IEEE.

70 Chen, P.-Y., Lin, B., Wang, I.-T. et al. (eds.) (2015). Mitigating effects of non-ideal synaptic device characteristics for on-chip learning. In: *2015 IEEE/ACM International Conference on Computer-Aided Design (ICCAD)*. Austin, TX: IEEE.

71 Jeong, Y., Kim, S., and Lu, W.D. (2015). Utilizing multiple state variables to improve the dynamic range of analog switching in a memristor. *Appl. Phys. Lett.* 107 (17): 173105.

72 Gokmen, T. and Haensch, W. (2020). Algorithm for training neural networks on resistive device arrays. *Front. Neurosci.* 14: 103.

73 Ambrogio, S., Narayanan, P., Tsai, H. et al. (2018). Equivalent-accuracy accelerated neural-network training using analogue memory. *Nature* 558 (7708): 60–67.

74 Zhang, Q., Wu, H., Yao, P. et al. (2018). Sign backpropagation: an on-chip learning algorithm for analog RRAM neuromorphic computing systems. *Neural Netw.* 108: 217–223.

75 Yang, T.-J. and Sze, V. (eds.) (2019). Design considerations for efficient deep neural networks on processing-in-memory accelerators. In: *2019 IEEE International Electron Devices Meeting (IEDM)*. San Francisco, CA: IEEE.

76 Hu, M., Strachan, J.P., Li, Z. et al. (eds.) (2016). Dot-product engine for neuromorphic computing: programming 1T1M crossbar to accelerate matrix-vector multiplication. In: *2016 53nd ACM/EDAC/IEEE Design Automation Conference (DAC)*. Austin, TX: IEEE.

77 Liu, B., Li, H., Chen, Y. et al. (eds.) (2014). Reduction and IR-drop compensations techniques for reliable neuromorphic computing systems. In: *2014 IEEE/ACM International Conference on Computer-Aided Design (ICCAD)*. San Jose, CA: IEEE.

78 Eryilmaz, S.B., Kuzum, D., Yu, S., and Wong, H.-S.P. (eds.) (2015). Device and system level design considerations for analog-non-volatile-memory based neuromorphic architectures. In: *2015 IEEE International Electron Devices Meeting (IEDM)*. San Francisco, CA: IEEE.

79 Woo, J. and Yu, S. (2019). Impact of selector devices in analog RRAM-based crossbar arrays for inference and training of neuromorphic system. *IEEE Trans. Very Large Scale Integr. (VLSI) Sys.* 27 (9): 2205–2212.

80 Yan, B., Liu, M., Chen, Y. et al. (eds.) (2019). On designing efficient and reliable nonvolatile memory-based computing-in-memory accelerators. In: *2019 IEEE International Electron Devices Meeting (IEDM)*. San Francisco, CA: IEEE.

81 He, Z., Lin, J., Ewetz, R. et al. (2019). Noise injection adaption: end-to-end reram crossbar non-ideal effect adaption for neural network mapping. In: *Proceedings of the 56th Annual Design Automation Conference 2019*. Las Vegas, NV: IEEE.

82 Jeong, Y., Zidan, M.A., and Lu, W.D. (2018). Parasitic effect analysis in memristor-array-based neuromorphic systems. *IEEE Trans. Nanotechnol.* 17 (1): 184–193.

83 Song, S., Miller, K.D., and Abbott, L.F. (2000). Competitive Hebbian learning through spike-timing-dependent synaptic plasticity. *Nat. Neurosci.* 3 (9): 919–926.

84 Chistiakova, M., Bannon, N.M., Chen, J.-Y. et al. (2015). Homeostatic role of heterosynaptic plasticity: models and experiments. *Front. Computat. Neurosci.* 9: 89.

85 Zucker, R.S. and Regehr, W.G. (2002). Short-term synaptic plasticity. *Ann. Rev. Phys.* 64 (1): 355–405.

86 Izhikevich, E.M. (2004). Which model to use for cortical spiking neurons? *IEEE Trans. Neural Netw.* 15 (5): 1063–1070.

87 London, M. and Hausser, M. (2005). Dendritic computation. *Annu. Rev. Neurosci.* 28: 503–532.

88 Pfeiffer, M. and Pfeil, T. (2018). Deep learning with spiking neurons: opportunities and challenges. *Front. Neurosci.* 12: 774.

89 Diehl, P.U., Neil, D., Binas, J. et al. (eds.) (2015). Fast-classifying, high-accuracy spiking deep networks through weight and threshold balancing. In: *International Joint Conference on Neural Networks (IJCNN)*. Killarney, Ireland: IEEE.

90 Huh, D. and Sejnowski, T.J. (eds.) (2018). Gradient descent for spiking neural networks. In: *Advances in Neural Information Processing Systems*. New York: Curran Associates, Inc.

91 Diehl, P.U. and Cook, M. (2015). Unsupervised learning of digit recognition using spike-timing-dependent plasticity. *Front. Comput. Neurosci.* 9: 99.

92 Ponulak, F. and Kasinski, A. (2010). Supervised learning in spiking neural networks with ReSuMe: sequence learning, classification, and spike shifting. *Neu. Comput.* 22 (2): 467–510.

93 Tavanaei, A. and Maida, A.J.N. (2019). BP-STDP: approximating backpropagation using spike timing dependent plasticity. *Neurocomputing* 330: 39–47.

94 Gerstner, W., Kempter, R., van Hemmen, J.L., and Wagner, H. (1996). A neuronal learning rule for sub-millisecond temporal coding. *Nature* 383 (6595): 76–78.

95 Bair, W. and Koch, C. (1996). Temporal precision of spike trains in extrastriate cortex of the behaving macaque monkey. *Neu. Comput.* 8 (6): 1185–1202.

96 Mostafa, H. (2018). Supervised learning based on temporal coding in spiking neural networks. *IEEE Trans. Neural Netw. Learn. Syst.* 29 (7): 3227–3235.

97 Thorpe, S. and Gautrais, J. (1998). Rank order coding. In: *Computational Neuroscience*, 113–118. Springer.

98 Tavanaei, A., Ghodrati, M., Kheradpisheh, S.R. et al. (2019). Deep learning in spiking neural networks. *Neural Netw.* 111: 47–63.

99 Deng, L., Wu, Y., Hu, X. et al. (2020). Rethinking the performance comparison between SNNS and ANNS. *Neural Netw.* 121: 294–307.

100 Jo, S.H., Chang, T., Ebong, I. et al. (2010). Nanoscale memristor device as synapse in neuromorphic systems. *Nano Lett.* 10 (4): 1297–1301.

101 Ambrogio, S., Balatti, S., Milo, V. et al. (eds.) (2016). Novel RRAM-enabled 1T1R synapse capable of low-power STDP via burst-mode communication and real-time unsupervised machine learning. In: *2016 IEEE Symposium on VLSI Technology*. Honolulu, HI: IEEE.

102 Zamarreno-Ramos, C., Camunas-Mesa, L.A., Perez-Carrasco, J.A. et al. (2011). On spike-timing-dependent-plasticity, memristive devices, and building a self-learning visual cortex. *Front. Neurosci.* 5: 26.

103 Pedretti, G., Bianchi, S., Milo, V. et al. (eds.) (2017). Modeling-based design of brain-inspired spiking neural networks with RRAM learning synapses. In: *2017 IEEE International Electron Devices Meeting (IEDM)*. San Francisco, CA: IEEE.

104 Guo, Y., Wu, H., Gao, B., and Qian, H. (2019). Unsupervised learning on resistive memory array based spiking neural networks. *Front Neurosci.* 13: 812.

105 Chang, C., Chen, P., Hudec, B. et al. (eds.) (2018). Interchangeable Hebbian and anti-Hebbian STDP applied to supervised learning in spiking neural network. In: *2018 IEEE International Electron Devices Meeting (IEDM)*. San Francisco, CA: IEEE.

106 Wang, W., Pedretti, G., Milo, V. et al. (2018). Learning of spatiotemporal patterns in a spiking neural network with resistive switching synapses. *Sci. Adv.* 4 (9): eaat4752.

107 Kim, S., Du, C., Sheridan, P. et al. (2015). Experimental demonstration of a second-order memristor and its ability to biorealistically implement synaptic plasticity. *Nano Lett.* 15 (3): 2203–2211.

108 Zhang, Y., He, W., Wu, Y. et al. (2018). Highly compact artificial memristive neuron with low energy consumption. *Small.* 14 (51): e1802188.

109 Yi, W., Tsang, K.K., Lam, S.K. et al. (2018). Biological plausibility and stochasticity in scalable VO_2 active memristor neurons. *Nat. Commun.* 9 (1): 4661.

110 Boybat, I., Le Gallo, M., Nandakumar, S.R. et al. (2018). Neuromorphic computing with multi-memristive synapses. *Nat. Commun.* 9 (1): 2514.

111 Ventra, M.D., Pershin, Y.V., and Chua, L.O. (2009). Circuit elements with memory: memristors, memcapacitors, and meminductors. *Proc. IEEE.* 97 (10): 1717–1724.

112 Wang, Z., Joshi, S., Savel'ev, S.E. et al. (2017). Memristors with diffusive dynamics as synaptic emulators for neuromorphic computing. *Nat. Mater.* 16 (1): 101–108.

113 Ohno, T., Hasegawa, T., Tsuruoka, T. et al. (2011). Short-term plasticity and long-term potentiation mimicked in single inorganic synapses. *Nat. Mater.* 10 (8): 591–595.

114 Kim, M.K. and Lee, J.S. (2018). Short-term plasticity and long-term potentiation in artificial biosynapses with diffusive dynamics. *ACS Nano.* 12 (2): 1680–1687.

115 Yang, Y., Chen, B., and Lu, W.D. (2015). Memristive physically evolving networks enabling the emulation of heterosynaptic plasticity. *Adv. Mater.* 27 (47): 7720–7727.

116 Sangwan, V.K., Lee, H.S., Bergeron, H. et al. (2018). Multi-terminal memtransistors from polycrystalline monolayer molybdenum disulfide. *Nature* 554 (7693): 500–504.

117 Fu, T., Liu, X., Gao, H. et al. (2020). Bioinspired bio-voltage memristors. *Nat. Commun.* 11 (1): 1861.

118 Indiveri, G., Linares-Barranco, B., Hamilton, T.J. et al. (2011). Neuromorphic silicon neuron circuits. *Front. Neurosci.* 5: 73.

119 Tuma, T., Pantazi, A., Le Gallo, M. et al. (2016). Stochastic phase-change neurons. *Nat. Nanotechnol.* 11 (8): 693–699.

120 Zhang, X., Wang, W., Liu, Q. et al. (2018). An artificial neuron based on a threshold switching memristor. *IEEE Electron. Device Lett.* 39 (2): 308–311.

121 You, Z. and Ramanathan, S. (2015). Mott Memory and Neuromorphic Devices. *Proc. IEEE* 103 (8): 1289–1310.

122 Pickett, M.D., Medeiros-Ribeiro, G., and Williams, R.S. (2013). A scalable neuristor built with Mott memristors. *Nat. Mater.* 12 (2): 114–117.

123 Chen, C., Yang, M., Liu, S. et al. (eds.) (2019). Bio-inspired neurons based on novel leaky-FeFET with ultra-low hardware cost and advanced functionality for all-ferroelectric neural network. In: *2019 Symposium on VLSI Technology*. Kyoto, Japan: IEEE.

124 Luo, J., Yu, L., Liu, T. et al. (eds.) (2019). Capacitor-less stochastic leaky-FeFET neuron of both excitatory and inhibitory connections for SNN with reduced hardware cost. In: *2019 IEEE International Electron Devices Meeting (IEDM)*. San Francisco, CA: IEEE.

125 Wang, Z., Joshi, S., Savel'ev, S. et al. (2018). Fully memristive neural networks for pattern classification with unsupervised learning. *Nat. Electron.* 1 (2): 137–145.

126 Wang, Z., Rao, M., Han, J.W. et al. (2018). Capacitive neural network with neuro-transistors. *Nat. Commun.* 9 (1): 3208.

127 Yan, B., Yang, Q., Chen, W. et al. (eds.) (2019). RRAM-based Spiking Nonvolatile Computing-In-Memory Processing Engine with Precision-Configurable In Situ Nonlinear Activation. In: *2019 Symposium on VLSI Technology*. Kyoto, Japan: IEEE.

128 Zhang, X., Wang, Z., Song, W. et al. (eds.) (2019). Experimental demonstration of conversion-based SNNs with 1T1R mott neurons for neuromorphic inference. In: *2019 IEEE International Electron Devices Meeting (IEDM)*. San Francisco, CA: IEEE.

129 Zhang, X., Lu, J., Wang, R., et al. (2021). Hybrid memristor-CMOS neurons for in-situ learning in fully hardware memristive spiking neural networks. *Sci. Bull.* 66 (16): 1624–1633.

130 Wan, W., Kubendran, R., Eryilmaz, S.B. et al. (eds.) (2020). 33.1 A 74 TMACS/W CMOS-RRAM neurosynaptic core with dynamically reconfigurable dataflow and in-situ transposable weights for probabilistic graphical models. In: *2020 IEEE International Solid-State Circuits Conference – (ISSCC)*. San Francisco, CA: IEEE.

131 Woźniak, S., Pantazi, A., Bohnstingl, T., and Eleftheriou, E. (2020). Deep learning incorporating biologically inspired neural dynamics and in-memory computing. *Nat. Mach. Intell.* 2 (6): 325–336.

132 Wu, T.F., Li, H., Huang, P.-C. et al. (2018). Hyperdimensional computing exploiting carbon nanotube FETs, resistive RAM, and their monolithic 3D integration. *IEEE J. Solid-State Circuits.* 53 (11): 3183–3196.

133 Li, X., Tang, J., Zhang, Q. et al. (2020). Power-efficient neural network with artificial dendrites. *Nat. Nanotechnol.* 15 (9): 776–782.

134 Moon, J., Ma, W., Shin, J.H. et al. (2019). Temporal data classification and forecasting using a memristor-based reservoir computing system. *Nat. Electron.* 2 (10): 480–487.

135 Romera, M., Talatchian, P., Tsunegi, S. et al. (2018). Vowel recognition with four coupled spin-torque nano-oscillators. *Nature* 563 (7730): 230–234.

136 Mahmoodi, M., Kim, H., Fahimi, Z. et al. (eds.) (2019). An analog neuro-optimizer with adaptable annealing based on 64 × 64 0T1R crossbar circuit. In: *2019 IEEE International Electron Devices Meeting (IEDM)*. San Francisco, CA: IEEE.

137 Kanerva, P. (2009). Hyperdimensional computing: an introduction to computing in distributed representation with high-dimensional random vectors. *Cognit. Comput.* 1 (2): 139–159.

138 Stuart, G.J. and Spruston, N. (2015). Dendritic integration: 60 years of progress. *Nat. Neurosci.* 18 (12): 1713–1721.

139 Sun, Z., Ambrosi, E., Bricalli, A., and Ielmini, D. (2018). Logic computing with stateful neural networks of resistive switches. *Adv. Mater.* 30 (38): e1802554.

140 Li, H., Wu, T.F., Rahimi, A. et al. (eds.) (2016). Hyperdimensional computing with 3D VRRAM in-memory kernels: device-architecture co-design for energy-efficient, error-resilient language recognition. In: *2016 IEEE International Electron Devices Meeting (IEDM)*. San Francisco, CA: IEEE.

141 Karunaratne, G., Le Gallo, M., Cherubini, G. et al. (2020). In-memory hyperdimensional computing. *Nat. Electron.* 3 (6): 327–337.

142 Payeur, A., Beique, J.C., and Naud, R. (2019). Classes of dendritic information processing. *Curr. Opin. Neurobiol.* 58: 78–85.

143 Larkum, M. (2013). A cellular mechanism for cortical associations: an organizing principle for the cerebral cortex. *Trends Neurosci.* 36 (3): 141–151.

144 Schemmel, J., Kriener, L., Müller, P., and Meier, K. (eds.) (2017). An accelerated analog neuromorphic hardware system emulating NMDA-and calcium-based non-linear dendrites. In: *2017 International Joint Conference on Neural Networks (IJCNN)*. Anchorage, AK: IEEE.

145 Pascanu, R., Mikolov, T., and Bengio, Y. (2013). On the difficulty of training recurrent neural networks. In: *Proceedings of the 30th International Conference on International Conference on Machine Learning-Volume 28*; Atlanta, GA, USA: JMLR.org, III–1310–III–1318.

146 Maass, W., Natschlager, T., and Markram, H. (2002). Real-time computing without stable states: a new framework for neural computation based on perturbations. *Neural Comput.* 14 (11): 2531–2560.

147 Du, C., Cai, F., Zidan, M.A. et al. (2017). Reservoir computing using dynamic memristors for temporal information processing. *Nat. Commun.* 8 (1): 2204.

148 Zhong, Y., Tang, J., Li, X. et al. (2021). Dynamic memristor-based reservoir computing for high-efficiency temporal signal processing. *Nat. Commun.* 12 (1): 408.

149 Bertschinger, N. and Natschlager, T. (2004). Real-time computation at the edge of chaos in recurrent neural networks. *Neural Comput.* 16 (7): 1413–1436.

150 Williame, J., Difini Accioly, A., Rontani, D. et al. (2019). Chaotic dynamics in a macrospin spin-torque nano-oscillator with delayed feedback. *Appl. Phys. Lett.* 114 (23).

151 Stieg, A.Z., Avizienis, A.V., Sillin, H.O. et al. (2012). Emergent criticality in complex turing B-type atomic switch networks. *Adv. Mater.* 24 (2): 286–293.

152 Izhikevich, E.M. (2000). Computing with oscillators. *Neural Netw.* 5255: 1–30.

153 Shukla, N., Tsai, W., Jerry, M. et al. (eds.) (2016). Ultra low power coupled oscillator arrays for computer vision applications. In: *2016 IEEE Symposium on VLSI Technology*. Honolulu, HI: IEEE.

154 Parihar, A., Shukla, N., Datta, S., and Raychowdhury, A. (2014). Exploiting synchronization properties of correlated electron devices in a non-Boolean computing fabric for template matching. *IEEE J. Emerg.Select. Topics Circuits Syst.* 4 (4): 450–459.

155 Hopfield, J.J. (1982). Neural networks and physical systems with emergent collective computational abilities. *Proc. Natl Acad. Sci. U S A.* 79 (8): 2554–2558.

156 Hu, S.G., Liu, Y., Liu, Z. et al. (2015). Associative memory realized by a reconfigurable memristive Hopfield neural network. *Nat. Commun.* 6 (1).

157 Zhou, Y., Wu, H., Gao, B. et al. (2019). Associative memory for image recovery with a high-performance memristor array. *Adv. Funct. Mater.* 29 (30).

158 Kirkpatrick, S., Gelatt, C.D. Jr., and Vecchi, M.P. (1983). Optimization by simulated annealing. *Science* 220 (4598): 671–680.

159 Shim, Y., Jaiswal, A., and Roy, K. (2017). Ising computation based combinatorial optimization using spin-Hall effect (SHE) induced stochastic magnetization reversal. *J. Appl. Phys.* 121 (19).

160 Shin, J.H., Jeong, Y.J., Zidan, M.A. et al. (eds.) (2018). Hardware acceleration of simulated annealing of spin glass by RRAM crossbar array. In: *2018 IEEE International Electron Devices Meeting (IEDM)*. San Francisco, CA: IEEE.

161 Cai, F., Kumar, S., Van Vaerenbergh, T. et al. (2020). Power-efficient combinatorial optimization using intrinsic noise in memristor Hopfield neural networks. *Nat. Electron.* 3 (7): 409–418.

162 Kumar, S., Strachan, J.P., and Williams, R.S. (2017). Chaotic dynamics in nanoscale NbO_2 Mott memristors for analogue computing. *Nature* 548 (7667): 318–321.

163 Zhang, W., Peng, X., Wu, H. et al. (eds.) (2019). Design guidelines of RRAM based neural-processing-unit: a joint device-circuit-algorithm analysis. In: *2019 56th ACM/IEEE Design Automation Conference (DAC)*. Las Vegas, NV: IEEE.

164 Lillicrap, T.P., Santoro, A., Marris, L. et al. (2020). Backpropagation and the brain. *Nat. Rev. Neurosci.* 21 (6): 335–346.

165 Liu, Z., Tang, J., Gao, B. et al. (2020). Neural signal analysis with memristor arrays towards high-efficiency brain–machine interfaces. *Nat. Commun.* 11 (1).

7

Neuromorphic Perceptual Systems with Emerging Devices

Ying Zhu, Changjin Wan, and Qing Wan

Nanjing University, School of Electronic Science & Engineering, Nanjing 210023, P. R. China

7.1 Background

In the 1980s, Hans Moravec, Rodney Brooks, Marvin Minsky, and others stated that, contrary to traditional assumptions, reasoning (which is high level in humans) requires very little computation, whereas sensorimotor skills (comparatively low level in humans) require enormous computational resources [1]. This is so-called the Moravec's paradox, which is the observation by artificial intelligence and robotics researchers. They also pointed out that we should expect the difficulty of reverse-engineering any human skill to be roughly proportional to the amount of time that skill has been evolving in animals. With great development in mechanics, sensors, automatic control systems, etc. machines like Boston Dynamics' dog-like robots have exhibited with great motor skills and can be applied in many scenarios, from documenting construction progress to monitoring remote environments. However, this level of sensorimotor skills is still far from that of the real animals that show superior adaptability, much lower energy consumption, integrity, universality, etc.

One key obstacle is due to the lack of approaches that enable the machinery to operate the same way as the animals. Recent efforts on achieving bioinspired sensing capabilities were deemed as the first step to achieve versatile sensorimotor skills. For instance, current progress on e-skin devices that resemble the haptic receptors could be regarded as a salutary lesson for the development of electronic receptors with other sense modalities. This has also resulted in the profound implications in the development of prosthetics [2–4], soft robotics or biorobotics [5–8], and wearable medical devices [9, 10]. One step was put forward by the most recent pursuit on the artificial sensory neuron, which aims at resembling the processing powers of the biological sensory neurons [11, 12]. This would no doubt propel the development on bioinspired sensorimotor technologies and may give rise to many paradigm breakers like neurorobotics with unprecedented awareness and intelligence.

In this chapter, the mimicking of perception by neuromorphic devices is illustrated in three parts. Firstly, we differentiate the concept of sensation and perception

Neuromorphic Devices for Brain-Inspired Computing: Artificial Intelligence, Perception, and Robotics, First Edition. Edited by Qing Wan and Yi Shi.

from biological and technological aspects and highlight the significance of realizing the artificial perception. Secondly, we provide a roadmap to pave the way for the realization of artificial perception by reviewing recent progress and endeavors on electronically emulating the related processes, taking a detailed look into the mechanisms and thoughts underlying these works, and discussing the possibility for building the integrated perceptual system. Thirdly, we address the open challenges for possible future advances in this emerging field.

7.2 Sensation and Perception

7.2.1 From Sensation to Perception

Organisms are sensorimotor systems [13]. Things in the world come into contact with our sensory surfaces, and we interact with them based on what the sensorimotor contact "affords" [13]. Our consciousness and knowledge begin with observation, and what we observe and how we interpret it lie at the very heart of sensation and perception. A sensation arises from the detection of physical/chemical energy from the environment and the conversion of them into neural signals by receptors (mechanoreceptors, thermoreceptors, nociceptors, etc.). Transduction of these signals to changes in membrane potential is believed to involve stimulation-activated ion channels in most receptors [14]. This change in membrane potential is called a receptor potential, which is a graded potential, increasing and decreasing with the intensity of the stimulus. This process allows our body to gather surrounding information, including tasting, hearing, seeing, smelling, and touching (Figure 7.1), and we can also feel things such as hunger and rumbling stomach. Sensory information is essential to interact and perceive the environment.

While perception is quite complex, it refers to the refining, organizing, and interpreting our sensations. It means the perception requires not only sensing but also the capabilities of memory, reasoning, learning, imagining, etc. The raw pattern of the external objects from the neural activity is called the proximal stimulus, which is then transmitted to the brain and processed. The resulting mental re-creation of

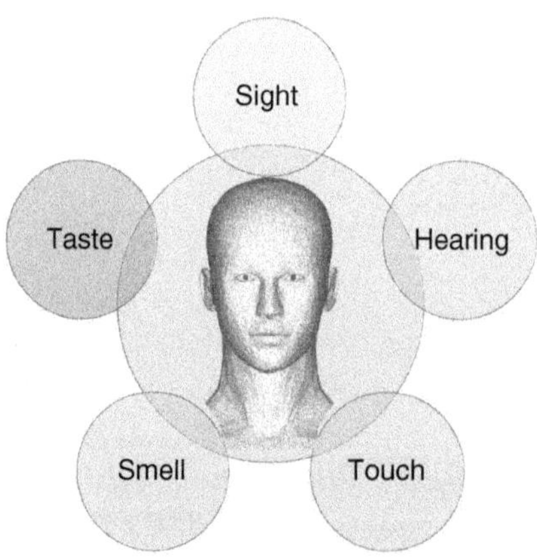

Figure 7.1 The five senses of human. The people face. Source: Adapted under the terms of the Pixabay license. Copyright 2020, Pixabay.

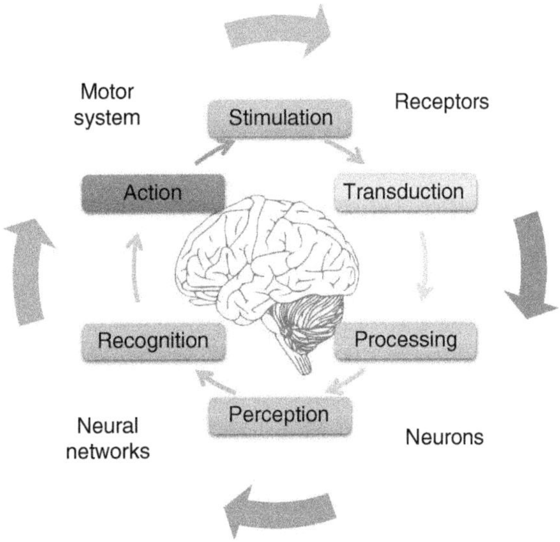

Figure 7.2 The perceptual process in the sensorimotor system. The human brain. Source: Adapted under the terms of the Pixabay license. Copyright 2020, Pixabay.

the distal stimulus is the percept. For example, the light from the shoe that enters a person's eye and stimulates the retina would trigger the proximal stimulus, and this process is called sensing. The image of the shoe reconstructed by the brain of the person is the percept.

There is a consensus that the perceptual process is made up of six steps [15]: environmental stimulus, transduction, processing, perception, recognition, and action (Figure 7.2), which is involved in comprehensive coordination of the neural and motor systems. In some sensory neurons, they respond directly to stimuli and can activate motor neurons to achieve muscle contraction. Such connections between sensory and motor neurons underlie motor reflex loops and several forms of involuntary behavior, including pain avoidance. As mentioned by R. S. Dahiya et al. [16], the tasks mediated via "sense of touch" can be grouped under three categories: manipulation (perception for action), exploration (action for perception), and reaction. In other words, perception is highly related to our voluntary/involuntary behaviors that interact with the environment.

Given this, sensation and perception provide the crucial capability to survive and initiate intelligence. For example, the "sense of touch" allows us to assess the contact parameters such as shape, surface texture, stiffness, temperature, etc. The realization of artificial perception would greatly enhance the adaptive capabilities of current electronic systems. Electronically implementing of perception would profoundly be of benefit to the development of such cutting-edge areas of the information such as robotics, cyborg, prosthetics, and neuromorphic engineering.

7.2.2 Pursing Artificial Perception by Neuromorphic Devices

However, previous efforts on bioinspired sensing much focused on how to achieve the biological sensing abilities (sensation) while overlooking the complex perceptual process (perception) that can greatly enhance these abilities. It may be due to the poor understanding of the biological perceptual process and the underestimation of the advantages of this process. For conventional robotic systems, a very tough issue is the low efficiency to process the massive amounts of sensory data, which in turn results in the low-level of sensorimotor skills. The biological system can refine and extract necessary features from the massive amounts of sensory data and process in a very efficient way. Our sensory system receives external stimulation around 1 Gb/s (estimated from the visual cell density in the retina). However, the human body sends 11 million bits/s to the brain through the refining of the sensory nervous system, yet the conscious mind seems to be able to extract the core information with a speed of only 50 bits/s to process [17]. By witnessing the great power of biological sensory processing, it is reasonable to translate this biological intelligence by electronic implementations. To fully release the potential of artificial perceptual systems, there are several challenges needed to be addressed.

Technologically, one main challenge is making electronic implementations to capture the essential functions of the neural system for perception. As demonstrated before, the perceptual process is highly complex and correlated to several subsystems of our body (e.g. peripheral nervous system and central nervous system). One possible way is to transform the conventional digital systems to adapt to biological processes. For example, M. M. Shulaker et al. realized a three-dimensional integrated circuit architecture with fine-grained and dense vertical connectivity between layers of computing, data storage, and input and output (in this instance, sensing) to mimic the olfactory sensory system [18]. Although this prototype can capture, store, and process massive amounts of sensory data, the whole network is highly defined with low plasticity and adaptability. The biological system can train its sensory system through the ever-changing sensory inputs and learn to identify, recognize, and interpret them based on the process called perceptual learning that can be deemed as the result of modification of the synaptic connections.

Alternatively, mimicking the biological sensory system from the bottom up might capture the very core of the biological neural system, i.e. "plasticity," and exceed the capabilities of current electronics. Given this, several groups have developed some electronic devices to mimic the sensory neuron functions and demonstrated potential applications in areas such as pattern recognition [19], prosthetics [2], and soft robots [20]. One main difference between these devices and previous bioinspired sensors is the introduction of the neuromorphic devices. As demonstrated in the previous chapters, the neuromorphic devices have a high degree of spatiotemporal processing power and different level of memory, which enable this new paradigm of sensor with the essential abilities as the sensory neuron (Figure 7.3). More importantly, the neuromorphic device is the basic element for building the neuromorphic networks, which can be potentially integrated with them for complex intelligent tasks. Furthermore, the addition of the parallel computing nature rendered by the neuromorphic component could be advantageous in integration of multiple sensory feedbacks to profoundly promote the sophistication of target systems [19].

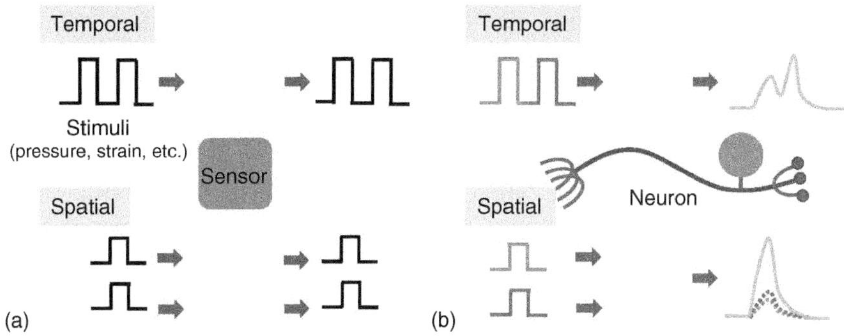

Figure 7.3 The temporal and spatial processing of (a) normal sensor and (b) sensory neuron.

7.3 Implementation of Artificial Perception

7.3.1 Building Block of Artificial Perception: Artificial Sensory Neuron

The sensory neuron could be deemed as a highly complex machine, which collects and transduces the sensory stimulations from receptors and conveys this information to the dorsal ganglia of the spinal cord where the cell body of the sensory neuron is located. In this way, this sensory information can be transduced, refined, and translated in the language of spike trains to the brain or spinal cord. The central nervous system then can process this information and make a suitable reaction. In such a loop, the sensory neuron plays a crucial role that is the interface between external stimulation and inner perception. Therefore, the implementation of the sensory neuron could be deemed as the first step in achieving artificial perception. However, the sensory neuron is highly complex in both functional and anatomical aspects, and therefore the simplification is needed. The sensory neuron can be divided into at least three parts (Figure 7.4): the sensor (receptor), the information channel (neural pathway), and the processor (cell body) for sensory transduction, information transmission, and preprocessing of sensory information (like action potential generation, signal integration, etc.). In this way, several artificial sensory neurons have been developed totally or partly based on this simplification.

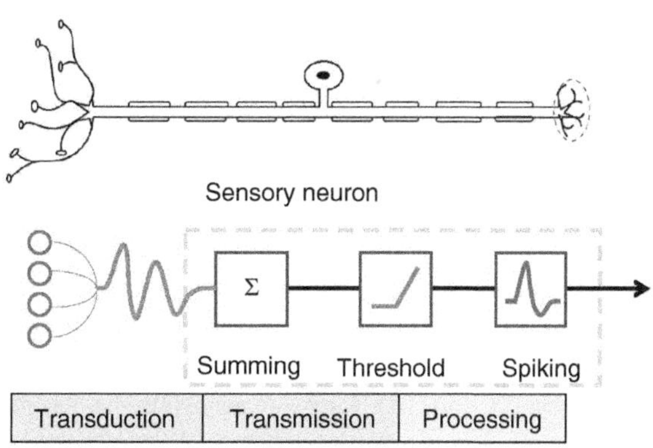

Figure 7.4 The simplification of the sensory neuron for the design of recent artificial sensory neurons. The neuron scheme. Source: Adapted under the terms of the Pixabay license. Copyright 2020, Pixabay.

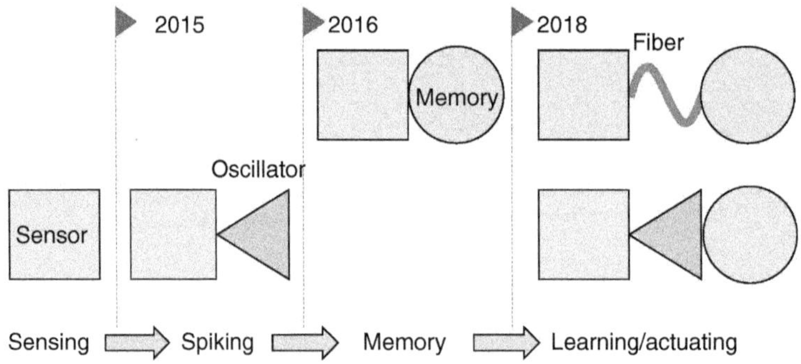

Figure 7.5 The evolution of the artificial sensory neuron. The square, triangle, circle, and wave denote the sensor, oscillator, memory, and nerve fiber.

Back in 2015, Z. Bao's group developed an organic digital mechanoreceptor, which well mimicked the pressure sensing and action potential generation functions of the biological tactile neuron [2] (Figure 7.5). After that, there are several groups who reported the integration of sensors with memory devices [21–23]. For example, Q. Sun et al. demonstrated the piezopotential-programmed multilevel nonvolatile memory, based on ion gel-gated field-effect transistors and piezoelectric nanogenerators [21]. Y. Zang et al. showed the dual-organic-transistor-based tactile-perception system could differentiate the touch speed [22]. These devices could be grouped by the artificial sensory memory [11], which represents the very basic function of the sensory neuron.

In 2018, researchers from Z. Bao and X. Chen's group proposed two kinds of artificial tactile neuron [12, 19], independently, which could be considered as the first batch of the artificial sensory neurons as they have resembled the essential components of the biological counterpart. Moreover, they demonstrated the potential for machine learning-based sensory processing. In the first model, the action potential generated from the organic digital mechanoreceptor was integrated and converted into postsynaptic currents by a synaptic transistor. The second model introduced the ionic conductor-based neural pathway for the transmission of information between the sensor and the synaptic transistor. This design would facilitate the construction of the soft robotic system as the whole device is stretchable, which is rendered by the soft ionic conductor. Based on the two models of artificial sensory neuron, the sensory data preprocessed by them were used for machine learning. As these devices can capture the spatiotemporal information of the stimulations [24], they provide more abundant features than the individual sensor well mimicking the function of the sensory neuron.

From then on, the neuromorphic devices were deemed as a key component for the artificial sensory neuron. This is because they can process sensory information just like the biological one [25]. More importantly, because the sensory neuron is the basic unit for the sensory nervous system, the artificial sensory neuron was then considered as the building block for the implementation of the artificial perception [26–29]. In Figure 7.6, we briefly list the typical artificial sensory neuron aimed for different sensing usages including haptic, nociceptive, olfactory, iconic,

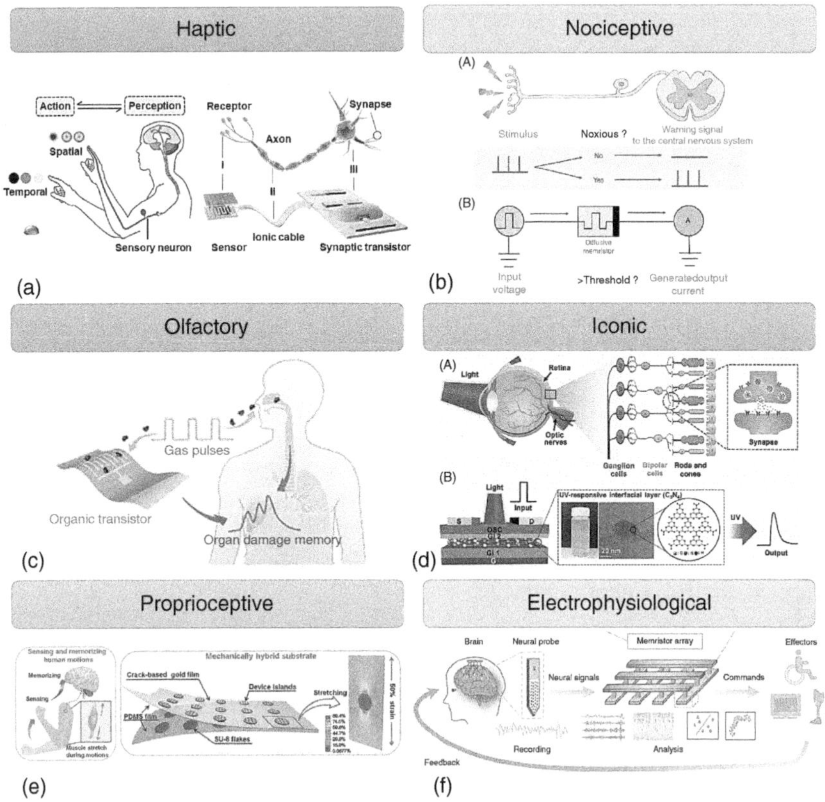

Figure 7.6 A summary of the artificial sensory neuron (or memory) with different sensory modalities: (a) haptic (mechano)neuron. Source: Adapted from Kim et al. [12]. (b) Nociceptive memory. Source: Yoon et al. [30]. (c) Olfactory memory. Source: Song et al. [31]. (d) Iconic neuron. Source: Park et al. [32]. (e) Proprioceptive type (muscular) neuron. Source: Liu et al. [33]. (f) Electrophysiological type (EEG) neural network. Source: Liu et al. [34].

proprioceptive, and electrophysiological artificial sensory neuron [35–38]. There are also some other reports based on these sensory modalities with different devices [32, 39–45], which have included almost all the five sensations, nociception, proprioception, and the sensing of electrophysiological signals. Hence, the integration of multisensory abilities is technically feasible, which would give rise to the completed artificial sensory system with biologically comparable perception. With the incorporation of additional sensory modalities that are not existed in human, like magnetic, UV/IR [46], etc. a supernatural sensory system would also be expected, which could be available for many more future application scenarios. Robots with multiple sensory modalities would be superior to human beings and would be more appropriate to be used in hazardous, dangerous, and even outer space environment.

7.3.2 Intelligent Tasks Based on Artificial Perception

To implement the artificial perception, a step forward is to apply these artificial sensory neurons for high-level intelligent activities like recognition, sorting, and

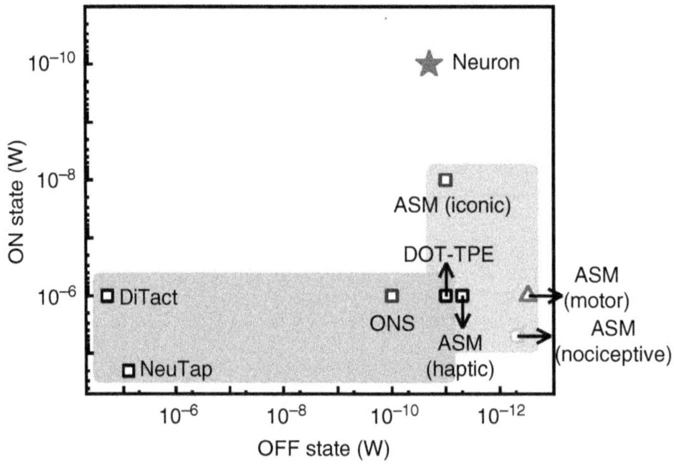

Figure 7.7 The power consumption for several kinds of artificial sensory neuron versus the biological neuron. Source: Wan et al. [11].

even motion control as demonstrated in Figure 7.2. In this case, the recent artificial sensory neurons have been explored in some intelligent tasks with the aid of software- or hardware-based neural networks. In contrast to the conventional way that senses, stores, and processes data in separated units, the artificial sensory networks can incorporate these processes into each sensory neuron, which could be an energy-efficient way. Although there is still a huge gap between technological and biological sensory neurons in terms of power consumption as shown in Figure 7.7, the rational design of device/system, utilization of proper materials, and employment of novel mechanisms would fill the gap in the future. However, the advantages of this technological route should be further verified by system-level implementations [47].

Up to date, intelligent tasks are mostly focused on two aspects: pattern recognition and prosthetics/robotics [48, 49]. The former is achieved based on the sensory data that is preprocessed by the artificial sensory neurons, and the latter aims to control the actuators, artificial and/or biological muscles, and therapeutic apparatus. This direction can also be conceptualized as in-sensor computing [46]. The ultimate goal of implementing artificial perception is to build electronic sensorimotor systems that can acquire humanlike sensorimotor skills with high energy efficiency and comparable sophisticated. Moreover, Table 7.1 has summarized the most recent progresses in this area [53].

7.3.2.1 Pattern Recognition Tasks

Generally, the feature data of a (tactile, visual, etc.) pattern is always fed up by peripheral equipment or directly generated by a computer. However, the artificial sensory neuron itself can obtain the pattern information through its sensing component and can be preprocessed through its neuromorphic component that integrates abundant spatiotemporal details and pattern features (Figure 7.8). This information then could be further processed by either software- or hardware-based neural networks

Table 7.1 The recent progresses on artificial sensory neuron.

Name	Sensory modality	Actuator and size	Feature action dimension	Recognition pattern	Synaptic emulations
Artificial afferent nerve [12]	Pressure	Insect leg ~3 cm	~1 mm	Braille pattern	Short-term potentiation
NeuTap [19]	Pressure	NA	NA	Braille pattern	Short-term potentiation; dendritic integration
Artificial somatic reflex arc [50]	Pressure	Electrochemical actuator ~2.5 cm	~1 mm	NA	Neural all-or-none law
Artificial optic neural synapse [51]	Light	NA	NA	Color-mixed pattern	Long-term potentiation and depression
Optoelectronic sensorimotor artificial synapse [20]	Light	Polymer actuator ~2 cm	~1 mm	NA	Short-term potentiation
Optoelectronic neuromorphic device [52]	Light	NA	NA	Illuminance gradation pattern	Visual synaptic functions; environment-adaptable perception
Base	Pressure and light	Myotubes ~1.5 mm	~5 μm	Multi-transparency pattern	Dendritic integration

Source: Adapted from Wan et al. [53].

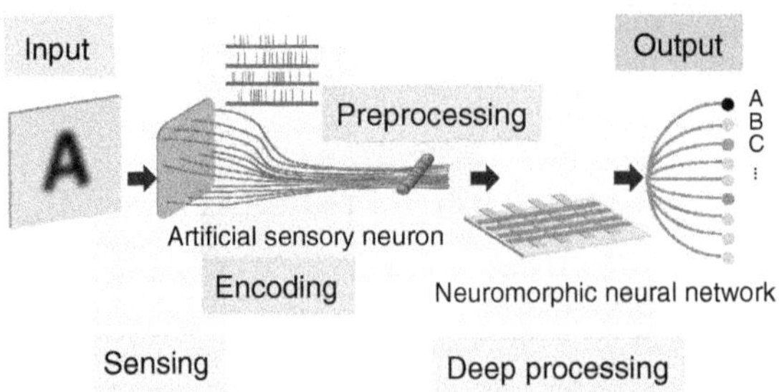

Figure 7.8 Information flow in an artificial perceptual system for pattern recognition.

that are feasible for implementing machine learning methods. For example, S. Seo et al. proposed an artificial optic neuron that incorporated synaptic and optical sensing functions [51]. Besides these device-level emulations, the synaptic devices have also been used to build artificial neural networks. In this way, the colored and color-mixed patterns detected and preprocessed by the artificial optic neurons can be directly fed into these networks that can perform pattern recognition tasks.

7.3.2.2 Prosthetics and Robotics Applications

On the other hand, the preprocessed or deep processed sensory information is also used for prosthetics and robotics applications like motion control and gesture adjustment [54, 55]. Animals' sensorimotor system is a sensing–memory–actuation cooperated system, in which animals can simultaneously monitor and memorize the corresponding motion information and perform these motions later [11]. With the electronic synapses that can refine, memorize, and adapt to some stimulation patterns, mimicking the plasticity function of the biological counterpart, the motion control would be more adaptive than conventional motors. For example, Y. Lee et al. proposed an artificial optic neuron based on a stretchable organic nanowire synaptic transistor, which can tune the bending angle of a polymer actuator according to the voltage pulse numbers [20]. Moreover, the artificial sensory neurons could innervate and control the motion of the biological or biohybrid substance/machinery/actuators (Table 7.1), which would be of profound implication for bioelectronics, prosthetics, and medicine. By integrating the artificial sensory and motor neurons with micro-actuators, the sensorimotor systems could be constructed from the bottom up, which could eventually realize the biological sensorimotor skills with low energy consumption and high sophistication in an integrated system. The parallel direction is the prosthetics, which has not yet achieved. It might due to the lack of effective, stable, and bidirectional interfaces for bridging the artificial and biological systems [27]. Also, the low-level similarity and compatibility of the current artificial sensory system are the main obstacles toward the practical applications.

7.3.3 A Roadmap Toward Neuromorphic Perceptual System

At the beginning of this Section 7.1, we have discussed about Moravec's paradox that is a hard nut to crack based on current technologies, and some paradigm breakers would be very much needed. With regard to this, it is believed that the future neuromorphic perceptual system could greatly extend the sensorimotor power of current

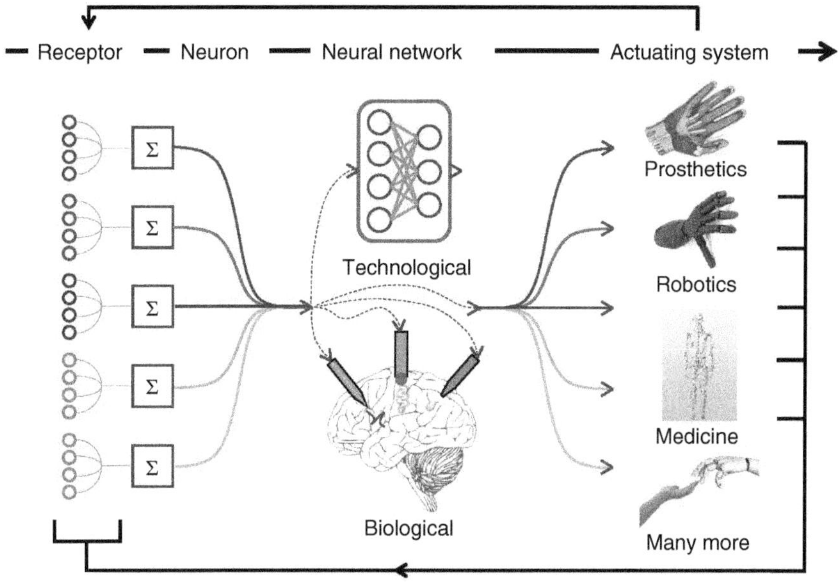

Figure 7.9 A roadmap for neuromorphic perceptual system. The human brain, hand, robotic hand, human shape medicine, and human/robotic hand. Source: Adapted under the terms of the Pixabay license. Copyright 2020, Pixabay.

machines, which would be a very promising candidate to solve the aforementioned question. Here, the aim is to propose a roadmap toward the neuromorphic perceptual system as shown in Figure 7.9.

There are at least four subsystems to mimic/replace the functions of the receptor, neuron, neural network, and actuating system. The current bionic sensors have well mimicked the function of the receptors and will be integrated and formed as the distributed sensing network to obtain all-sided, seamless, high-throughput sensory information. This information is then preprocessed by the artificial sensory neuron. This would require the artificial sensory neuron to filter, refine, and memorize this information (may not be eternal). More importantly, the artificial sensory neuron is required to extract the spatiotemporal information of the target as concisely as possible based on the preprocessed procedure [43, 56, 57]. In this case, the synaptic device would play an important role in constructing this neuron, and other functional elements like oscillator and energy suppliers that are responsible for signaling and powering, respectively, are also very necessary [24].

Above the primary sensory network is a more complex computing system that could be both technological and biological neural networks. To connect with a technological neural network, the sensory information could be further processed and applied for intelligent tasks. More importantly, with the input of massive sensory information, the neural network could be highly trained, which can in turn improve the intelligent level of the network. On the other hand, the sensory information obtained and preprocessed by the artificial sensory neurons could also connect with the biological tissues with the aid of brain–machine interfaces [28, 55, 58]. These interfaces could be potential, chemical, optogenetic types, enabling both artificial transmission of the biosignals between prosthesis and nervous system and stimulation and recording of these signals with mechanically compliant, implantable neural interfaces [27]. As the information is processed in a biologically similar way, the connection with biological tissues would provide novel solutions for a wide range of applications such as sense and motor restoration. Finally, the output from either technological or biological neural network could be sent to actuating systems with regard to the target applications.

7.4 Challenges and Perspectives

7.4.1 Challenges

The mimicry of biological perception involved in sensing, processing, and actuating has been achieved, although it is far from complete as many challenges should be addressed before achieving the advantages. One major challenge of artificial perception could arise from the poor understanding of the biological process. The nervous systems are highly complex, and the mechanism of intelligence is still unclear for scientists. Although we would improve the artificial one with deeper understanding, a trend in ideas believes that the brain-level intelligence might be achieved by resembling the elements, structures, and functions of the biological nervous systems [59]. In this case, developing functional devices with higher biological accuracy and integrating them with structures closer to the biological systems would also help to improve this artificial system. Following this trend, many more bionic sensors, neuromorphic devices, and other functional devices with biological similarity should be developed, and structural engineering should be taken into account when assembling them.

Another important challenge comes from universality. The digital logic systems can be transplanted, integrated with, and extended to almost any other digital logic systems, as they are built on the same binary language and universal interface standards. The artificial perceptual system based on neuromorphic device is still at the very early stage. Moreover, devices and systems developed recently are based on various devices that are rooted in various rules. In the future, universal protocols and standards should be developed for the fabrication/manufacturing/integration of these devices/systems, and it would be compatible with current digital systems. The large-scale, low-cost, and reproducible technologies should be developed for practical applications.

Besides, there are many scientific and engineering challenges. For example, devices that can mimic the biological functions in more accurate manners are still very much needed. The emerging devices like ionotronics that could share similar ionic dynamics as the biological counterparts would have the great potential to be the core building block. Technologically, the devices/systems with artificial perception are better to fabricate in soft form, which can be compatible with biological systems. There are also a number of general engineer problems such as efficient power management, cost control, performance optimization, etc. This also requires a convergence of multiple disciplines including materials science, microelectronics, nanoscience, mechanical engineering, data science, and more.

7.4.2 Conclusions and Perspectives

Despite its infancy, the neuromorphic perceptual systems have already implicated in a broad range of applications like bionics, prosthetics, robotics, and artificial intelligence. The long-term goal is to build a sensorimotor system, which aims to solve the puzzle in implementing highly sophisticated sensorimotor skills by electronics. The machines endowed with these sensorimotor skills are expected to work, interact, and care with humans much closer and safer than today's "ancestors." Furthermore, the sensory system and motor system are believed to be the input and output of an intelligent system, while the current software-based artificial systems are separated from the two systems. This greatly increases the data transmission burden and data redundancy, resulting in an unprecedented energy budget in the artificial intelligence (AI) era.

Future trends in the large number, multimodal, and multipoint integrated systems that incorporate the sensing/memory/processing would be continued and speeded up as the essential components in accessing the complex sensory information, and the interfaces with hardware neural networks are almost available now. In the near future, this would give rise to some low-level perception-based applications like intelligent sensory feedback for robotics and (partial) sense restoration for disabled people. This trend would also be of benefit to precision medicine by collecting timely, all-around, and precise parameters for health assessment. For example, a near-skin (inner and outer) environmental assessment patch could be developed to access sensory data in a biologically accurate manner and to provide accurate and timely feedback (e.g. therapy, drug release, and other medical care).

We are confident that the neuromorphic perceptual system would greatly improve current technologies in cyborgs, robotics, human–machine systems, prosthetic systems, and more. It is reasonable to replicate the biological intelligence by electronic building blocks, which may also be expected to extend the capability of the conventional digital systems with regard to cognitive, emotional, social, and psychological aspects that have not been well replicated currently. We believe the neuromorphic perceptual system is a promising paradigm breaker for autonomous artificial intelligence, and it would undoubtedly shed light on future advances concerning various translational implementations.

References

1 Moravec, H.P. (1988). *Mind Children: The Future of Robot and Human Intelligence*. Cambridge: Harvard University Press.

2 Tee, B.C.-K., Chortos, A., Berndt, A. et al. (2015). A skin-inspired organic digital mechanoreceptor. *Science* 350 (6258): 313–316.

3 Chortos, A., Liu, J., and Bao, Z. (2016). Pursuing prosthetic electronic skin. *Nat. Mater.* 15: 937.

4 Tabot, G.A., Dammann, J.F., Berg, J.A. et al. (2013). Restoring the sense of touch with a prosthetic hand through a brain interface. *Proc. Natl. Acad. Sci.* 110 (45): 18279–18284.

5 Shin, S.R., Migliori, B., Miccoli, B. et al. (2018). Electrically driven microengineered bioinspired soft robots. *Adv. Mater.* 30 (10): 1704189.

6 Park, S.-J., Gazzola, M., Park, K.S. et al. (2016). Phototactic guidance of a tissue-engineered soft-robotic ray. *Science* 353 (6295): 158–162.

7 Martinez, R.V., Fish, C.R., Chen, X., and Whitesides, G.M. (2012). Elastomeric origami: programmable paper-elastomer composites as pneumatic actuators. *Adv. Funct. Mater.* 22 (7): 1376–1384.

8 Kim, Y.S., Liu, M., Ishida, Y. et al. (2015). Thermoresponsive actuation enabled by permittivity switching in an electrostatically anisotropic hydrogel. *Nat. Mater.* 14: 1002.

9 Son, D., Lee, J., Qiao, S. et al. (2014). Multifunctional wearable devices for diagnosis and therapy of movement disorders. *Nat. Nanotechnol.* 9: 397.

10 Lee, H., Choi, T.K., Lee, Y.B. et al. (2016). A graphene-based electrochemical device with thermoresponsive microneedles for diabetes monitoring and therapy. *Nat. Nanotechnol.* 11: 566.

11 Wan, C., Cai, P., Wang, M. et al. (2019). Artificial sensory memory. *Adv. Mater.*: 1902434.

12 Kim, Y., Chortos, A., Xu, W. et al. (2018). A bioinspired flexible organic artificial afferent nerve. *Science* 360 (6392): 998–1003.

13 Harnad, S. (2017). Chapter 2 – To cognize is to categorize: cognition is categorization. In: *Handbook of categorization in cognitive science*, 2e (eds. H. Cohen and C. Lefebvre), 21–54. San Diego: Elsevier.

14 French, A.S. and Torkkeli, P.H. (2012). Chapter 36 – Sensory receptors and mechanotransduction. In: *Cell Physiology Source Book*, 4e (ed. N. Sperelakis), 633–647. San Diego: Academic Press.

15 May, M. and Chudler, E.H. (2009). *Sensation and Perception*. Dordrecht: Facts On File, Incorporated.

16 Ravinder, S. and Dahiya, M.V. (2013). *Robotic Tactile Sensing*, 248. Netherlands: Springer.

17 Markowsky, G. (2017). Information theory. Available from: https://www.britannica.com/science/information-theory/Physiology (accessed 28 June 2021).

18 Shulaker, M.M., Hills, G., Park, R.S. et al. (2017). Three-dimensional integration of nanotechnologies for computing and data storage on a single chip. *Nature* 547: 74.

19 Wan, C., Chen, G., Fu, Y. et al. (2018). An artificial sensory neuron with tactile perceptual learning. *Adv. Mater.* 30 (30): 1801291.

20 Lee, Y., Oh, J.Y., Xu, W. et al. (2018). Stretchable organic optoelectronic sensorimotor synapse. *Sci. Adv.* 4 (11): eaat7387.

21 Sun, Q., Ho, D.H., Choi, Y. et al. (2016). Piezopotential-programmed multilevel nonvolatile memory as triggered by mechanical stimuli. *ACS Nano* 10 (12): 11037–11043.

22 Zang, Y., Shen, H., Huang, D. et al. (2017). A dual-organic-transistor-based tactile-perception system with signal-processing functionality. *Adv. Mater.* 29 (18): 1606088.

23 Zhu, B., Wang, H., Liu, Y. et al. (2016). Skin-inspired haptic memory arrays with an electrically reconfigurable architecture. *Adv. Mater.* 28 (8): 1559–1566.

24 Wan, C., He, Y., Jiang, S. et al. (2020). Emerging devices for biologically accurate neuron. *ACS Appl. Electron. Mater.* 2 (2): 389–397.

25 Molina-Lopez, F., Gao, T.Z., Kraft, U. et al. (2019). Inkjet-printed stretchable and low voltage synaptic transistor array. *Nat. Commun.* 10 (1): 2676.

26 Zhu, M., He, T., and Lee, C. (2020). Technologies toward next generation human machine interfaces: from machine learning enhanced tactile sensing to neuromorphic sensory systems. *Appl. Phys. Rev.*. 7 (3): 031305.

27 Li, P., Anwar Ali, H.P., Cheng, W. et al. (2020). Bioinspired prosthetic interfaces. *Adv. Mater. Technol.* 5 (3): 1900856.

28 Yang, J.C., Mun, J., Kwon, S.Y. et al. (2019). Electronic skin: recent progress and future prospects for skin-attachable devices for health monitoring, robotics, and prosthetics. *Adv. Mater.* 31 (48): 1904765.

29 Xiao, T.P., Bennett, C.H., Feinberg, B. et al. (2020). Analog architectures for neural network acceleration based on non-volatile memory. *Appl. Phys. Rev.* 7 (3): 031301.

30 Yoon, J.H., Wang, Z., Kim, K.M. et al. (2018). An artificial nociceptor based on a diffusive memristor. *Nat. Commun.* 9 (1): 417.

31 Song, Z., Tong, Y., Zhao, X. et al. (2019). A flexible conformable artificial organ-damage memory system towards hazardous gas leakage based on a single organic transistor. *Mater. Horiz.* 6 (4): 717–726.

32 Park, H.-L., Kim, H., Lim, D. et al. (2020). Retina-inspired carbon nitride-based photonic synapses for selective detection of UV light. *Adv. Mater.* 32 (11): 1906899.

33 Liu, Y., Liu, Z., Zhu, B. et al. (2017). Stretchable motion memory devices based on mechanical hybrid materials. *Adv. Mater.* 29 (34): 1701780.

34 Liu, Z., Tang, J., Gao, B. et al. (2020). Neural signal analysis with memristor arrays towards high-efficiency brain–machine interfaces. *Nat. Commun.* 11 (1): 4234.

35 Yang, X., Xiong, Z., Chen, Y. et al. (2020). A self-powered artificial retina perception system for image preprocessing based on photovoltaic devices and memristive arrays. *Nano Energy* 78: 105246.

36 Zhou, F., Zhou, Z., Chen, J. et al. (2019). Optoelectronic resistive random access memory for neuromorphic vision sensors. *Nat. Nanotechnol.* 14 (8): 776–782.

37 Chen, Q., Zhang, Y., Liu, S. et al. (2020). Switchable perovskite photovoltaic sensors for bioinspired adaptive machine vision. *Adv. Intell. Syst.* 2 (9): 2000122.

38 Wei, H., Ni, Y., Sun, L. et al. (2021). Flexible electro-optical neuromorphic transistors with tunable synaptic plasticity and nociceptive behavior. *Nano Energy* 81: 105648.

39 Lv, Z., Wang, Y., Chen, J. et al. (2020). Semiconductor quantum dots for memories and neuromorphic computing systems. *Chem. Rev.* 120 (9): 3941–4006.

40 Jiang, C., Li, Q., Sun, N. et al. (2020). A high-performance bionic pressure memory device based on piezo-OLED and piezo-memristor as luminescence-fish neuromorphic tactile system. *Nano Energy.* 77: 105120.

41 Wu, C., Kim, T.W., Park, J.H. et al. (2020). Self-powered tactile sensor with learning and memory. *ACS Nano* 14 (2): 1390–1398.

42 Feng, G., Jiang, J., Zhao, Y. et al. (2020). A sub-10 nm vertical organic/inorganic hybrid transistor for pain-perceptual and sensitization-regulated nociceptor emulation. *Adv. Mater.* 32 (6): 1906171.

43 Li, Y., Lu, J., Shang, D. et al. (2020). Oxide-based electrolyte-gated transistors for spatiotemporal information processing. *Adv. Mater.* 32 (47): 2003018.

44 Wu, Z., Lu, J., Shi, T. et al. (2020). A habituation sensory nervous system with memristors. *Adv. Mater.* 32 (46): 2004398.

45 Tan, H., Tao, Q., Pande, I. et al. (2020). Tactile sensory coding and learning with bio-inspired optoelectronic spiking afferent nerves. *Nat. Commun.* 11 (1): 1369.

46 Zhou, F. and Chai, Y. (2020). Near-sensor and in-sensor computing. *Nat. Electron.* 3 (11): 664–671.

47 Tan, H., Zhou, Y., Tao, Q. et al. (2021). Bioinspired multisensory neural network with crossmodal integration and recognition. *Nat. Commun.* 12 (1): 1120.

48 Wang, M., Yan, Z., Wang, T. et al. (2020). Gesture recognition using a bioinspired learning architecture that integrates visual data with somatosensory data from stretchable sensors. *Nat. Electron.* 3 (9): 563–570.

49 Wei, H., Shi, R., Sun, L. et al. (2021). Mimicking efferent nerves using a graphdiyne-based artificial synapse with multiple ion diffusion dynamics. *Nat. Commun.* 12 (1): 1068.

50 He, K., Liu, Y., Wang, M. et al. (2020). An artificial somatic reflex arc. *Adv. Mater.* 32 (4): 1905399.

51 Seo, S., Jo, S.-H., Kim, S. et al. (2018). Artificial optic-neural synapse for colored and color-mixed pattern recognition. *Nat. Commun.* 9 (1): 5106.

52 Kwon, S.M., Cho, S.W., Kim, M. et al. (2019). Environment-adaptable artificial visual perception behaviors using a light-adjustable optoelectronic neuromorphic device array. *Adv. Mater.* 31 (52): 1906433.

53 Wan, C., Cai, P., Guo, X. et al. (2020). An artificial sensory neuron with visual-haptic fusion. *Nat. Commun.* 11 (1): 4602.

54 Yokota, T., Nakamura, T., Kato, H. et al. (2020). A conformable imager for biometric authentication and vital sign measurement. *Nat. Electron.* 3 (2): 113–121.

55 You, I., Mackanic, D.G., Matsuhisa, N. et al. (2020). Artificial multimodal receptors based on ion relaxation dynamics. *Science* 370 (6519): 961–965.

56 He, Y., Nie, S., Liu, R. et al. (2019). Spatiotemporal information processing emulated by multiterminal neuro-transistor networks. *Adv. Mater.* 31 (21): 1900903.

57 Wang, W., Pedretti, G., Milo, V. et al. (2018). Learning of spatiotemporal patterns in a spiking neural network with resistive switching synapses. *Sci. Adv.* 4 (9): eaat4752.

58 Boutry, C.M., Negre, M., Jorda, M. et al. (2018). A hierarchically patterned, bioinspired e-skin able to detect the direction of applied pressure for robotics. *Sci. Robot.* 3 (24): eaau6914.

59 Tiejun Huang, L.S., Tang, H., Pan, G. et al. (2016). Research on multimedia technology 2015 – advances and trend of brain-like computing. *J. Image Graph.* 11: 1411–1424.

Index

a

active metal conductive filament mechanism 138
Ag/$CH_3NH_3PbI_{(3-x)}Cl_x$/FTO structure 150–151
$Ag_5In_5Sb_{60}Te_{30}$ 127
AgInSbTe (AIST)-based chalcogenide material 136
Ag nanoclusters 97, 190
Ag/PEDOT:PSS/Ta memristor 150, 151
Ag/SiO_2/Pt device 9
Al/$CH_3NH_3PbBr_3$/BCCP/ITO 154
all-spin artificial neural network (ASANN) 61–62
all-spin spiking neural network (ASSNN) 62
analog-like synaptic device 52–54
analog-to-digital converters (ADCs) 179
animals' sensorimotor system 226
anion migration related filaments 8, 11–14
ANN computing 36–39
application-specific integrated circuits (ASICs) 174
artificial auditory device 113–114
artificial intelligence (AI) 17, 91, 125, 158, 173
artificial intelligence of things (AIoT) 125
artificial neural device 57–61
artificial neural networks (ANNs) 7, 51, 91, 125
artificial neurons based on TS effects 136–139
artificial perception 221
artificial perceptual system for pattern recognition 226
artificial sensory neuron 221–225
artificial synapses based on MS behaviors 129–136
artificial tactile device 109–112
artificial tactile neuron 222
artificial vision device 112–113
artificial visualization systems 166–167
as-fabricated heterosynaptic device 98
associative memory based injection locking 73–74
Atkinson-Shiffrin multi-store memory model 94
atomic layer deposition (ALD) 2
Au/Li_xMoS_2/Au memristor 96, 97

b

backpropagation-STDP (BP-STDP) 189
Bienenstock–Cooper–Munro (BCM) 23, 24, 26
big data 18
bi-layered HfO_2/TiO_x cognitive memristor 93, 94
binarized neural networks (BNNs) 185
binary magnetic tunnel junction 48–49
biological neural networks (BioNNs) 27, 173
biological neurons 28
biological synapses 19

Neuromorphic Devices for Brain-Inspired Computing: Artificial Intelligence, Perception, and Robotics, First Edition. Edited by Qing Wan and Yi Shi.

bioNNs 174
biophysical memristive neurons 34–36
bioplausible memristive neurons 31–34
biorealistic memristive synapses 22
bio-voltage diffusive memristor 190
bit slicing scheme 185
bottom electrode (BE) 11, 12, 34, 93, 131, 150, 154
Braindrop 176
brain-inspired multi-functional perception learning system 95
brain-inspired neuromorphic computing 125
brain-like neuromorphic devices 91

c

calmodulin-dependent kinase II 96
capacitively coupled neuro-transistors 114
2048-cell TiN/TaO_x/HfO_2/TiN memristor array 183
chalcogenide materials
- artificial neurons based on TS effects 136–139
- artificial synapses based on MS behaviors 129–136
- hardware neural networks 139–143
- ovonic memory switching (OMS) 126–129
- ovonic threshold switching (OTS) 126–129

charge-trapping-based memristors 155–157
chemical-biosensors 101–102
chemical vapor deposition (CVD) 3
CMOS-based in-situ nonlinear activation (ISNA) modules 195
CMOS-based neuromorphic computing 174–176
C_{60} nanoparticles 158
C_3N_4 nanodots 112
complementary metal-oxide-semiconductor (CMOS) 39, 72, 150
computing-in-memory (CIM) hardware 174
conditioned stimulus (CS) 108, 113
conductive filaments (CFs) 8, 41, 97
confined-type phase-change memory 130
convolutional neural networks (CNNs) 37, 125, 139, 173
$CsPbBr_3$ QDs 164, 165

d

deep belief networks (DBNs) 173
deep neural networks (DNNs) 173
Delta rule 37, 38
dendrite neuron and dendritic algorithm 106–109
dendritic computing 196–202, 205
device-to-application simulation system 178
digital-analog converters (DACs) 179
2,7-dioctyl[1]benzothieno[3,2-*b*][1]benzothiophene (C8-BTBT) 164
DNNs based on synaptic devices
- architecture and algorithm optimization 185
- array demonstrations 179–182
- chip and system implementations 183–185
- device performance requirements 178–179

dopamine 102
dual-gate IZO-based EDL transistor 104
dual-organic-transistor-based tactile-perception system 222
dynamic pseudo-memcapacitors (DPMs) 192
dynamic random-access memory (DRAM) 176

e

EDL transistors (EDL-Ts) 99
EGT based neuromorphic transistors 103–105
electrical-double-layer (EDL) modulation 99

electrically programmable read-only memory (EPROM) 126
electrochemical doping/de-doping 99
electrochemical metallization cells (ECM) 8
electrochemical metallization memristors (ECMMs) 93
electrochemical random-access memory (ECRAM) 176
electrochemical transistors (ECTs) 99
electro-iono-photoactive 2D chalcogenide neuristors 113
electrolyte-gate transistors (EGTs) 99, 158–162
electron energy-loss spectra (EELS) 15, 16
electronic artificial neurons and synapses 125
ethyl viologen diperchlorate [$EV(ClO_4)_2$]/ triphenylamine-containing polymer 152
excitatory postsynaptic current (EPSC) 20
excitatory postsynaptic potential (ePSP) 18

f
fault-tolerant algorithms 186
ferroelectric devices 125
ferroelectric field-effect transistors (FeFETs) 162–163, 176
ferromagnetic (FM) layers 48
field-induced nucleation model 128
filamentary-type resistive switching 8–14
  anion migration related filaments 11–14
  cation migration related filaments 8–11
flash memory 176
flexible floating-gate synaptic transistor 158
flexible gate-tunable resistive switching 190
flexible memristors 95
flexible oxide-based neuromorphic transistor 103
floating-gate transistors 157–159
free layer (FL) 48
fully-connected networks (FCNs) 173

g
gate coupling effects 98
$Ge_1Sb_2Te_4$ 127
$Ge_1Sb_4Te_7$ 127
$Ge_2Sb_2Te_5$ 127, 129
$Ge_3Sb_2Te_6$ 127
$Ge_8Sb_2Te_{11}$ 127
$Ge_{10}Si_{12}As_{30}Te_{48}$ 126
GeTe-Sb_2Te_3 pseudo-binary alloy system 126, 127
GeTe/Sb_2Te_3 superlatice-like layers 130
graphene 3
graphene quantum dots (GQDs)/organic hybrid systems 156
graphics processing units (GPUs) 174

h
halide perovskites 150, 154, 155
haptic type (mechano)neuron 223
hardware neural networks 139–143
heavy mental (HM) 57
Hebb's rule 188
heterosynaptic depression 98
heterosynaptic facilitation 98
heterosynaptic plasticity 97–98, 188
HH neurons 29, 34
highly interconnected neuromorphic architecture (HINA) 161
high resistance state (HRS) 4, 8, 48, 56, 58, 93, 126, 185
high-resolution transmission electron microscopy (HRTEM) 12, 33
Hodgkin–Huxley (HH)
  axon 35
  model 28, 29
  neuron model 188
homosynaptic plasticity 96, 98
Hopfield neural network (HNN) 139, 203–204

human senses 218
hyperdimensional (HD) computing 196, 201

i
IBM's TrueNorth 176
ι-carrageenan 150
iconic neuron 223
Ielmini model 129
IF neurons 29
indium–zinc–oxide (IZO) 20, 100
inhibited postsynaptic current (IPSC) 20
inhibitory postsynaptic potential (IPSP) 18
injection locking 73
inorganic halide perovskite quantum dots (IHP QDs) 165
in-plane magnetic anisotropy (IMA) 48
integrate-fire (IF) model 28, 56
intelligent tasks 204, 220, 223–226
Intel's Loihi 176
interface-type resistive switching
  n-type semiconductor 15
  spectromicroscopy 16
  TEM and EELS 16
Internet of Things (IoT) 18
ion coupling 99
ion gel-gated field-effect transistors 222
ionic/electronic hybrid electrolyte gated transistors (EGTs) 99
ion migration-based memristors 153–155
ion-sensitive field-effect transistors 102
ions mobility 10–11
IZO-based EDL transistor 103

k
1Kb HfO_x-based 1T1R artificial synapse array 202
160Kb 1T1R memristor array 181
64k-cell PCM synaptic array 142
Kirchhoff laws 7

l
Landau-Lifshitz-Gilbert (LLG) equation 54
4-layer Bayesian neural network 181
4-layer three-dimensional (3D) vertical RRAM (VRAM)/FinFET kernel 196
leaky-integrate-and-fire (LIF) 28, 29, 31, 32, 36, 54, 65, 188
Li_xMoS_2-based synaptic device 97
local graded potential (LGP) 28
locally competitive algorithm (LCA) 38
logic-in-memory circuits 196
long-short term network (LSTM) 139
long term depression (LTD) 23, 131
long-term memristive synapses 23–27
long-term plasticity (LTP) 18, 19
long term potentiation (LTP) 23, 24, 131
long-term potentiation/depression (LTP/LTD) 98, 188
low resistance state (LRS) 8, 48, 58, 93, 126
low resistive state (LRS) 4, 151
low-voltage indium-zinc-oxide homojunction electric-double-layer transistor 102

m
magnetic devices 125
magnetic random-access memory (MRAM) 176
magnetic skyrmions 62, 64
magnetic tunnel junction (MTJ)
  all-spin artificial neural network 61–62
  all-spin spiking neural network 62
  analog-like synaptic device 52–54
  artificial neural device 57–61
  spiking neural device 54–57
  stochastic binary synaptic device 51–52
  write/read operation
    binary magnetic tunnel junction 48–49

multi-level spintronic memristor 49–51
Manhattan rule 38
matrix-vector multiplications (MVMs) 176
mechanoreceptors 218
membrane potential (MP) 27, 28, 31, 54, 105, 137, 218
memristive bioinspired devices 17
  memristive neurons 27–36
  memristive synapses 18–27
memristive device 1, 8
  materials 2–4
  metal/insulator/metal structure 1–2
memristive neural networks
  ANN computing 36–39
  SNN computing 39–41
memristive neurons 27–36
  biophysical memristive neurons 34–36
  bioplausible memristive neurons 31–34
memristive synapses 18–27
  long-term memristive synapses 23–27
  short-term memristive synapses 20–23
memristor-based CIM macros 183
memristor based neuromorphic devices 92–96
memristors 125
metal filament conducting-based memristors 150–152
metal/insulator/metal (MIM) 1–2
metaplasticity 92
metavalent bonding 129
mimicked multi-store memory model 94
MIT cell 34
Modified National Institute of Standards and technology (MNIST) 37
monolayer molybdenum disulfide (MoS_2) 98
Moore's law 91
multi-gate neuromorphic transistor 105–106
multi-gate neuromorphic transistor for pH-sens 103
multi-gate P3HT electrochemical transistors (ECTs) 109
multi-gate synergic modulation 105
multilayer perceptron (MLP) 125, 139
multi-level spintronic memristor 49–51
multi-memristive synapse design 190
multiple $Au/Li_xMoS_2/Au$ memristors 97
multiplication-addition-permutation (MAP) 196
multi-terminal memtransistor 98, 190
multi-terminal neuromorphic memristors
  heterosynaptic plasticity 97–98
  memristor based neuromorphic devices 92–96
  multi-terminal memtransistor 98
  synaptic competition and cooperation 96–97
multi-terminal neuromorphic transistors
  artificial auditory device 113–114
  artificial tactile device 109–112
  artificial vision device 112–113
  chemical-biosensors 101–102
  dendrite neuron and dendritic algorithm 106–109
  EGT based neuromorphic transistors 103–105
  multi-gate neuromorphic transistor 105–106
  multi-gate neuromorphic transistor for pH-sens 103
  neuron MOS (νMOS) transistor 99
  neuron transistors for chemical-biosensor 102
  pattern recognition 109
multi-terminal synaptic transistor structure 115

n

Na^+ channels 27
nanogranular SiO_2 based electrolyte 102
nanogranular SiO_2 electrolyte film 103
nanogranular SiO_2 films 100

nanoparticle organic memory field-effect transistors (NOMFET) 158
Nb_2O_5-based Mott devices 192
NbO_x-based Mott memristors 202
neural networks 54
neuromorphic computing 2
  background introduction 173–174
  based synchronization 77
  CMOS-based 174–176
  dendritic computing 196–202
  DNNs based on synaptic devices 178–187
  Hopfield neural network (HNN) 203–204
  hyperdimensional (HD) computing 196, 201
  motivation for 174
  oscillatory neural networks (ONNs) 203
  principles of memristors 176–177
  reservoir computing 202–203
  SNNs based on neuromorphic devices 187–195
neuromorphic perceptual systems 227
  artificial perception 219–221
  challenges and perspectives 228–229
  sensation and perception 218–219
neuromorphic tactile processing system (NeuTap) 110
neuron MOS (νMOS) transistor 99
neuron transistors for chemical-biosensor 102
nociceptive memory 223
nociceptors 218
nonvolatile resistive switching 4, 7, 8, 31
normal sensor 221
n-type semiconductor 15
nucleation-dominant GeTe-Sb_2Te_3 pseudo-binary alloy system 127

o

Ohm and Kirchhoff laws 7, 18
olfactory memory 223
on-chip memory (SRAM) 183
one-selector-one-resistor (1S1R) 179
one-transistor-one-memristive device (1T1M) 37, 41
one-transistor-one-resistor (1T1R) 179
optoelectronic transistors 164–165
organic digital mechanoreceptor 222
organic neuromorphic devices
  artificial visualization systems 166–167
  tactile-perception systems 167–169
  three-terminal devices 157–165
  two-terminal devices 149–157
organometallic halide perovskites 154
oscillators 72
oscillatory neural networks (ONNs) 203
ovonic memory switching (OMS) 126
ovonic threshold switching (OTS) 126

p

paired-pulse depression (PPD) 20, 21
paired-pulse facilitation (PPF) 20–22, 92
passive memristive device (0T1R) 179
pattern recognition 109, 226
Pavlov classical conditioning 113
Pavlov's learning behaviors 108
$(PEA)_2PbBr_4$ 155
PCMO 16
2-PCM synapse 133
Pd/HfO_2/Ta memristor arrays 192
P-doped nanogranular SiO_2 films 100
PEDOT+PSS electrochemical transistor 109
PEO-EV$(ClO_4)_2$/triphenylamine-based polyimide (TPA-PI) system 153
PEO-LiCF_3SO_3/poly[2-methoxy-5-(2-ethylhexyloxy)-1,4-phenylenevinylene] (MEH-PPV) system 153
perovskite quantum dot (QD) 113
perpendicular magnetic anisotropy (PMA) 48, 49
persistent photoconductivity (PPC) effect 113
phase-change heterostructure (PCH) 136
phase-change memory (PCM) 176

phase-change storage disc technology 126
phase locked loop (PLL) 72
photoemission electron microscopy (PEEM) 13
piezoelectric nanogenerators (PENG) 110, 111, 222
piezopotential-programmed multilevel non-volatile memory 222
piezotronic artificial synapse 110
pinned layer (PL) 48
plasma-enhanced chemical vapor deposition (PECVD) 100
plasticity-related proteins (PRPs) 96
PMOS transistor 60
polyacrylonitrile (PAN) film 164
polycrystalline monolayer molybdenum disulfide (MoS_2) 190
polydopamine-based wearable memristor device 95
polyethylene oxide (PEO)/terpyridyl-iron polymer (TPy-Fe)/ITO structure 153
poly (vinyl alcohol)-graphene oxide (PVA-GO) hybrid nanocomposite 95
poly(isoindigo-co-bithiophene) [P(IID-BT)] 163
poly(vinylidenefluoride-co-trifluoro-ethylene) [P(VDF-TrFE)] 162
poly (vinylidenefluoride-trifluoro-ethylene) [P(VDF-TrFE)] film 162
poly(3,4-ethylenedioxythiophene):poly-styrene sulfonate (PEDOT:PSS) 34, 150
poly[(1-vinylpyrrolidone)-co-(2-ethyl-dimethylammonioethyl methacrylate ethyl sulfate)] [P(VP-EDMAEMAES)] 163
postsynaptic neurons (POSTNs) 114
post-tetanic potentiation (PTP) 19, 20
prestructural ordering (incubation) method 130
presynaptic neurons (PRENs) 114
principles of memristors 176
programmable in-memory HD computing system 196
programmable metallization cells (PMCs) 8
prosthetics 217
prosthetics and robotics applications 226
proton-conducting graphene oxide coupled neuron transistors 108
Pt/Ag/SiO_x:Ag/Ag/Pt diffusive memristor 192
Pt/GeSe/Pt device 137
Pt/Ta_2O_5/TaO_x/Pt capacitor 192
Pt/TaO_x/AlO_y/Al dynamic memristor device 202
pulsed laser deposition (PLD) 3
pulses trigger depression (PPD) 22

r

receptor potential 218
rectified linear unit (ReLU) 40, 139
recurrent neural networks (RNNs) 67, 75–77, 173, 202
redox reaction-based memristors 152–153
reduced instruction set computing (RISC) 183
remote supervised method (ReSuMe) 189
reprogrammable CMOS-RRAM CIM system 183
reprogrammable neuromorphic computing hardware 183
reservoir computing (RC) 47, 202–203
reservoir computing based STO 74–75
RESET operation 4
resistive random-access memory (RRAM) 176
resistive switching (RS)
  filamentary-type 8–14
  interface-type resistive switching 14–17
  I–V characteristics 5
  nonvolatile resistive switching 7
  volatile resistive switching 5–7

resistive switching (RS) behavior 92
resistive switching devices 125
restricted Boltzmann machine (RBM) 142

S

scanning electron micrograph (SEM) 38
scanning transmission X-ray microscopy (STXM) 13
Schottky barrier 15
$Sc_{0.2}Sb_2Te_3$ material 130
self-organized atomic switch networks 203
sensation and perception 218–219
sense amplifiers (SAs) 48, 183, 185
sense of touch 219
sensorimotor system 166, 167, 219, 224, 226, 229
sensory memory (SM) 94, 222
sensory neuron 110, 167, 217, 219–223
short-term facilitation/depression (STF/STD) 188
short-term/long-term synaptic plasticities (STP/LTP) 92
short-term memristive synapses 20–23
short-term plasticity (STP) 18, 19, 23
sign-based backpropagation (SBP) training method 186
silicon-on-insulator (SOI) wafers 102
SiO_xN_y:Ag diffusive memristor 190
skyrmion-based electronics 63
skyrmion-based neuromorphic computing
  overview 62–64, 71
  skyrmion-based neuron devices 65–67
  skyrmion-based reservoir computing 67–68
  skyrmion-based stochastic computing 68–69
  skyrmion-based synapse devices 64–65
skyrmion-based neuron devices 65–67
skyrmion-based reservoir computing 67–68
skyrmion-based stochastic computing 68–69
skyrmion-based synapse devices 64–65
skyrmion-based unconventional computing 70
skyrmion Hall effect (SkHE) 64
skyrmionics 63, 65
skyrmion reshuffler device 69
SNN computing 39–41
SNNs based on neuromorphic devices 187–195
  learning rules and memory principles 188–189
  synaptic devices and neuronal devices 189–192
30 × 10 × 10 SNN with hybrid in-situ learning 192
(2-4) soft-/biorobotics 217
soma 106
spectromicroscopy 16
spike-based training algorithms 189
spike rate dependent plasticity (SRDP) 25, 136
spike-timing dependent plasticity (STDP) 24, 25, 52, 92, 131, 142, 188
spiking neural device 54–57
spiking neural network (SNN) 51, 125, 140, 173
spiking rate dependent plasticity (SRDP) 92
spin-orbit torque-magnetic tunnel junction (SOT-MTJ) 56
spin-torque nano-oscillators 203
spin torque oscillators (STO)
  advantages 77
  associative memory based injection locking 73–74
  device 64
  neuromorphic computing based synchronization 77
  overview 72–73
  recurrent neural network based on delayed feedback 75–77
  reservoir computing based 74–75
spin transfer torque (STT) effect 48
spintronic devices 47
spintronic memristors 51, 62

spintronic neuromorphic devices 47
Stanford's Neurogrid 176
starch-based bio-polysaccharide electrolyte gated indium-gallium-zinc oxide (IGZO) phototransistor 112
starch-based flexible biomemristors 95
static random-access memory (SRAM) 176
stochastic binary synaptic device 51–52
stochastic computing (SC) 64
Street View House Numbers (SVHN) dataset 202
stretchable organic nanowire synaptic transistor (s-ONWST) 166
STT-MTJ 49, 51, 52
Stuck-At-Fault (SAF) 187
synapse 18
synaptic cell programming 180
synaptic filtering 92
synaptic plasticity 188
synaptic weight 64

t

tactile-perception systems 167–169
tactile sensation 109
$Ta/EV(ClO_4)_2$ 153
TaO_x-based bilayer memristors 190
$Te_{48}As_{30}Si_{12}Ge_{10}$ chalcogenide 128
$Te_{81}Ge_{15}Sb_2S_2$ 126
tensor processing units (TPUs) 174
Tera/Giga operations per second (T/GOPS) 176
tetrafluoroethylene-perfluoro-3,6-dioxa-4-methyl-7-octenesulfonic acid copolymer (PFI) 154
thermoreceptors 218
three-dimensional integrated circuit architecture 220
3D XPoint™ 126
three-terminal organic neuromorphic devices
  electrolyte-gate transistors (EGTs) 158–162
  ferroelectric field-effect transistors (FeFETs) 162–163
  floating-gate transistors 157–158
  optoelectronic transistors 164–165
threshold switching (TS) devices 136
threshold switching memristor (TSM) 191
"Tiki-Taka" training algorithm 186
TiO_2-based device 13
top electrode (TE) 11
trans-impedance amplifiers (TIAs) 179
transition metal chalcogenides (TMDs) 3
transmission electron microscopy (TEM) 9
  interface-type resistive switching 14–17
  $Pt/TiO_2/Pt$ device 12
  Ti/PCMO junction 16
1T1R memristor-based synapses 190
128 × 8 1T1R memristor array 181
two-dimensional (2D) planar geometry 98
two-terminal organic neuromorphic devices
  charge-trapping-based memristors 155–157
  ion migration-based memristors 153–155
  metal filament conducting-based memristors 150–152
  redox reaction-based memristors 152–153

u

ultra-low-power SNN processor 142
unconditioned stimulus (US) 108, 112, 113
UZH's ROLLS 176

v

valence change mechanism (VCM) 11, 93
valence change memory memristors (VCMMs) 93

vector-matrix multiplication (VMM) 7, 18, 139
volatile resistive switching 5–7
voltage controlled oscillators (VCOs) 72
voltage-gated calcium (Ca^{2+}) channels 105

W

wearable medical devices 217
weighted summation 7
Werbos's backpropagation algorithm 173
WO_x memristors 202

X

X-ray absorption spectroscopy (XAS) 13
X-ray diffraction (XRD) 13
X-ray PEEM (XPEEM) 14
$30 \times 10 \times 10$ SNN with hybrid in-situ learning 192

Z

ZJU's Darwin 176